REUNITING THE ANTIPODES – CONSTRUCTIVE AND NONSTANDARD VIEWS OF THE CONTINUUM

SYNTHESE LIBRARY

STUDIES IN EPISTEMOLOGY,

LOGIC, METHODOLOGY, AND PHILOSOPHY OF SCIENCE

VOLUME 306

REUNITING THE ANTIPODES – CONSTRUCTIVE AND NONSTANDARD VIEWS OF THE CONTINUUM

Symposium Proceedings, San Servolo,
Venice, Italy, May 16–22, 1999

Edited by

PETER SCHUSTER
Ludwig-Maximilians-Universität, München, Germany

ULRICH BERGER
University of Wales, Swansea, United Kingdom

and

HORST OSSWALD
Ludwig-Maximilians-Universität, München, Germany

KLUWER ACADEMIC PUBLISHERS
DORDRECHT / BOSTON / LONDON

A C.I.P. Catalogue record for this book is available from the Library of Congress.

ISBN 1-4020-0152-5

Published by Kluwer Academic Publishers,
P.O. Box 17, 3300 AA Dordrecht, The Netherlands.

Sold and distributed in North, Central and South America
by Kluwer Academic Publishers,
101 Philip Drive, Norwell, MA 02061, U.S.A.

In all other countries, sold and distributed
by Kluwer Academic Publishers,
P.O. Box 322, 3300 AH Dordrecht, The Netherlands.

Printed on acid-free paper

Printed in the Netherlands.

TABLE OF CONTENTS

In memoriam

Detlef Laugwitz

(1932–2000)

FOREWORD

The introduction of differentials by Leibniz and Newton might be considered as the first appearance of infinitesimals in the mathematics of modern times, famous also because these objects were soon derided by Berkeley as ghosts of a highly unsure existence. Infinitesimals, i.e., nonvanishing positive 'numbers' smaller than any given fraction, had still been very useful for Euler and Cauchy; they became ruled out only later on, in the wake of the putatively rigorous tendencies which took over mathematical analysis during the 19th century. Since, the best known renewal of that venerable concept has been carried out in the preceding four decades by Abraham Robinson and others, who designed this nonstandard analysis to deal with infinitely small and infinitely large entities in a truly rigorous manner. Because of making a rather unrestricted use of classical logic and set theory and, in particular, of the axiom of choice, Robinson's theory in its full-fledged form has widely been suspected to be nonconstructive from the outset. In addition, the nonstandard idea of discretising the continuum seems to be even less compatible with the intuitionistic concept of a continuum in the true sense of the word, than with the classical atomistic notion.

The distance between constructive and nonstandard mathematics, however, is actually much smaller than it appears to be. Indications for this are that nonstandard practice often looks rather constructive, and that very small numbers unknown to vanish are indispensable to distinguish constructive mathematics from its traditional counterpart. At least from any naive point of view, it is therefore no wonder that constructive mathematics eventually proved its capability to tackle also relatively abstract objects such as infinitesimals. This progress cannot be thought of without the revival of constructive thinking since the 1960es, initiated by the work of Errett Bishop, Per Martin-Löf, and others as well as by the development of digital computers, which has eventually lead to today's pragmatic way in which constructive mathematics sees itself. Some far-reaching approaches to constructive nonstandard mathematics have indeed been undertaken quite recently, whence time was ripe for the first meeting dedicated simultaneously to constructive and nonstandard mathematics—and, of course, to the reunion of these seeming antipodes.

Consisting of peer-reviewed articles written on the occasion of such an event, this volume offers views of the continuum from various standpoints. Including historical and philosophical issues, the topics of the contributions range from the foundations, the practice, and the applications of constructive and nonstandard mathematics, to the interplay of these areas and the development of a unified theory.

Further Talks. The following talks were given at the conference but will be published elsewhere.

PETR ANDREYEV (Nizhnii Novgorod State University, Russia), *Definable standardness predicates in Internal Set Theory*

DOUGLAS BRIDGES (University of Canterbury, Christchurch, New Zealand), *Constructive investigations of functions of bounded variations*

NIGEL J. CUTLAND (University of Hull, England), *Constructive aspects of nonstandard methods in fluid mechanics*

DIRK VAN DALEN (Rijksuniversiteit Utrecht, The Netherlands), *Indecomposable subsets of the continuum*

JENS ERIK FENSTAD (University of Oslo, Norway), *Computability theory over the nonstandard reals*

JAMES HENLE (Smith College, Northhampton, Massachusetts), *Non-nonstandard analysis: category, measure, and integration*

CHRIS IMPENS (University of Gent, Belgium), *Some thoughts on nonstandard geometry*

HAJIME ISHIHARA (Japan Advanced Institute of Science and Technology), *A note on the Gödel-Gentzen translation*

H. JEROME KEISLER (University of Wisconsin, Madison), *Nonstandard methods in o-minimal structures*

M. ALI KHAN (John Hopkins University, Baltimore, Maryland), *Modelling 'neglibility' in mathematical economics: an application of Loeb spaces*

P. EKKEHARD KOPP (University of Hull, England), *Hyperfinite discretisations and convergence in option pricing models*

XIAOAI LIN (National University of Singapore), *On the almost independence of correspondences on Loeb space*

WILHELMUS A. J. LUXEMBURG (California Institute of Technology, Pasadena), *A Schauder type theorem for internal linear operators*

PER MARTIN-LÖF (University of Stockholm, Sweden), *Nonstandard type theory*

JOAN R. MOSCHOVAKIS (Occidental College, Los Angeles; University of Athens, Greece), *The intuitionistic continuum as an extension of the classical one*

JUHA OIKKONEN (University of Helsinki, Finland), *Some geometric ideas related to Brownian motion*

HERVÉ PERDRY (Université de Franche-Comte, Besançon, France), *Computing in the constructive henselisation of a valued field*

HANS PLOSS (Universität Wien, Austria), *On the rearrangement of series*

MICHAEL REEKEN (Universität Wuppertal, Germany), *Discretising the continuum*

HERMANN RENDER (Universität Duisburg, Germany), *Borel measure extensions defined on sub-σ-algebras*

GIOVANNI SAMBIN (Università di Padova, Italy), *Real numbers in formal topology*

PETER SCHUSTER (Universität München, Germany), *Elementary choiceless constructive analysis*

BAS SPITTERS (Katolieke Universiteit Nijmegen, The Netherlands), *A constructive converse of the mean value theorem*

YENENG SUN (National University of Singapore), *Asymptotic, hyperfinite, and continuum models*

WIM VELDMAN (Katolieke Universiteit Nijmegen, The Netherlands), *On some sets that are not positively Borel*

HANS VERNAEVE (University of Gent, Belgium), *Reducing distributions to hyperreal functions*

MANFRED WOLFF (Universität Tübingen, Germany), *On the approximation of operators and their spectra*

Further Participants

STEFANO BARATELLA (Università di Trento, Italy)
JOSEF BERGER (Universität München, Germany)
JENS BLANCK (University of Wales, Swansea)
JONAS ELIASSON (University of Uppsala, Sweden)
FERRUCCIO GUIDI (Università di Padova, Italy)
GÖRAN HAMRIN (University of Uppsala, Sweden)
MARTIN KÜBLER (Universität München, Germany)
JUHA RUOKOLAINEN (University of Helsinki, Finland)

Acknowledgements. This meeting was particularly sponsored by the *Volkswagen-Stiftung*, Hannover, which generously contributed to the travel and accommodation expenses of many participants. Additional financial support originated in the *Istituto Italiano per gli Studi Filosofici*, Naples and Venice, and in the *Münchner Universitätsgesellschaft*, Munich.

Local logistic assistance was provided by the *Venice International University* staff and, especially, by Umberto Curi and Raffeala Toffolo; a helping hand often came from Josef Berger, Giovanni Curi, Martin Kübler, Monika Seisenberger, and Julia Zink.

The preparation of the final version of this volume had not been possible without the technical assistance by Uwe Lück. We also wish to thank Ingrid Krabbenbos, Jolanda Voogd, and Rudolf Rijgersberg from *Kluwer Academic Publishers* for their patience during the publication process, and Dokie van Dalen for both the idea and the permission to include the group photograph taken by her.

Last but not least, the editors are most grateful to all the participants, authors, and anonymous referees, without whose respective contributions this volume had hardly emerged.

PETER SCHUSTER (Ludwig-Maximilians-Universität München)

ULRICH BERGER (University of Wales Swansea)

HORST OSSWALD (Ludwig-Maximilians-Universität München)

Munich and Swansea.

REUNITING THE ANTIPODES— CONSTRUCTIVE AND NONSTANDARD VIEWS OF THE CONTINUUM

NONSTANDARD CONSTRUCTION OF STABLE TYPE EUCLIDEAN RANDOM FIELD MEASURES

Sergio Albeverio

Institut für Angewandte Mathematik, Universität Bonn, D-53115 Bonn, Germany
BiBoS-Research Centre, D-33615 Bielefeld, Germany
CERFIM, Locarno, Switzerland; Acc. Arch., USI, Switzerland
albeverio@uni-bonn.de

Jiang-Lun Wu

Institut für Angewandte Mathematik der Universität Bonn, D-53115 Bonn, Germany
Department of Mathematics, University of Wales Swansea,
Singleton Park, Swansea SA2 8PP, United Kingdom
J.L.Wu@Swansea.ac.uk

Keywords: Stable type Euclidean random field measures, hyperfinite flat integral formula, large deviation, nonstandard analysis.

Mathematical Subject Classification: 28C20, 60G60, 81T25, 60F10, 03H05

Abstract: A nonstandard construction of stable type Euclidean random fields via hyperfinite flat integrals and stable white noise is given. Moreover, a brief account on an extension of Cutland's flat integral formula for (centered) Gaussian measures on the Hilbert space l_2 to the case of Banach spaces $l_p, 1 \leq p < \infty$, is presented.

Introduction

The aim of this paper is to derive a nonstandard flat integral representation for certain stable type Euclidean random field measures. In the case of Gaussian Euclidean random field measures, this was done in [3] where a flat integral formula for Nelson's free field measure has been given. In his seminal paper [6], Cutland studied the nonstandard flat integral representation of Wiener measure on the classical Wiener space $C_0[0,1]$, which gives a nonstandard justification of Donsker's (heuristic) "flat integral". He then used such representation to give a fairly simple

P. Schuster, U. Berger and H. Osswald (eds.),
Reuniting the Antipodes – Constructive and Nonstandard Views of the Continuum, 1–18.

and intuitive nonstandard proof of a Schilder's large deviation principle for the Wiener measure. Furthermore, in [7, 8, 9, 10], he extended such investigations to various (centered) Gaussian measures, which provides a shorter (and nonstandard) version of the large deviation principle discussed in general, for centered Gaussian measures on separable Banach spaces, in section III.3.4 of [11]. Let us also mention the interesting work [19] by Osswald, where he presents a further nonstandard construction of Brownian motion in abstract Wiener spaces based on [6]. In the last section of this paper, we will present shortly an extension of Cutland's work [9] on flat integral representation for measures on l_2 to the case of the Banach spaces $l_p, 1 \leq p < \infty$.

The study of large deviations for (non Gaussian) Euclidean random field measures seems delicate but possible. Also, one may expect to be able to discuss the scaling limits for such random field measures. To this purpose, a flat integral formula is apparently useful. Our strategy is to find an appropriate (hyperfinite) lattice setting and to construct certain lattice measures via inverse Fourier transform, then utilizing the fact that the stable Euclidean random field measures are induced by Euclidean random fields independent at each point to make a product measure of all such lattice measures, and finally to use Loeb measure structure to get the flat integral formula. This idea has been further utilized in [4] to investigate a functional integral realization for the class of Euclidean random field models for constructive quantum field theory developed in recent years in all space-time dimensions by Albeverio, Gottschalk and Wu.

In this paper, we take for granted the familiarity with the preliminaries on nonstandard analysis and the Loeb measure construction presented e.g. in [1], [5] and [18].

1. EUCLIDEAN RANDOM FIELD MEASURES

Let $\mathcal{D} := C_0^\infty(\mathbb{R}^d)$ be the vector space consisting of all C^∞-smooth functions on $\mathbb{R}^d$ ($d \in \mathbb{N}$) with compact support endowed with Schwartz topology and $\mathcal{D}'$ its (topological) dual space. Let $\mathcal{B}$ denote the Kolmogorov σ-algebra on $\mathcal{D}'$ generated by cylindrical sets of $\mathcal{D}'$ (which coincides with the topological σ-algebras generated by the strong or weak topologies of $\mathcal{D}'$). Let $p \in (0, 2]$ be arbitrarily fixed. From [13],

$$f \in \mathcal{D} \mapsto e^{-\int_{\mathbb{R}^d} |f(x)|^p dx} \in \mathbb{R}(\subset \mathbb{C})$$

is a characteristic functional on the nuclear space $\mathcal{D}$. By the well-known Bochner-Minlos' theorem (see e.g. [13]), there exists a unique probability measure μ on $(\mathcal{D}', \mathcal{B})$ such that

$$C(f) := \int_{\mathcal{D}'} e^{i<f,\omega>} d\mu(\omega) = e^{-\int_{\mathbb{R}^d} |f(x)|^p dx}, \quad f \in \mathcal{D}. \tag{1.1}$$

Moreover, there is a Euclidean random field[1] $F : \mathcal{D} \times (\mathcal{D}', \mathcal{B}, \mu) \to \mathbb{R}$ determined by $F(f,\omega) =< f, \omega >, f \in \mathcal{D}, \omega \in \mathcal{D}'$. We call F a stable random field and its probability law μ a stable random field measure.

Now let $m > 0$ if $d = 1, 2$ and $m \geq 0$ is $d \geq 3$ and let $\alpha \in (0,1)$. Then the following stochastic pseudo-differential equation

$$(-\Delta + m^2)^\alpha X = F$$

induces a stable type Euclidean random field $X : \mathcal{D} \times (\mathcal{D}', \mathcal{B}, \mu_X) \to \mathbb{R}$ via $X(f,\omega) =< f, \omega >, f \in \mathcal{D}, \omega \in \mathcal{D}'$, where $\mu_X(D) := \mu((\Delta + m^2)^{-\alpha} D), D \in \mathcal{B}$, whose characteristic functional

$$C_X(f) := \int_{\mathcal{D}'} e^{i<f,\omega>} d\mu_X(\omega) = e^{-\int_{\mathbb{R}^d} |(\Delta+m^2)^{-\alpha} f(x)|^p dx}, \quad f \in \mathcal{D}. \tag{1.2}$$

Let us point out that if $p = 2$, μ and μ_X are Gaussian measures on $\mathcal{D}'$ supported by certain Sobolev spaces with negative indices (while μ_X is just Nelson's free field measure if $\alpha = \frac{1}{2}$, which was already studied using methods of nonstandard analysis in [1] and [3]). Also F introduced here is an interesting special case of infinitely divisible (Euclidean) random fields discussed e.g. in [2] (in the terminology of [13], an infinitely divisible random field is called "a generalized random process with independent value at every point"). Our main aim here is to give a representation formula for (the non Gaussian measures) μ and μ_X. Since there is no inverse Fourier transform for probability measures on ∞-dimensional spaces, we will realize our aim by using nonstandard analysis.

Similar methods will also be used to discuss Gaussian measures on l_p for $1 \leq p < \infty$, see Section 3.

2. NONSTANDARD CONSTRUCTION OF μ AND μ_X

Let us first give a hyperfinite representation of $\mathcal{D}'$ by following [14, 15]. Fix a polysaturated nonstandard model. Let $N \in {}^*\mathbb{N} \setminus \mathbb{N}$ be arbitrarily

[1] By "Euclidean", we mean that the probability law is invariant under the (proper) Euclidean transformation group.

fixed and $\delta := \frac{1}{N}$, an infinitesimal. We set

$$T := \{-N, -N + \frac{1}{N}, ..., -\frac{1}{N}, 0, \frac{1}{N}, ..., N - \frac{1}{N}, N\} \subset {}^*\mathbb{R}$$

and $\mathcal{L} := T^d \equiv \underbrace{T \times \cdots \times T}_{d\ times} \subset {}^*\mathbb{R}^d, d \in \mathbb{N}$. Let ${}^*\mathbb{R}^{\mathcal{L}}$ stands for the internal space of all internal functions from $\mathcal{L}$ into ${}^*\mathbb{R}$. We set

$$< f, g > := \sum_{t \in \mathcal{L}} \delta^d f(t) g(t), \quad f, g \in {}^*\mathbb{R}^{\mathcal{L}}.$$

Definition 2.1 *(Keßler [15]) $f \in {}^*\mathbb{R}^{\mathcal{L}}$ is called S-continuos whenever g is infinitesimal in ${}^*\mathcal{D}(K)$ for some compact set $K \subset \mathbb{R}^d$ implies that $< f, g >$ is infinitesimal in ${}^*\mathbb{R}$. Moreover, $f \in {}^*\mathbb{R}^{\mathcal{L}}$ is said to be $\mathcal{D}'$-nearstandard if $< f, \cdot > |_{{}^*\mathcal{D}(K)}$ is S-continuous for any compact set $K \subset \mathbb{R}^d$.*

$< f, \cdot >$ being linear on ${}^*\mathbb{R}^{\mathcal{L}}$ for $f \in {}^*\mathbb{R}^{\mathcal{L}}$, the necessary and sufficient condition for $< f, \cdot > |_{{}^*\mathcal{D}(K)}$ to be S-continuous for any compact set $K \subset \mathbb{R}^d$ is that $< f, g >$ is finite whenever g is finite in ${}^*\mathcal{D}(K)$. Thus, $f \in {}^*\mathbb{R}^{\mathcal{L}}$ is $\mathcal{D}'$-nearstandard if $< f, g >$ is finite for any compact set $K \subset \mathbb{R}^d$ and for any $\mathcal{D}(K)$-finite $g \in {}^*\mathcal{D}(K)$ (where $g \in {}^*\mathcal{D}(K)$ is said to be $\mathcal{D}(K)$-finite if the internal suprema $sup_{x \in {}^*K} |g^{(n)}(x)|, n \in \mathbb{N}$, are finite).

We denote by $Ns({}^*\mathbb{R}^{\mathcal{L}})$ the totality of $\mathcal{D}'$-nearstandard functions. We define the (weak) standard part mapping $st : Ns({}^*\mathbb{R}^{\mathcal{L}}) \to \mathcal{D}'$ via duality:

$$< st(f), g > = {}^\circ(< f, {}^*g >), \quad \forall g \in \mathcal{D}.$$

$< st(f), \cdot >$ defines a distribution essentially because of the definition of the linear induction limit topology. The standard part mapping is continuous on each $\mathcal{D}(K)$ and hence on $\mathcal{D}$. On the other hand, from [14], every standard distribution $g \in \mathcal{D}'$ has a hyperfinite representation $f \in Ns({}^*\mathbb{R}^{\mathcal{L}}) : st(f) = g$. Therefore $st[Ns({}^*\mathbb{R}^{\mathcal{L}})] = \mathcal{D}'$.

Let us now turn to the construction of μ and μ_X. We begin to argue formally. In the hyperfinite lattice setting, we have $f = (f_t)_{t \in \mathcal{L}} \in {}^*\mathbb{R}^{\mathcal{L}}$ and $g = (g_t)_{t \in \mathcal{L}} \in {}^*\mathbb{R}^{\mathcal{L}}$ as hyperfinite sequences (or vectors). Since μ is the probability distribution of "a generalized random process with independent value at every point", we have $\mu = \prod_{t \in \mathcal{L}} \mu_t$, where $\mu_t := Proj_t \mu, t \in \mathcal{L}$, the marginal probability distribution of μ. Taking a hint from (1.1), we have for any $(f_t)_{t\in\mathcal{L}} \in {}^*\mathbb{R}^{\mathcal{L}} \cap {}^*\mathcal{D}$ (i.e., the hyperfinite segment of ${}^*f$ for $f \in \mathcal{D}$),

$$\int_{{}^*\mathbb{R}^{\mathcal{L}}} e^{i\sum_{t\in\mathcal{L}}\delta^d f_t q_t} \prod_{t\in\mathcal{L}} d\mu_t(q_t) = e^{-\sum_{t\in\mathcal{L}}\delta^d |f_t|^p}$$

namely

$$\prod_{t\in\mathcal{L}} \int_{{}^*\mathbb{R}} e^{i\delta^d f_t q_t} d\mu_t(q_t) = \prod_{t\in\mathcal{L}} e^{-\delta^d |f_t|^p}$$

which further implies that

$$\int_{{}^*\mathbb{R}} e^{i\delta^d f_t q_t} d\mu_t(q_t) = e^{-\delta^d |f_t|^p} .$$

Setting $\mu_t^\delta(\cdot) := \mu_t(\delta^d \cdot)$, then

$$\int_{{}^*\mathbb{R}} e^{i f_t q_t} d\mu_t^\delta(q_t) = e^{-\delta^d |f_t|^p} .$$

Remarking that the above equality is a one-dimensional Fourier transform, one can take inverse Fourier transform to get the following expression for the density of μ_t (i.e. the Radon-Nikodym derivative with respect to one-dimensional Lebesgue measure)[2]

$$\frac{d\mu_t^\delta(q_t)}{dq_t} = \frac{1}{2\pi}\int_{{}^*\mathbb{R}} e^{-i f_t q_t} e^{-\delta^d |f_t|^p} df_t\, , \quad (f_t)_{t\in\mathcal{L}} \in {}^*\mathbb{R}^{\mathcal{L}} \cap {}^*\mathcal{D}$$

where dq_t and df_t stand for one-dimensional Lebesgue measure. Clearly this paves a way for us to construct μ.

Let $(\Omega, \mathcal{A}(\Omega), P)$ be a given internal probability space. The associated Loeb space is denoted by $(\Omega, \mathcal{A}_L(\Omega), P_L)$. Let $\{\eta_t(\omega) : \omega \in \Omega\}_{t\in\mathcal{L}}$ be an internal family of independent, identically distributed ${}^*\mathbb{R}$-valued random variables on $(\Omega, \mathcal{A}(\Omega), P)$, each $\eta_t : \Omega \to {}^*\mathbb{R}$ has (internal) density h given by

$$h(q_t) = \frac{1}{2\pi}\int_{{}^*\mathbb{R}} e^{-(i f_t q_t + \delta^d |f_t|^p)} df_t,\ (f_t)_{t\in\mathcal{L}} \in {}^*\mathbb{R}^{\mathcal{L}} \cap {}^*\mathcal{D}, (q_t)_{t\in\mathcal{L}} \in {}^*\mathbb{R}^{\mathcal{L}}$$

---

[2]Actually this inverse Fourier transform can be computed explicitly, for instance, by using the formulae of items 82 and 83 on page 25 of F. Oberhettinger: *Fourier Transforms of Distributions and Their Inverses. A Collection of Tables.* Academic Press, New York, London, 1973. But we do not need it here.

so that the inverse Fourier transform of h is

$$\int_{^*\mathbb{R}} e^{if_tq_t}h(q_t)dq_t = e^{-\delta^d|f_t|^p}, \ (f_t)_{t\in\mathcal{L}} \in {}^*\mathbb{R}^{\mathcal{L}} \cap {}^*\mathcal{D}, (q_t)_{t\in\mathcal{L}} \in {}^*\mathbb{R}^{\mathcal{L}}$$

and the probability law of η_t is

$$P\{\omega \in \Omega : \eta_t(\omega) \in B\} = \int_B h(q_t)dq_t, \quad B \in \mathcal{A}({}^*\mathbb{R})$$

where $\mathcal{A}({}^*\mathbb{R})$ denotes the internal algebra of internal subsets of ${}^*\mathbb{R}$. Namely, $\{\eta_t(\omega) : \omega \in \Omega\}_{t\in\mathcal{L}}$ is an internal family of independent ${}^*\mathbb{R}$-valued stable random variables of index p on $(\Omega, \mathcal{A}(\Omega), P)$. In particular, when $p = 1$, each η_t is a random variable with Cauchy distribution. Moreover, for any $t \in \mathcal{L}$,

$$\int_\Omega e^{i\theta\eta_t(\omega)}dP(\omega) = e^{-\delta^d|\theta|^p}, \quad \theta \in {}^*\mathbb{R}.$$

Now let us define

$$\eta_A(\omega) := \sum_{t\in A} \eta_t(\omega), \quad A \in \mathcal{A}(\mathcal{L}).$$

The mapping $\eta : \mathcal{A}(\mathcal{L}) \to {}^*\mathbb{R}$ is said to be *S-stable white noise* on $(\mathcal{L}, \mathcal{A}(\mathcal{L}), \delta^d)$ with respect to the given $(\Omega, \mathcal{A}(\Omega), P)$. Let $|A|$ denote the internal cardinal number of $A \in \mathcal{A}(\mathcal{L})$. One can verify the following result

Proposition 2.2 *The standard family $\{{}^\circ\eta_A(\omega) : A \in \mathcal{A}_L(\mathcal{L}), {}^\circ(|A|\delta^d) < \infty\}$ is an one-dimensional ($\mathbb{R}$-valued) infinitely divisible (stable) random process indexed by the family of the sets $\{A \in \mathcal{A}_L(\mathcal{L}), {}^\circ(|A|\delta^d) < \infty\}$ with characteristic function*

$$\int_\Omega e^{i\lambda\, {}^\circ\eta_A(\omega)}dP_L(\omega) = e^{-{}^\circ(|A|\delta^d)|\lambda|^p}, \lambda \in \mathbb{R}, \ A \in \mathcal{A}_L(\mathcal{L}), {}^\circ(|A|\delta^d) < \infty.$$

We call η stable white noise on the Loeb space $(\mathcal{L}, \mathcal{A}_L(\mathcal{L}), (\delta^d)_L)$ with respect to the Loeb space $(\Omega, \mathcal{A}_L(\Omega), P_L)$. If $p = 2$, η is just a Gaussian white noise.

Proof. It suffices to compute the internal characteristic function of η. In fact, for $\lambda \in \mathbb{R}$, $A \in \mathcal{A}_L(\mathcal{L})$, by independence of $\eta_t, t \in \mathcal{L}$, we have

$$\int_\Omega e^{i\lambda\,\eta_A(\omega)}dP(\omega) = \prod_{t\in A}\int_\Omega e^{i\lambda\,\eta_t(\omega)}dP(\omega) = \prod_{t\in A} e^{-\delta^d|\lambda|^p} = e^{-(|A|\delta^d)|\lambda|^p}.$$

This shows that the standard part ${}^\circ\eta_A(\omega)$ exists P_L-a.s. for any $A \in \mathcal{A}_L(\mathcal{L})$ with ${}^\circ(|A|\delta^d) < \infty$, and the standard family $\{{}^\circ\eta_A(\omega) : A \in \mathcal{A}_L(\mathcal{L}), {}^\circ(|A|\delta^d) < \infty\}$ fulfils the assertion. ∎

Now let us define

$$\mathsf{F}(f,\omega) := \sum_{t\in\mathcal{L}} \delta^d f(t)\eta_t(\omega), \quad f \in {}^*\mathcal{D} \cap {}^*\mathbb{R}^{\mathcal{L}}$$

and

$$\mathsf{X}_t(\omega) := \sum_{t'\in\mathcal{L}} \delta^d G(t-t')\eta_{t'}(\omega), \quad t \in \mathcal{L}$$

$$\mathsf{X}(f,\omega) := \sum_{t\in\mathcal{L}} \delta^d f(t)\mathsf{X}_t(\omega) = < f, \mathsf{X}_\cdot(\omega) >, \quad f \in {}^*\mathcal{D} \cap {}^*\mathbb{R}^{\mathcal{L}}$$

where

$$G(x) := (2\pi)^{-d} \int_{\mathbb{R}^d} \frac{\cos(kx)dk}{(|k|^2+m^2)^\alpha}$$

is the Green's function of $(-\Delta + m^2)^\alpha$. We have

Proposition 2.3 *$\mathsf{F}(\cdot,\omega)$ and $\mathsf{X}(\cdot,\omega)$ are S-continuous on ${}^*\mathcal{D}$ for P_L-almost all $\omega \in \Omega$. Moreover, let $\tilde{f}$ be a hyperfinite lifting of $f \in \mathcal{D}$, then*

$${}^\circ\mathsf{F}(\tilde{f},\omega) = \int_{\mathcal{L}} f(t)d\,{}^\circ\eta_t(\omega), \quad {}^\circ\mathsf{X}(\tilde{f},\omega) = \int_{\mathcal{L}} (-\Delta+m^2)^{-\alpha} f(t)d\,{}^\circ\eta_t(\omega)$$

for P_L-a.a. $\omega \in \Omega$. Hence, $\{{}^\circ\mathsf{F}(\tilde{f},\omega) : f \in \mathcal{D}), \omega \in \Omega\}$ and $\{{}^\circ\mathsf{X}(\tilde{f},\omega) : f \in \mathcal{D}), \omega \in \Omega\}$ are the stable random field F and the stable type Euclidean random field X on the Loeb space $(\Omega, \mathcal{A}_L(\Omega), P_L)$, respectively.

Proof. Concerning S-continuity, by linearity, it suffices to verify that for $f \in \mathcal{D}$, $\mathsf{F}(\tilde{f},\omega)$ and $\mathsf{X}(\tilde{f},\omega)$ are finite for P_L-almost all $\omega \in \Omega$, where $\tilde{f} \in Ns({}^*\mathbb{R}^{\mathcal{L}})$ is any (hyperfinite) lifting of f. Namely we need to show the nearstandardness of F and X. Remarking that in general the (absolute) moments of $\eta_t, t \in \mathcal{L}$ may not exist (as for example in case of $p=1$, each η_t is Cauchy distributed which has no moment), one can not apply the moment estimate techniques employed e.g. in [3] for dealing with the Gaussian case ($p=2$). But we can use Fourier transform to carry out the internal characteristic functions of F and X. In fact, for any $\lambda \in \mathbb{R}$, by independence of $\eta_t, t \in \mathcal{L}$,

$$\int_\Omega e^{i\lambda\mathsf{F}(\tilde{f},\omega)}dP(\omega) = e^{-|\lambda|^p\sum_{t\in\mathcal{L}}\delta^d|\tilde{f}(t)|^p}$$

$$\int_\Omega e^{i\lambda \mathsf{X}(\tilde f,\omega)}dP(\omega) = e^{-|\lambda|^p \sum_{t\in\mathcal{L}} \delta^d |(-\Delta_\delta+m^2)^{-\alpha}\tilde f(t)|^p}$$

where Δ_δ stands for the hyperfinitely discrete Laplace operator on the lattice $\mathcal{L}$ and $(-\Delta_\delta + m^2)^{-\alpha}\tilde f(t) = \sum_{t'\in\mathcal{L}} \delta^d G(t-t')\tilde f(t'), t \in \mathcal{L}$. Since $f \in \mathcal{D}$,we have

$$^\circ(\sum_{t\in\mathcal{L}} \delta^d |\tilde f(t)|^p) < \infty, \quad ^\circ\sum_{t\in\mathcal{L}} \delta^d |(-\Delta_\delta + m^2)^{-\alpha}\tilde f(t)|^p < \infty$$

which imply that the standard part $^\circ\mathsf{F}(\tilde f,\omega)$ and $^\circ\mathsf{X}(\tilde f,\omega)$ exist (and are finite) P_L-a.s for any hyperfinite lifting $\tilde f$ of $f \in \mathcal{D}$. The rest statements of the proposition are then straightforward. ∎

Now for $A \in \mathcal{A}(^*I\!R^{\mathcal{L}})$, set

$$\Gamma(A) := P\{\omega \in \Omega : \delta^d\eta_\cdot(\omega) \in A\}, \quad \Lambda(A) := P\{\omega \in \Omega : \mathsf{X}_\cdot(\omega) \in A\}.$$

By independence of $\eta_t, t \in \mathcal{L}$, we have

$$\begin{aligned}&\Gamma(A)\\ = &\prod_{t\in\mathcal{L}} P\{\omega : \delta^d\eta_t(\omega) \in Proj_t A)\}\\ = &(2\pi\delta^d)^{-|\mathcal{L}|}\int_A\left[\int_{^*I\!R^{\mathcal{L}}\cap{}^*\mathcal{D}} e^{-i\sum_{t\in\mathcal{L}}\delta^d f_t q_t - \sum_{t\in\mathcal{L}}\delta^d|f_t|^p}\prod_{t\in\mathcal{L}} df_t\right]\prod_{t\in\mathcal{L}} dq_t\end{aligned}$$

and

$$\begin{aligned}&\Lambda(A)\\ = &P\{\omega : \sum_{t\in\mathcal{L}}\delta^d G(\cdot - t)\eta_t(\omega) \in A\}\\ = &\prod_{t\in\mathcal{L}} P\{\omega : \delta^d\eta_t(\omega) \in Proj_t((G)^{-1}A)\}\\ = &\kappa\int_A\left[\int_{^*I\!R^{\mathcal{L}}\cap{}^*\mathcal{D}} e^{-i\sum_{t\in\mathcal{L}}\delta^d\{[(-\Delta_\delta+m^2)^{-\alpha}f_t]q_t - |f_t|^p\}}\prod_{t\in\mathcal{L}} df_t\right]\prod_{t\in\mathcal{L}} dq_t\end{aligned}$$

where $\kappa := (2\pi\delta^d)^{-|\mathcal{L}|}det((G)^{-1}), (G) := (G(t-t'))_{t,t'\in\mathcal{L}}$ is an $|\mathcal{L}| \times |\mathcal{L}|$-matrix induced by G, and $(G)^{-1}$ the inverse matrix of (G).

On the other hand, we remark that the space of all paths $\mathsf{F}(\cdot,\omega)$ and $\mathsf{X}(\cdot,\omega)$ determined respectively by

$$\mathsf{F}(f,\omega) = \sum_{t\in\mathcal{L}}\delta^d f(t)\eta_t(\omega), \quad \mathsf{X}(f,\omega) = \sum_{t\in\mathcal{L}}\delta^d f(t)\mathsf{X}_t(\omega), \quad f \in {}^*I\!R^{\mathcal{L}} \cap {}^*\mathcal{D}$$

is the collection of all internal S-continuous mappings from ${}^*\mathbb{R}^{\mathcal{L}} \cap {}^*\mathcal{D}$ to ${}^*\mathbb{R}$, i.e., $\mathsf{F}(\cdot,\omega), \mathsf{X}(\cdot,\omega) \in Ns({}^*\mathbb{R}^{\mathcal{L}})$. For $A \in \mathcal{A}(Ns({}^*\mathbb{R}^{\mathcal{L}}))$, we define

$$\hat{A} := \left\{ q \in {}^*\mathbb{R}^{\mathcal{L}} : \left(f \in {}^*\mathbb{R}^{\mathcal{L}} \cap {}^*\mathcal{D} \to \sum_{t \in \mathcal{L}} \delta^d f(t) q(t) \in {}^*\mathbb{R} \right) \in A \right\}.$$

The proof of the following result is similar to that of Lemma 3.6 of [3]

Proposition 2.4 *The mapping* $\hat{}: A \to \hat{A}$ *is injective. Moreover,* $\hat{A} \in \mathcal{A}({}^*\mathbb{R}^{\mathcal{L}})$ *if and only if* $A \in \mathcal{A}(Ns({}^*\mathbb{R}^{\mathcal{L}}))$.

The following result is the alternative formulation of our Proposition 2.3 in terms of measures which gives a hyperfinite flat integral representation of μ and μ_X

Theorem 2.5 *We have*

$$\mu(D) = \Gamma_L((\widehat{st^{-1}(D)})), \quad \mu_X(D) = \Lambda_L((\widehat{st^{-1}(D)}))$$

where $D \in \mathcal{B}$ *with* $st^{-1}(D) := \{q \in Ns({}^*\mathbb{R}^{\mathcal{L}}) : st(q) \in D\}$.

The flat integral construction of μ and μ_X permits a study of the asymptotic behaviours, e.g. large deviations and scaling limits, of certain Euclidean random fields, with applications to critical phenomena in statistical physics (for the case $d = 1$ and μ a perturbed Gaussian measure, see e.g. [20]; see also for this area of investigations e.g. [12]). We intend to investigate this topic in our forthcoming research.

3. A REMARK ON A FLAT INTEGRAL REPRESENTATION OF GAUSSIAN MEASURES ON l_p, $1 \leq p < \infty$

In this final section, we present an extension of Cutland's flat integral formula for (centered) Gaussian measures on l_2 (cf. [9]) to the case of Banach spaces $l_p, 1 \leq p < \infty$.

Let B be a Banach space with B' its (topological) dual space, and let $< \cdot, \cdot >$ denote the dual pairing between B' and B. It is known, e.g.

from Proposition IV.2.8 in [21], that the characteristic functional of a Gaussian (Radon) measure μ on B is of the form

$$C_\mu(f) := \int_B e^{i<f,x>} d\mu(x) = e^{i<f,m>-\frac{1}{2}<f,Rf>}, \quad f \in B' \qquad (1.3)$$

where $m \in B$ is the mean of μ and $R : B' \to B$ is the covariance operator which is symmetric and positive. Conversely, suppose μ is a Radon measure on B with the characteristic functional given by (1.3) with $m \in B$ and $R : B' \to B$ being symmetric and positive, then μ is a Gaussian measure on B with mean m and covariance operator R.

It is worthwhile to notice that for an infinite-dimensional Banach space B, not every functional of the form (1.3) is the characteristic functional of a Gaussian measure on B. A well-known example is that the functional

$$C(h) = \exp\{-\frac{1}{2}(h,h)_H\}, \qquad h \in H$$

is a functional of the form given by (1.3), where H is an infinite dimensional Hilbert space with inner product $(\cdot,\cdot)_H$, but it is not the characteristic functional of a Radon measure on H. Therefore, finding necessary and sufficient conditions on $m \in B$ and on $R : B' \to B$ such that a functional of the form (1.3) is the characteristic functional of a Gaussian measure on B is an important and interesting problem. Clearly, $m \in B$ in (1.3) can be arbitrary, so the problem of description of Gaussian measures is equivalently reduced to the problem of decribing the class of symmetric, positive operators $R : B' \to B$ for which the functional

$$C(f) = \exp\{-\frac{1}{2} < f, Rf >\}, \quad f \in B'$$

is a characteristic functional. Thus, without loss of generality, from now on we only deal with centered Gaussian measures (i.e., with mean $m = 0$) and by "a Gaussian measure" we shall always mean a centered Gaussian measure.

In the case of a (general) Hilbert space H, it is known, see e.g. [21], that a necessary and sufficient condition on $R : H \to H$ such that $C_\mu(f) = e^{-\frac{1}{2}(f,Rf)_H}$ is a characteristic functional of a Gaussian (Radon) measure μ on H is that R is a symmetric, positive nuclear operator. Furthermore, if H is separable, then it is well-known (see, e.g. [16]), that a necessary and sufficient condition is that $R : H \to H$ is a Hilbert-Schmidt operator. In the case of general Banach spaces, however, the above problem of description of Gaussian measures in terms of characteristic functionals is still open (this is related to some problems of the geometric theory of Banach spaces).

In the case of Banach sequence spaces $l_p, 1 \leq p < \infty$, where

$$l_p = \{x = (x_1, x_2, \cdots) \in \mathbb{R}^{\mathbb{N}} : \|x\|_p := (\sum_{n \in \mathbb{N}} \|x_n\|^p)^{\frac{1}{p}} < \infty\},\ 1 \leq p < \infty,$$

the following result, due to Vakhania (see Theorem V.5.6 of [21]), provides a description of Gaussian measures on l_p. Let $\{e_n\}_{n \in \mathbb{N}}$ denote the standard basis of the space $\mathbb{R}^{\mathbb{N}}$ and let $p \in [1, \infty)$. Let us identify, as usual, the dual space l'_p with $l_{p'}$, where $p' := \frac{p}{p-1}$ with the convention that $\frac{1}{0} = \infty$ (In case $p = 1$, the dual space is l_∞ defined by $l_\infty = \{x = (x_1, x_2, \cdots) \in \mathbb{R}^{\mathbb{N}} : \|x\|_\infty := \sup_{n \in \mathbb{N}} |x_n| < \infty\}$).

Theorem 3.1 *([21]) An operator $R : l_{p'} \to l_p$ is the covariance operator of a Gaussian measure μ on l_p iff it is symmetric, positive and*

$$\sum_{n \in \mathbb{N}} < e_n, Re_n >^{\frac{p}{2}} < \infty.$$

In fact, μ is uniquely determined by an infinite, symmetric, nonnegative definite matrix $(< e_k, Re_n >)_{k,n \in \mathbb{N}}$. We call it the *covariance matrix* of μ.

Remark 3.2 *There are some important classical Banach spaces which are related to the sequence spaces $l_p, 1 \leq p < \infty$. For $p = 2$ we have the standard example of a separable, real Hilbert space (every separable, real Hilbert space being isomorphic to l_2). For each $p \in [1, \infty)$, the spaces $L_p([0,1])$ as well as $C[0,1]$, have a Schauder basis and can be represented as l_p. Another example of a space isomorphic with l_p is the function space H_p, $p \in [1, \infty)$, of all analytic functions on the unit disc of the complex plane endowed with the norm defined by $\|f\|_p := (\int |f(x+iy)|^p dxdy)^{\frac{1}{p}}$. Also, $l_p, 1 \leq p < \infty$, are isomorphic to some Orlicz sequence spaces and Lorentz sequence spaces, see e.g. [17]. The study of Gaussian measures on l_p obviously and automatically gives information also on Gaussian measures on Banach spaces isomorphic with l_p $(1 \leq p < \infty)$.*

Let $p \in [1, \infty)$ be arbitrarily fixed and μ be a Gaussian measure on l_p whose covariance matrix $(< e_m, Re_n >)_{m,n \in \mathbb{N}}$ is diagonal with $\sigma_n^2 :=< e_n, Re_n > \ > 0$ for all $n \in \mathbb{N}$. For each $n \in \mathbb{N}$, let μ_n be the Gaussian measure on $\mathbb{R}$ with variance σ_n^2, i.e., the n-th coordinate x_n is distributed as $\mathcal{N}(0, \sigma_n^2)$ (with $\mathcal{N}(0, c)$ standing for a Gaussian random variable with mean zero and covariance c) under μ. Thus, μ_n has a density(with respect to the Lebesgue measure dt on $\mathbb{R}$) $g_n(t) = (2\pi\sigma_n^2)^{-\frac{1}{2}} \exp\{-\frac{t^2}{2\sigma_n^2}\}$ so that $d\mu_n(t) = g_n(t)dt$. Clearly, the infinite product measure $\prod_{n\in\mathbb{N}} \mu_n$ is a Gaussian measure on $\mathbb{R}^{\mathbb{N}}$(in the sense that every coordinate projection image of it is Gaussian, i.e., $\prod_{n\in\mathbb{N}} \mu_n$ is a Gaussian measure with respect to the vector space determined by all the coordinate projections. See, e.g. [21] for the general definition of Gaussian Radon measure with respect to a topological vector space). Furthermore, we have the following result, which indicates that l_p is a support of $\prod_{n\in\mathbb{N}} \mu_n$ and the coordinate components of $x = (x_n)_{n\in\mathbb{N}}$ are mutually independent $\mathcal{N}(0, \sigma_n^2)$-random variables under μ.

Proposition 3.3

$$\prod_{n\in\mathbb{N}} \mu_n(l_p) = 1; \quad \mu = \prod_{n\in\mathbb{N}} \mu_n.$$

Proof. To show the first equality, we note that $\forall m \in \mathbb{N}$,

$$\begin{aligned}
\mathbf{E}_{\prod_{n\in\mathbb{N}} \mu_n}\left(\sum_{n=1}^{m} |x_n|^p\right) &= (2\pi\sigma_n^2)^{-\frac{1}{2}} \int_{\mathbb{R}} |x_n|^p \exp\{-\frac{x_n^2}{2\sigma_n^2}\} dx_n \\
&= \left[(2\pi)^{-\frac{1}{2}} \int_{\mathbb{R}} |t|^p e^{-\frac{t^2}{2}} dt\right] \sum_{n=1}^{m} (\sigma_n^2)^{\frac{p}{2}} \\
&= c_p \sum_{n=1}^{m} \sigma_n^p
\end{aligned}$$

where c_p is the p-th order absolute moment of the normal distribution $\mathcal{N}(0,1)$ on $\mathbb{R}$. Thus

$$\mathbf{E}_{\prod_{n\in\mathbb{N}} \mu_n}\left(\sum_{n\in\mathbb{N}} |x_n|^p\right) = \lim_{m\to\infty} \mathbf{E}_{\prod_{n\in\mathbb{N}} \mu_n}\left(\sum_{n=1}^{m} |x_n|^p\right) = c_p \sum_{n\in\mathbb{N}} \sigma_n^p < \infty.$$

Therefore, $\prod_{n\in\mathbb{N}} \mu_n(l_p) = 1$.

As for the proof of the second equality, we note that $\prod_{n\in\mathbb{N}} \mu_n$ is also a Gaussian measure on l_p, thus it suffices to show that the associated covariance matrices of $\prod_{n\in\mathbb{N}} \mu_n$ and μ are the same. The covariance matrix of $\prod_{n\in\mathbb{N}} \mu_n$ is calculated as follows

$$\mathbf{E}_{\prod_{n\in\mathbb{N}} \mu_n}(x_j^2) = < e_j, Re_j >, \quad j \in \mathbb{N}$$

$$\mathbf{E}_{\prod_{n\in\mathbb{N}}\mu_n}(x_j x_k) = 0, \quad j \neq k, j, k \in \mathbb{N}$$

and thus the covariance matrix of $\prod_{n\in\mathbb{N}}\mu_n$ is $(\delta_{jk} < e_j, Re_k >)_{j,k\in\mathbb{N}}$, which is the same as that of μ. Therefore we obtain $\mu = \prod_{n\in\mathbb{N}}\mu_n$. ∎

Now we can define the action functional of μ on l_p by

$$I(x) := \frac{1}{2}\sum_{n\in\mathbb{N}} \frac{x_n^2}{\sigma_n^2} \in \mathbb{R}\cup\{\infty\}.$$

Consequently, we are able to define a corresponding Hilbert space, i.e., the Cameron-Martin subspace of l_p, associated with μ by

$$H := \{x \in \mathbb{R}^{\mathbb{N}} : I(x) < \infty\}$$

with the inner product $(x,y)_H := \sum_{n\in\mathbb{N}} \frac{x_n y_n}{\sigma_n^2}$.

Proposition 3.4 *(i) $H \subset l_p$; (ii) $\|\cdot\|_p$ is a measurable norm on H in the sense of Gross (see, e.g., [16]).*

Proof. (i) First we remark that if $\forall n \in \mathbb{N}, a_n > 0$, and $\sum_{n\in\mathbb{N}} a_n < \infty$, then $\forall\alpha > 1, \sum_{n\in\mathbb{N}}(a_n)^\alpha \leq (\sum_{n\in\mathbb{N}} a_n)^\alpha < \infty$. Thus $\forall x \in H$, we have $\sum_{n\in\mathbb{N}} \frac{x_n^2}{\sigma_n^2} < \infty$. Therefore

$$\sum_{n\in\mathbb{N}} (\frac{x_n^2}{\sigma_n^2})^{\frac{p+1}{2}} \leq (\sum_{n\in\mathbb{N}} \frac{x_n^2}{\sigma_n^2})^{\frac{p+1}{2}} < \infty$$

since $\forall p > 1, \frac{p+1}{2} \geq 1$. On the other hand, we have

$$\sum_{n\in\mathbb{N}} (\sigma_n^2)^{\frac{p(p+1)}{2}} \leq (\sum_{n\in\mathbb{N}} (\sigma_n^2)^{\frac{p}{2}})^{p+1} < \infty.$$

Now from the Hölder inequality $\|fg\|_p \leq \|f\|_q\|g\|_r$ for $1 \leq p, q, r \leq \infty$ and $\frac{1}{p} = \frac{1}{q} + \frac{1}{r}$, by taking $q = p+1$ and $r = p(p+1)$, we obtain that

$$\|x\|_p = (\sum_{n\in\mathbb{N}} |x_n|^p)^{\frac{1}{p}} = \left[\sum_{n\in\mathbb{N}} \left(\frac{x_n^2}{\sigma_n^2}\right)^{\frac{p}{2}} \sigma_n^p\right] \leq \|x\|_H \left[\sum_{n\in\mathbb{N}} \sigma_n^p\right]^{\frac{1}{p}} < \infty.$$

Hence $H \subset l_p$.

(ii) By (i), we get

$$\|x\|_p \leq \|x\|_H \left[\sum_{n\in\mathbb{N}} \sigma_n^p\right]^{\frac{1}{p}}, \qquad x \in H$$

which shows that $\|\cdot\|_p$ is weaker than $\|\cdot\|_H$ on H. On the other hand, it is clear that the restriction of μ to H, denoted by $\mu|_H$, is the *standard Gauss measure* (see e.g. [16] for the definition) on H, namely,

$$\mu|_H(x \in H : \mathcal{P}x \in F) = (2\pi)^{-\frac{(\dim \mathcal{P}H)}{2}} \int_F \exp\{-\frac{1}{2}\|x\|_H^2\}dx$$

where $\mathcal{P}$ is an arbitrary finite dimensional orthogonal projection of H, F is a Borel subset of $\mathcal{P}H$ and dx is the $\dim(\mathcal{P}H)$-dimensional Lebesgue measure on $\mathcal{P}H$. Moreover, μ is the (unique) σ-additive extension of $\mu|_H$ to l_p. Thus, by the known Dudley-Feldman-LeCam's theorem (see e.g. p.91 of [16]), $\|\cdot\|_p$ is a measurable norm on H. ∎

Following the lines of [9], the space l_p is naturally represented in ${}^*\mathbb{R}^N$ for any hyperfinite $N \in {}^*\mathbb{N} \setminus \mathbb{N}$ as follows. An element $X = (X_1, X_2, \cdots, X_N) \in {}^*\mathbb{R}^N$ is called p-nearstandard if

$$\sum_{n\in\mathbb{N}} {}^\circ|X_n|^p \approx \sum_{n=1}^{N} |X_n|^p < \infty.$$

(Where $a \approx b$ means that the difference $a - b$ is infinitesimal.) We denote by $ns_p({}^*\mathbb{R}^N)$ the collection of all p-nearstandard elements of ${}^*\mathbb{R}^N$. Furthermore, we define the standard part of $X \in ns_p({}^*\mathbb{R}^N)$ by

$${}^\circ X := st_p(X) = ({}^\circ X_n)_{n\in\mathbb{N}} \in l_p.$$

Remark 3.5 *The relationship between p-nearstandard on* ${}^*\mathbb{R}^N$ *and the usual* $\|\cdot\|_p$*-nearstandard (namely, nearstandard in the* $\|\cdot\|_p$*-topology, see e.g. [1]) on* l_p *is as follows.* $X \in {}^*\mathbb{R}^N$ *is p-nearstandard in the above sense if* $X|_N := (\hat{X}_n)_{n\in{}^*\mathbb{N}} \subset {}^*l_p$ *is* $\|\cdot\|_p$*-nearstandard, where*

$$\hat{X}_n := \begin{cases} X_n, & 1 \le n \le N; \\ 0, & n > N. \end{cases}$$

Also, an equivalent characterization of $X \in ns_p({}^*\mathbb{R}^N)$ *is* $\sum_{n=1}^N |X_n|^p < \infty$ *and* $\sum_{n=M}^N |X_n|^p \approx 0$ *for all hyperfinite* $M \in {}^*\mathbb{N} \setminus \mathbb{N}$ *with* $M < N$.

Definition 3.6 *We define the nonstandard counterpart of the action functional I on* ${}^*\mathbb{R}^N$ *by*

$$J(X) := \frac{1}{2}\sum_{n=1}^{N} \frac{X_n^2}{\sigma_n^2}$$

where $\{e_n\}_{1\leq n\leq N}$ is a hyperfinite initial segment of the *-version of the standard basis $\{e_n\}_{n\in\mathbb{N}}$ given above.

We have the following relationship between I and its nonstandard counterpart J:

Proposition 3.7 *(i) If $J(X)$ is finite, then $X \in ns_p({}^*\mathbb{R}^N)$; (ii) Moreover, $I({}^\circ X) \leq {}^\circ J(X)$, $X \in ns_p({}^*\mathbb{R}^N)$; (iii) if $X = {}^*x|_N$ for $x \in l_p$, then $J(X) \approx I(x)$.*

Proof. (i) $\forall m \in {}^*\mathbb{N} \cap [1, N]$, we have

$$\begin{aligned}(\sum_{n=m}^{N} |X_n|^p)^{\frac{1}{p}} &= \left[\sum_{n=m}^{N} \left(\frac{X_n^2}{\sigma_n^2}\right)^{\frac{p}{2}} \sigma_n^p\right]^{\frac{1}{p}} \\ &\leq \left(\sum_{n=1}^{N} \frac{X_n^2}{\sigma_n^2}\right)^{\frac{1}{2}} \left(\sum_{n=m}^{N} \sigma_n^p\right)^{\frac{1}{p}} = \sqrt{2}J(X)^{\frac{1}{2}} \left(\sum_{n=m}^{N} \sigma_n^p\right)^{\frac{1}{p}}.\end{aligned}$$

Let $m = 1$, then $\sum_{n=1}^{N} |X_n|^p$ is finite. On the other hand, let $m \in {}^*\mathbb{N} \setminus \mathbb{N}$ and $m \leq N$, then $\sum_{n=m}^{N} |X_n|^p \approx 0$ since $J(X)$ is finite and $\sum_{n=m}^{N} \sigma_n^p \approx 0$. Hence, $X \in ns_p({}^*\mathbb{R}^N)$.

(ii) and (iii) are proved by the same argument as in the proof of Lemma 3.3 of [9]. ∎

Let Γ be the internal probability measure on ${}^*\mathbb{R}^N$ given by $\Gamma = \prod_{n=1}^{N} {}^*\mu_n$. Namely, the covariance of Γ is the hyperfinite diagonal matrix $(\sigma_n^2)_{1\leq n\leq N}$. Then we have the following hyperfinite flat integral formula

$$\Gamma(A) = \kappa \int_A \exp\{-J(X)\} dX = \kappa \int_A \exp\{-\frac{1}{2}\sum_{n=1}^{N} \frac{X_n^2}{\sigma_n^2}\} \prod_{n=1}^{N} dX_n$$

for all *Borel subsets $A \in {}^*\mathbb{R}^N$, where $dX = \prod_{n=1}^{N} dX_n$ is the *Lebesgue measure on ${}^*\mathbb{R}^N$ and $\kappa := \prod_{n=1}^{N} (2\pi\sigma_n^2)^{-\frac{1}{2}}$ is the normalization constant. We have the following result

Proposition 3.8 *X is p-nearstandard for Γ_L-a.a. $X \in {}^*\mathbb{R}^N$.*

Proof. We remark that ${}^\circ X_n \sim \mathcal{N}(0, \sigma_n^2)$ for each $n \in \mathbb{N}$, thus

$$\mathbf{E}_{\Gamma_L}(\sum_{n\in\mathbb{N}} {}^\circ|X_n|^p) = \lim_{m\to\infty} \mathbf{E}_{\Gamma_L}(\sum_{n=1}^{m} {}^\circ|X_n|^p) \leq c_p \sum_{n\in\mathbb{N}} \sigma_n^p < \infty$$

and

$$\mathbf{E}\left[{}^\circ(\sum_{n=1}^{N}|X_n|^p) - \sum_{n\in\mathbb{N}}{}^\circ|X_n|^p\right] = \lim_{m\to\infty}\mathbf{E}_{\Gamma_L}{}^\circ(\sum_{n=m}^{N}|X_n|^p) = 0.$$

Hence, for Γ_L-a.a. $X\in{}^*\mathbb{R}^N$, $\sum_{n\in\mathbb{N}}{}^\circ|X_n|^p = {}^\circ(\sum_{n=1}^{N}|X_n|^p) < \infty$. ■

We have thus obtained the following main result of this section, which gives a flat integral formula for Gaussian measures on l_p.

Theorem 3.9 *Suppose μ is a Gaussian measure on l_p with covariance operator $R : l_{p'} \to l_p$ which induces an infinite, diagonal matrix $(\sigma_n^2)_{n\in\mathbb{N}}$ with $\sum_{n\in\mathbb{N}}\sigma_n^p < \infty$. Then*

$$\mu(A) = \Gamma_L(st_p^{-1}(A)), \quad A\subset l_p$$

where $st_p^{-1}(A) = \{X\in{}^*\mathbb{R}^N : st_p(X)\in A\}$.

The hyperfinite flat integral formula for μ can be used to prove a Cameron-Martin-type formula and a Schilder's large deviation principle for μ on l_p. Recall that we had defined the Cameron-Martin subspace of l_p associated with μ as the Hilbert space $H = \{x\in\mathbb{R}^N : I(x) < \infty\}$. A direct calculation then gives

Proposition 3.10 *Let $z\in H$ and define $\mu^z(A) := \mu(A-z)$ for $A\subset l_p$. Then $\mu^z \ll \mu$ and*

$$\frac{d\mu^z}{d\mu}(x) = \exp\{\sum_{n\in\mathbb{N}}\frac{x_n z_n}{\sigma_n^2} - I(z)\}, \qquad \mu - a.s.\ \ x\in l_p.$$

Let $\varepsilon > 0$, we define $\mu_\varepsilon(A) := \mu(\varepsilon^{-1}A)$ for $A\subset l_p$. By using Hölder inequality and mimicking the proofs of Theorems 4.1 and 4.2 in [9], one can show the following Schilder's large deviation principle for $\{\mu_\varepsilon : \varepsilon > 0\}$.

Theorem 3.11 *(i)For every open set $G\subset l_p$,*

$$\underline{\lim}_{\varepsilon\to 0}\varepsilon^2\log\mu_\varepsilon(G) \geq -\inf I(G);$$

(ii) for every closed set $F\subset l_p$,

$$\overline{\lim}_{\varepsilon\to 0}\varepsilon^2\log\mu_\varepsilon(F) \leq -\inf I(F).$$

Acknowledgments

Stimulating discussions with Nigel J. Cutland and Horst Osswald and the financial support from DFG through the project "Stochastische Analysis und Systeme mit unendlich vielen Freiheitsgraden" are gratefully acknowledged. The organizers and Volkswagen-Stiftung should be thanked for the invitation resp. the financial support to the Venice Symposium "Reuniting the Antipodes: Constructive and Nonstandard Views of the Continuum".

References

[1] Albeverio, S., Fenstad, J.E., Høegh–Krohn R. and Lindstrøm T. (1986). *Nonstandard Methods in Stochastic Analysis and Mathematical Physics.* New York: Academic Press.

[2] Albeverio, S. and Wu, J.-L. (1995). Euclidean random fields obtained by convolution from generalized white noise. J. Math. Phys. **36**, 5217–5245.

[3] Albeverio, S. and Wu, J.-L. (1996). A mathematical flat integral realization and a large deviation result for the free Euclidean field. Acta Appl. Math. **45**, 317–348.

[4] Albeverio, S. and Wu, J.-L. (2000). On hyperfinite integral representation of Euclidean random field measures. Bonn SFB 256-preprint.

[5] Cutland, N.J. (1983). Nonstandard measure theory and its applications. Bull. London Math. Soc. **15**, 529–589.

[6] Cutland, N.J. (1987). Infinitesimals in action. J. London Math. Soc. **35**, 202–216.

[7] Cutland, N.J. (1989). The Brownian bridge as a flat integral. Math. Proc. Cambridge Phil. Soc. **106**, 343–354.

[8] Cutland, N.J. (1990). An action functional for Lévy Brownian motion. Acta Appl. Math. **18**, 487–495.

[9] Cutland, N.J. (1991). On large deviations in Hilbert space. Proc. Edinburgh Math. Soc. **34**, 487–495.

[10] Cutland, N.J. (1990). Nonstandard representation of Gaussian measures. *White Noise – Mathematics and Application* (eds. Hida, T., Kuo, H.-H., Potthoff, J. and Streit, L.). Singapore: World Scientific, pp73–92.

[11] Deuschel, J.-D. and Stroock, D.W. (1989). *Large Deviations.* San Diego: Academic Press.

[12] Fritz, J., Lebowitz, J.L. and Szász, D. (Editors) (1981). *Random Fields. Rigorous Results in Statistical Mechanics and Quantum Field Theory* (Vol. I, II). Colloquia Mathematica Societatis János Bolyai **27**, , Amsterdam: North-Holland Publishing Co..

[13] Gelfand, I.M. and Vilenkin, N.Ya. (1964). *Generalized Functions, IV. Some Applications of Harmonic Analysis.* New York: Academic Press.

[14] Keßler, C. (1984). Nonstandard methods in random fields. Bochum: Dissertation.

[15] Keßler, C. (1988). On hyperfinite representation of distributions. Bull. London Math. Soc. **20**, 139–144.

[16] Kuo, H.-H. (1975). *Gaussian Measures in Banach Spaces.* Lect. Notes Math. **463**. Berlin: Springer-Verlag.

[17] Lindenstrauss, J. and Tzafriri, L. (1977). *Classical Banach Spaces. I. Sequence Spaces.* Berlin: Springer-Verlag.

[18] Lindstrøm, T. (1988). An invitation to nonstandard analysis. *Nonstandard Analysis and Its Applications* (ed. by Cutland, N.J.). Cambridge: Cambridge University Press, pp1–105.

[19] Osswald, H. (2000). Infinitesimal in abstract Wiener spaces. To appear in *Stochastic Processes, Physics and Geometry: New Interplays. A Volume in Honor of Sergio Albeverio.* Proceedings of the Leipzig International Conference on Infinite Dimensional (Stochastic) Analysis and Quantum Physics, the Canadian Mathematical Society (CMS) Conference Proceeings Series.

[20] Rosen, J. and Simon, B. (1975). Global support properties of stationary ergodic processes. Duke Math. J. **42**, 51–55.

[21] Vakhania, N.N., Tarieladze, V.I. and Chobanyan, S.A. (1987). *Probability Distributions on Banach Spaces.* Dordrecht: D. Reidel Publishing Company.

THE CONTINUUM IN SMOOTH INFINITESIMAL ANALYSIS

John L. Bell
Department of Philosophy, University of Western Ontario
London, Ontario N6A 3K7, Canada
jbell@julian.uwo.ca

Keywords: continuum, smooth infinitesimal analsis, constructive analysis

Abstract: In this paper an investigation is made of the properties of the continuum in smooth infinitesimal analysis: it is shown that it differs in certain important respects from its counterpart in constructive analysis.

As presented in [1], *smooth infinitesimal analysis*, SIA, is a theory formulated within higher-order intuitionistic logic and based on the following axioms:

Axioms for the continuum, or smooth real line R. These are the usual axioms for a(n) (intuitionistic) field expressed in terms of two operations $+$ and $\cdot$, and two distinguished elements $0, 1$.

Axioms for the strict order relation $<$ on R. These are:

1 $a < b$ and $b < c$ implies $a < c$.

2 $\neg(a < a)$.

3 $a < b$ implies $a + c < b + c$ for any c.

4 $a < b$ and $0 < c$ implies $a \cdot c < b \cdot c$.

5 either $0 < a$ or $a < 1$.

6 $a \neq b$[1] implies $a < b$ or $b < a$.

[1]Here $a \neq b$ stands for $\neg a = b$. It should be pointed out that axiom 6 is omitted in some presentations of SIA, e.g. those in [3] and [4].

P. Schuster, U. Berger and H. Osswald (eds.),
Reuniting the Antipodes – Constructive and Nonstandard Views of the Continuum, 19–24.

The relation $\leq$ on $\mathbf{R}$ is defined by $a \leq b \Leftrightarrow \neg b < a$. The open interval (a, b) and closed interval $[a, b]$ are defined as usual, viz. $(a, b) = \{ x : a < x < b \}$ and $[a, b] = \{ x : a \leq x \leq b \}$; similarly for half-open, half-closed, and unbounded intervals.

Write Δ for the subset $\{ x : x^2 = 0 \}$ of $\mathbf{R}$; we use the letter ε as a variable ranging over Δ. Elements of Δ are called (nilsquare) *infinitesimals* or *microquantities*. Since, clearly, $0 \in \Delta$, Δ may be regarded as an *infinitesimal neighbourhood of* 0. Δ is subject to the

Microaffineness Principle. *For any map* $g : \Delta \to \mathbf{R}$ *there exist unique* $a, b \in \mathbf{R}$ *such that, for all* ε, *we have*

$$g(\varepsilon) = a + b \cdot \varepsilon.$$

Notice that then $a = g(0)$.

From these three axioms it follows that the continuum in SIA differs in certain key respects from its counterpart in *constructive analysis* CA, which is furnished with an elegant axiomatization in [2].

To begin with, the third basic property of the strict ordering relation $<$ in CA, given as axiom R2(3) on p. 102 of [2], and which may be written

(*) $$\neg(x < y \vee y < x) \to x = y$$

is incompatible with the axioms of SIA. For (*) implies

(**) $$\forall x \neg(x < 0 \vee 0 < x) \to x = 0.$$

But in SIA we have by Exercise 1.6 and Thm. 1.1 (i) of [1],

$$\forall x \in \Delta \neg(x < 0 \vee 0 < x) \wedge \Delta \neq \{0\},$$

which clearly contradicts (**).

Thus in CA the set Δ of infinitesimals would be degenerate (i.e., identical with $\{0\}$), while the nondegeneracy of Δ in SIA is one of its characteristic features.

Next, call a binary relation S on $\mathbf{R}$ *stable* if it satisfies

$$\forall x \forall y (\neg\neg x S y \to x S y).$$

In CA, the equality relation is stable, a fact which again follows from principle R2(3) referred to above. But in SIA it is not stable, for, as shown in Thm. 1.1 (ii) of [1], there we have $\forall x \in \Delta \neg\neg x = 0$. If $=$ were stable, it would follow that $\forall x \in \Delta\, x = 0$, in other words, that Δ is degenerate, which is not the case in SIA.

Axiom 6 of SIA, together with the transitivity and irreflexivity of $<$, implies that $<$ is stable. This may be seen as follows. Suppose $\neg\neg a < b$. Then certainly $a \neq b$, since $a = b \rightarrow \neg a < b$ by irreflexivity. Therefore $a < b$ or $b < a$. The second disjunct together with $\neg\neg a < b$ and transitivity gives $\neg\neg a < a$, which contradicts $\neg a < a$. Accordingly we are left with $a < b$. As can be deduced from assertion 8 on p. 103 of [2], the stability of $<$ implies *Markov's principle*, which is not affirmed in CA.[2]

A subset $A \subseteq \mathbf{R}$ is *indecomposable* if it admits only trivial partitionings, that is, if $A = U \cup V$ and $U \cap V = \emptyset$, then $U = \emptyset$ or $V = \emptyset$. Clearly A is indecomposable iff any map $f : A \rightarrow 2 = \{0, 1\}$ is constant.

In SIA one also assumes the

Constancy Principle. *If $A \subseteq \mathbf{R}$ is any closed interval on $\mathbf{R}$, or $\mathbf{R}$ itself, and $f : A \rightarrow \mathbf{R}$ satisfies $f(a + \varepsilon) = f(a)$ for all $a \in A$ and $\varepsilon \in \Delta$, then f is constant.*

As shown in Thm. 2.1 of [1], it follows in SIA from the Constancy Principle that $\mathbf{R}$ *itself and each of its closed intervals is indecomposable.* From this we can deduce that in SIA *all intervals in* $\mathbf{R}$ *are indecomposable.* To do this we employ the following

Lemma. Suppose that A is an inhabited subset of $\mathbf{R}$ satisfying

(*) for any $x, y \in A$ there is an indecomposable set B such that

$$\{x, y\} \subseteq B \subseteq A.$$

Then A is indecomposable.

Proof. Suppose A satisfies (*) and $A = U \cup V$ with $U \cap V = \emptyset$. Since A is inhabited, we may choose $a \in A$. Then $a \in U$ or $a \in V$. Suppose $a \in U$; then if $y \in V$ there is an indecomposable B for which $\{a, y\} \subseteq B \subseteq A = U \cup V$. It follows that $B = (B \cap U) \cup (B \cap V)$, whence $B \cap U = \emptyset$ or $B \cap V = \emptyset$. The former possibility is ruled out by the fact that $a \in B \cap U$, so $B \cap V = \emptyset$, contradicting $y \in B \cap V$. Therefore $y \in V$ is impossible; since this is the case for arbitrary y, we conclude

[2] In versions of SIA that omit axiom 6 neither the stability of $<$, nor Markov's principle, can be derived.

that $V = \emptyset$. Similarly, if $a \in V$, then $U = \emptyset$, so that A is indecomposable as claimed.

We use this lemma to show that the open interval $(0, 1)$ is indecomposable; similar arguments work for arbitrary intervals. In fact, if $\{x, y\} \subseteq (0, 1)$, it is easy to verify that

$$\{x, y\} \subseteq [\tfrac{xy}{x+y}, \tfrac{1-xy}{2-x-y}] \subseteq (0, 1).$$

Thus, in view of the indecomposability of closed intervals, $(0, 1)$ satisfies condition (*) of the lemma, and so is indecomposable.

Aside from certain infinitesimal subsets to be discussed below, in SIA indecomposable subsets of $\mathbf{R}$ correspond to connected subsets of $\mathbf{R}$ in classical analysis, that is, to intervals. In particular, any puncturing of $\mathbf{R}$ is *decomposable*, for it follows immediately from Axiom 6 that

$$\mathbf{R} - \{a\} = \{\, x : x > a \,\} \cup \{\, x : x < a \,\}.$$

Similarly, the set $\mathbf{R} - \mathbf{Q}$ of irrational numbers is decomposable as

$$\mathbf{R} - \mathbf{Q} = [\{\, x : x > 0 \,\} - \mathbf{Q}] \cup [\{\, x : x < 0 \,\} - \mathbf{Q}\}.$$

This is in sharp contrast with the situation in *intuitionistic analysis* IA, that is, CA augmented by Kripke's scheme, the continuity principle, and bar induction. For it is shown in [5] that in IA not only is any puncturing of $\mathbf{R}$ indecomposable, but that this is even the case for the set of irrational numbers (further indecomposabilty results for IA may be found in [6].) This would seem to indicate that in some sense the continuum in SIA is considerably less "syrupy"[3] than its counterpart in IA.

It can be shown that the various "infinitesimal" subsets of $\mathbf{R}$ introduced in [1] are indecomposable. For example, the indecomposability of Δ can be established as follows. Suppose $f : \Delta \to \{0, 1\}$. Then by Microaffineness there are unique $a, b \in \mathbf{R}$ such that $f(\varepsilon) = a + b \cdot \varepsilon$ for all ε. Now $a = f(0) = 0$ or 1; if $a = 0$, then $b \cdot \varepsilon = f(\varepsilon) = 0$ or 1, and clearly $b \cdot \varepsilon \neq 1$. So in this case $f(\varepsilon) = 0$ for all ε. If on the other hand $a = 1$, then $1 + b \cdot \varepsilon = f(\varepsilon) = 0$ or 1; but $1 + b \cdot \varepsilon = 0$ would imply $b \cdot \varepsilon = -1$ which is again impossible. So in this case $f(\varepsilon) = 1$ for all ε. Therefore f is constant and Δ indecomposable.

[3] It should be emphasized that this phenomenon is a consequence of axiom 6: it cannot necessarily be affirmed in versions of SIA not including this axiom.

In SIA *nilpotent infinitesimals* are defined to be the members of the sets

$$\Delta_k = \{ x \in \mathbf{R} : x^{k+1} = 0 \},$$

for $k = 1, 2, \ldots$, each of which may be considered an infinitesimal neighbourhood of 0. These are subject to the

Micropolynomiality Principle. *For any $k \geq 1$ and any $g : \Delta_k \to \mathbf{R}$, there exist unique $a, b_1, \ldots, b_k \in \mathbf{R}$ such that for all $\delta \in \Delta_k$ we have*

$$g(\delta) = a + b_1\delta + b_2\delta^2 + \cdots + b_k\delta^k.$$

Micropolynomiality implies that no Δ_k coincides with $\{0\}$.

An argument similar to that establishing the indecomposability of Δ does the same for each Δ_k. Thus let $f : \Delta_k \to \{0, 1\}$; Micropolynomiality implies the existence of $a, b_1, \ldots, b_k \in \mathbf{R}$ such that $f(\delta) = a + \zeta(\delta)$, where $\zeta(\delta) = b_1\delta + b_2\delta^2 + \cdots + b_k\delta^k$. Notice that $\zeta(\delta) \in \Delta_k$, that is, $\zeta(\delta)$ is nilpotent. Now $a = f(0) = 0$ or 1; if $a = 0$ then $\zeta(\delta) = f(\delta) = 0$ or 1, but since $\zeta(\delta)$ is nilpotent it cannot =1. Accordingly in this case $f(\delta) = 0$ for all $\delta \in \Delta_k$. If on the other hand $a = 1$, then $1 + \zeta(\delta) = f(\delta) = 0$ or 1, but $1 + \zeta(\delta) = 0$ would imply $\zeta(\delta) = -1$ which is again impossible. Accordingly f is constant and Δ_k indecomposable.

The union $\mathbf{D}$ of all the Δ_k is the *set of nilpotent infinitesimals*, another infinitesimal neighbourhood of 0. The indecomposability of $\mathbf{D}$ follows immediately by applying the Lemma above.

The next infinitesimal neighbourhood of 0 is the closed interval $[0, 0]$, which, as a closed interval, is indecomposable. It is easily shown that $[0, 0]$ includes $\mathbf{D}$, so that it does not coincide with $\{0\}$.

It is also easily shown, using axioms 2 and 6, that $[0, 0]$ coincides with the set

$$\mathbf{I} = \{ x \in \mathbf{R} : \neg\neg x = 0 \}.$$

So $\mathbf{I}$ is indecomposable. (In fact the indecomposability of $\mathbf{I}$ can be proved independently of axioms 1–6 through the general observation that, if A is indecomposable, then so is the set $A^\star = \{ x : \neg\neg x \in A \}$.)

Finally, we observe that the sequence of infinitesimal neighbourhoods of 0 generates a strictly ascending sequence of decomposable subsets containing $\mathbf{R} - \{0\}$, namely:

$$\mathbf{R} - \{0\} \subset (\mathbf{R} - \{0\}) \cup \{0\} \subset (\mathbf{R} - \{0\}) \cup \Delta_1 \subset (\mathbf{R} - \{0\}) \cup \Delta_2 \subset \ldots$$
$$(\mathbf{R} - \{0\}) \cup \mathbf{D} \subset (\mathbf{R} - \{0\}) \cup [0, 0].$$

Note

It was hearing Dirk van Dalen's stimulating talk at the conference that got me thinking about the topic discussed in this paper. I would like to thank Peter Schuster for encouraging me to write it, and the referee for helpful suggestions.

References

[1] Bell, J.L., *A Primer of Infinitesimal Analysis.* Cambridge University Press, 1998.

[2] Bridges, D.S., *Constructive mathematics: a foundation for computable analysis.* Theoretical Computer Science **219** (1999), 95–109.

[3] Kock, A., *Synthetic Differential Geometry.* Cambridge University Press, 1981.

[4] McLarty, C., *Elementary Categories, Elementary Toposes.* Oxford University Press, 1992.

[5] van Dalen, D., *How connected is the intuitionistic continuum?* Journal of Symbolic Logic **62** (1997), 1147–1150.

[6] van Dalen, D., *From Brouwerian counterexamples to the creating subject.* Studia Logica **62** (1999), 305–314.

CONSTRUCTIVE UNBOUNDED OPERATORS

Douglas Bridges
Department of Mathematics and Statistics, University of Canterbury
Private Bag 4800, Christchurch, New Zealand
d.bridges@math.canterbury.ac.nz

Hajime Ishihara
School of Information Science, Japan Advanced Institute of Science and Technology
Tatsunokuchi, Ishikawa 923-12, Japan
ishihara@jaist.ac.jp

Keywords: constructive, unbounded operator, sequentially continuous.

Abstract: The existence of points outside the domain of an unbounded linear mapping between normed spaces is investigated constructively.

Introduction

In this paper we begin a constructive investigation of unbounded operators by studying a number of questions about their domains. We intend this to be the start of a programme to investigate systematically the constructive theory of unbounded operators, and incidentally to counter, once and for all, some common misconceptions about the viability of such a theory (see [10] and [5]).

Since constructive analysis uses intuitionistic, rather than classical, logic, even at the base level of the theory of unbounded operators there are problems that are classically trivial but constructively significant. For example, the obvious contradiction argument, based on the closed graph theorem, merely shows that the domain of a closed unbounded operator is not complete; it does not enable us to construct a point of the initial Banach space at which the operator is not defined. In the general case we have only managed a partial, weak solution to this

P. Schuster, U. Berger and H. Osswald (eds.),
Reuniting the Antipodes – Constructive and Nonstandard Views of the Continuum, 25–37.

problem; but in the special case of a closed unbounded operator T on a Hilbert space, such that both T and T^* are densely defined, we can find $\xi \in H$ such that $T\xi$ is not defined.

Our setting is Bishop's constructive analysis (**BISH**—see [2] or [3]), which we can regard as mathematics carried out with intuitionistic logic ([14], [15]). At one stage we add Church's thesis, to enable us to prove a stronger result in recursive constructive analysis, which is simply one model of BISH. Further information about BISH and other varieties of constructive mathematics can be found in [8], [1], and [18].

Let T be a linear mapping between normed spaces X and Y. We say that T is

- **not bounded** if it is contradictory that it be bounded;
- **unbounded** if there exists a sequence (x_n) in X such that $\lim_{n\to\infty} x_n = 0$ and $\forall n\ (\|Tx_n\| > 1)$.

There is a constructive distinction between these two concepts: whereas the second clearly implies the first, the converse implication depends on **Markov's Principle**,

If (a_n) *is a binary sequence such that* $\neg\forall n\,(a_n = 0)$, *then* $\exists n\,(a_n = 1)$,

This principle, which represents an unbounded search, cannot be derived within, but is consistent with, Heyting arithmetic—that is, Peano arithmetic with intuitionistic logic; see [8], pages 137–138. For future reference, note that Markov's Principle is equivalent to the statement

$$\forall x \in \mathbf{R}\,(\neg\,(x = 0) \Rightarrow x \neq 0)\,,$$

where, in the context of a normed space X, $x \neq 0$ means $\|x\| > 0$. We shall return to the distinction between *bounded* and *not unbounded* at the end of the paper. We say that T is **strongly extensional** if

$$\forall x, x' \in X\ \left(Tx \neq Tx' \Rightarrow x \neq x'\right).$$

Markov's Principle implies that every linear mapping between normed spaces is strongly extensional. Without Markov's Principle the best we can prove constructively is that every linear mapping of a Banach space into a normed space is strongly extensional; see [6], Corollary 2.

1. CONSTRUCTING POINTS OUTSIDE DOMAINS

In this section we examine the problem of finding points outside the domain of an unbounded linear mapping between normed spaces. Our first result is connected with the **limited principle of omniscience (LPO)**,

> *If* (a_n) *is a binary sequence, then either* $a_n = 0$ *for all* n *or else there exists* n *such that* $a_n = 1$,

a weak form of the law of excluded middle that cannot be derived in intuitionistic logic (and that is false, even with classical logic, in the recursive model of BISH—see Chapter 3 of [8]).

Recall that every normed linear space X can be embedded as a dense subset of a Banach space $\widehat{X}$, the **completion** of X.

Proposition 1 *Let* T *be a strongly extensional unbounded linear mapping between normed spaces* X, Y. *Then for each binary sequence* (a_n) *there exists* $x \in \widehat{X}$ *such that if* $x \in X$, *then*

$$\forall n \ (a_n = 0) \vee \exists n \ (a_n = 1) .$$

PROOF. We may assume that (a_n) has at most one term equal to 1. For each positive integer n choose a unit vector $x_n \in X$ such that $\|Tx_n\| > n^2$. Then $\sum_{n=1}^{\infty} a_n \|Tx_n\|^{-1} x_n$ converges to a sum x in the Banach space $\widehat{X}$, by comparison with $\sum_{n=1}^{\infty} n^{-2}$. Suppose that $x \in X$. Then either $Tx \neq 0$ or $\|Tx\| < 1$. In the first case, as T is strongly extensional, $x \neq 0$ and so there exists n such that $a_n = 1$. In the second, suppose that there exists N such that $a_N = 1$; then $Tx = \|Tx_N\|^{-1} Tx_N$ and so $\|Tx\| = 1$, a contradiction; whence $a_n = 0$ for all n. Q.E.D.

Ishihara [11] has shown that in recursive constructive mathematics—constructive mathematics plus Church's thesis—every linear mapping on a Banach space is sequentially continuous and therefore not unbounded; this result also holds in the intuitionistic model of constructive mathematics ([18], page 354, 2.8). Since BISH is consistent with classical (that is, traditional) mathematics, we cannot hope to prove that recursive/intuitionistic continuity result within BISH. What we can prove, however, is that the existence of unbounded linear mappings on a Banach space is essentially nonconstructive.

Corollary 2 *If there exists a strongly extensional unbounded linear mapping on a Banach space, then LPO holds.*

PROOF. This is an immediate consequence of Proposition 1. Q.E.D.

A linear mapping $T : X \to Y$ between normed spaces is said to be **closed** if its graph is a closed subset of $X \times Y$; that is, if

$$(x_n \to x \in X \text{ and } Tx_n \to y \in Y) \Rightarrow y = Tx.$$

The constructive **closed graph theorem** says that if X is complete and the graph of T is both closed and separable, then T is sequentially continuous ([12], Corollary 2). The separability hypothesis is not needed in classical analysis, where the conclusion is the stronger one that T is bounded.

Proposition 3 *If T is an unbounded closed linear mapping, with a separable graph, of a normed space X into a Banach space Y, then* $\neg\left(\widehat{X} = X\right)$.

PROOF. If $\widehat{X} = X$, then the closed graph theorem shows that T is sequentially continuous, a contradiction. Q.E.D.

Let H be a Hilbert space, and T a linear mapping of a dense subspace of H into H; then we say that T is a **densely defined operator on** H. The **adjoint** T^* of a densely defined operator T on H is defined as in classical mathematics. Thus the domain of T^* comprises those $x \in H$ for which there exists $y \in H$ (which is then uniquely defined by x) such that

$$\langle y, z\rangle = \langle x, Tz\rangle \quad (z \in H),$$

and for such an x we write $T^*x = y$. In contrast to the classical situation, T^* may not be defined even if T is a *bounded* operator defined on the whole space H; see [9].

When dealing with unbounded linear operators on linear subsets of a Hilbert space, we can obtain substantial improvements upon Corollary 2. For the first of these we need the constructive **uniform boundedness theorem** (a contrapositive of the usual classical version of that result):

> *If (A_n) is a sequence of bounded linear mappings from a Banach space X into a normed space Y, and (x_n) is a sequence of unit vectors in X such that $\|A_n x_n\| \to \infty$ as $n \to \infty$, then there exists $x \in X$ such that $\|A_n x\| \to \infty$ as $n \to \infty$* ([16], page 61).

Proposition 4 *If H is a Hilbert space, and T is an unbounded densely defined operator on H with an adjoint, then there exists an element ξ of H such that $T\xi$ is undefined.*

PROOF. Let (x_n) be a sequence converging to 0 in H such that $\|Tx_n\| \to \infty$. Applying the uniform boundedness theorem to the linear functionals $x \mapsto \langle x, Tx_n \rangle$, we obtain a unit vector $\xi \in H$ such that $|\langle \xi, Tx_n \rangle| \to \infty$ as $n \to \infty$. If $T\xi$ is defined, then $\langle \xi, Tx_n \rangle = \langle T^*\xi, x_n \rangle \to 0$, a contradiction. Hence $T\xi$ is not defined. Q.E.D.

If T is an unbounded densely defined operator on a Hilbert space H, what can we say about its adjoint? It is closed (see [13]), but need not be unbounded; in fact, it can be 0 even though T is densely defined (see [17]). Classically, T^* is unbounded: for if it were bounded, we could extend it by continuity to a bounded operator on H whose adjoint would be bounded; whence T would have a bounded extension. When H is separable, the same result holds constructively, but, of course, with a direct proof.

Proposition 5 *If T is a densely defined operator on a separable Hilbert space H such that T^* is unbounded, then T is unbounded.*

PROOF. Since it is a dense subset of a separable space, the domain of T is separable; so there exists an increasing sequence (V_n) of finite–dimensional subspaces whose union is dense in that domain and therefore in H. Let P_n be the projection of H on V_n. Given $K > 0$, choose a unit vector $\xi \in H$ such that $\|T^*\xi\| > K + 1$. Then choose n such that

$$\|T^*\xi - P_n T^*\xi\| = \rho(T^*\xi, V_n) < 1$$

and therefore $\|P_n T^*\xi\| > K$. Setting

$$z = \|P_n T^*\xi\|^{-1} P_n T^*\xi,$$

we see that $P_n z = z$, $\|z\| = 1$, and

$$K < \|P_n T^*\xi\| = \langle P_n T^*\xi, z \rangle = \langle \xi, T P_n z \rangle = \langle \xi, Tz \rangle \leq \|Tz\|. \qquad \text{Q.E.D.}$$

Corollary 6 *Let T be an unbounded densely defined operator on a Hilbert space, such that T^* is densely defined. Then there exists $\xi \in H$ such that $T\xi$ is undefined.*

PROOF. Use the preceding two results. Q.E.D.

Returning to the context of Banach spaces, we show how to improve Corollary 2 under Church's thesis (that is, in the recursive model of BISH).

To each binary Turing machine $\mathcal{M}$ —that is, Turing machine with input alphabet $\{0,1\}$ and tape alphabet $\{0,1,\mathbf{B}\}$, where $\mathbf{B}$ denotes the blank symbol—there corresponds a unique computable partial function $\varphi : \mathbf{N} \to \mathbf{N}$. The value $\varphi(n)$, if defined, is obtained as the natural number whose unary form is the output when $\mathcal{M}$ is run with the unary form of n as input. (See [4], Ch. 3.)

Proposition 7 *Assuming Church's Thesis, let T be an unbounded linear mapping of a normed space X into a normed space Y. Then there exists a sequence (x_n) in $\widehat{X}$ such that*

$$\neg\forall n\ (Tx_n \text{ is defined}) .$$

PROOF. Let $\mathcal{M}_0, \mathcal{M}_1, \ldots$ be an effective enumeration of all binary Turing machines, and for each n let φ_n be the partial function from $\mathbf{N}$ to $\mathbf{N}$ computed by $\mathcal{M}_n$. Define a binary sequence $(a_{n,k})_{n,k=0}^{\infty}$ as follows:

$$a_{n,k} = \begin{cases} 1 & \text{if } \mathcal{M}_n \text{ computes } \varphi_n(n) \text{ in exactly } k \text{ steps} \\ 0 & \text{otherwise.} \end{cases}$$

By Proposition 1, for each n there exists $x_n \in \widehat{X}$ such that if Tx_n is defined, then we can decide whether $\varphi_n(n)$ is defined. So if Tx_n is defined for each n, then we can decide the restricted Halting Problem, which is absurd ([4], (4.2)). Q.E.D.

Here is another proof of the last result, based on **Specker's Theorem**:

> *Under Church's thesis, there exists a strictly increasing sequence (q_n) of rational numbers in $[0,1]$ that is eventually bounded away from any given real number: in other words, for each $x \in \mathbf{R}$ there exist $\delta > 0$ and N such that $|x - q_n| \geq \delta$ for all $n \geq N$.* ([8], Ch. 3, (3.1)).

Assuming Church's Thesis, let (r_n) be a decreasing sequence in the open interval $(0,1)$ that is eventually bounded away from any given real number. Let (p_n) be an enumeration of all ordered pairs (α, β) of rational numbers with $\alpha < \beta$. Choose a sequence (x_n) converging to 0 in X, such that $\|Tx_n\| = 1 + r_n$ for each n. For each n define a binary sequence $(a_{n,k})_{k=0}^{\infty}$ such that

$$\begin{aligned} a_{n,k} = 0 &\ \Rightarrow\ \forall j \leq k\ (\|Tx_j\| < p_{n,2}) , \\ a_{n,k} = 1 &\ \Rightarrow\ \exists j \leq k\ (\|Tx_j\| > p_{n,1}) , \end{aligned}$$

where $p_n = (p_{n,1}, p_{n,2})$. By Proposition 1, there exists $\xi_n \in X$ such that if $T\xi_n$ is defined, then

$$\forall k \; (\|Tx_k\| < p_{n,2}) \vee \exists k \; (\|Tx_k\| > p_{n,1}) \,.$$

If $T\xi_n$ is defined for each n, then $\inf_{k\geq 1} \|Tx_k\|$ exists, by the least–upper–bound principle ([3], page 37, (4.3)); whence the sequence (r_n) converges, a contradiction.

2. UNBOUNDEDNESS ON SUBSPACES

It is trivial to prove that an unbounded linear mapping $T : X \to Y$ between normed spaces that is bounded on a subspace S of X is unbounded off S. It is not trivial to prove the corresponding result when T is only known to be sequentially continuous on S (as might be the case when the restriction of T to S is the inverse of a bounded linear mapping [12]); indeed, we have had to add the hypothesis that S be **located**, in the sense that

$$\rho(x, S) = \inf \{\|x - s\| : s \in S\}$$

exists for each $x \in X$.

Proposition 8 *Let T be an unbounded linear mapping of a normed space X into a normed space Y, and let S be a complete located subspace of X on which T is sequentially continuous. Then there exists a sequence (x_n) in X such that for each n, $\rho(x_n, S) > 0$, $\lim_{n\to\infty} x_n = 0$, and $\|Tx_n\| > 1$.*

PROOF. It will suffice to show that for any sequence (x_n) that converges to 0 in X and satisfies $\|Tx_n\| > 1$ for each n, there exists N such that $\rho(x_N, S) > 0$. Without loss of generality, we may assume that $\|x_n\| < 1/n$ for each n. By Lemma (3.8), page 92 of [3], for each n there exists $s_n \in S$ such that if $x_n \neq s_n$, then $\rho(x_n, S) > 0$. Define an increasing binary sequence (λ_n) such that

$$\begin{aligned} \lambda_n = 0 \quad &\Rightarrow \quad \forall k \leq n \; (\|x_k - s_k\| < 1/n \,\wedge\, \|Tx_k - Ts_k\| < 1/n) \,, \\ \lambda_n = 1 - \lambda_{n-1} \quad &\Rightarrow \quad x_n \neq s_n . \end{aligned}$$

(Note that if $Tx_n \neq Ts_n$, then $x_n \neq s_n$.) We may assume that $\lambda_2 = 0$. It will suffice to show that $\lambda_m = 1$ for some m: for then there must exist $n \leq m$ such that $\lambda_n = 1 - \lambda_{n-1}$, so that $x_n \neq s_n$ and therefore $\rho(x_n, S) > 0$.

If $\lambda_n = 0$, set $\zeta_n = 0$. If $\lambda_n = 1 - \lambda_{n-1}$, set $\zeta_k = s_{n-1}$ for all $k \geq n$; then

$$\begin{aligned}\|\zeta_n - \zeta_{n-1}\| &= \|s_{n-1}\| \\ &\leq \|x_{n-1} - s_{n-1}\| + \|x_{n-1}\| \\ &< \tfrac{1}{n-1} + \tfrac{1}{n-1} \\ &= \tfrac{2}{n-1}.\end{aligned}$$

It now follows that

$$\|\zeta_m - \zeta_n\| < \tfrac{2}{n-1} \quad (m \geq n).$$

So (ζ_n) is a Cauchy sequence in S and therefore converges to a limit $\zeta \in S$. Since T is sequentially continuous on S, we have $\lim_{n\to\infty} T\zeta_n = T\zeta$. Either $T\zeta \neq 0$ or $\|T\zeta\| < 1/2$. In the first case we have $T\zeta_m \neq 0$, so that $\zeta_m \neq 0$ and therefore $\lambda_m = 1$, for all sufficiently large m. In the second case we can choose m such that $\|T\zeta_k\| < 1/2$ for all $k \geq m$. Suppose that $\lambda_m = 0$. If there exists $n > m$ such that $\lambda_n = 1 - \lambda_{n-1}$, then $\zeta = s_{n-1}$, and so (as $n \geq 3$)

$$\|T\zeta\| \geq \|Tx_{n-1}\| - \|Tx_{n-1} - Ts_{n-1}\| > 1 - \tfrac{1}{n-1} > \tfrac{1}{2},$$

a contradiction. Hence $\lambda_n = 0$ for all $n \geq m$, and therefore for all n It follows that

$$\|s_n\| \leq \|x_n\| + \|x_n - s_n\| \leq \tfrac{2}{n} \to 0 \text{ as } n \to \infty$$

and that

$$\|Ts_n\| \geq \|Tx_n\| - \|Tx_n - Ts_n\| > 1 - \tfrac{1}{n} \geq \tfrac{1}{2}$$

for each $n \geq 2$, which contradicts the sequential continuity of T on S. Hence, in fact, we must have $\lambda_m = 1$. Q.E.D.

Proposition 8 is much easier to prove when X is a Hilbert space. In that case we let P be the projection of X onto S, and consider any sequence (x_n) that converges to 0 in X and satisfies $\|Tx_n\| > 1$ for each n. Since $Px_n \to 0$ and T is sequentially continuous on S, we have $TPx_n \to 0$; so

$$\|T(I-P)x_n\| \geq \|Tx_n\| - \|TPx_n\| > 1$$

for all sufficiently large n, where $(I - P)\, x_n \in S\perp$.

Corollary 9 *Let X be a normed space, Y a Banach space, S a complete located subspace of X, and T a closed unbounded linear mapping on X that has a separable graph. Then there exists a sequence (x_n) converging to 0 such that $\rho(x_n, S) > 0$ and $\|Tx_n\| > 1$ for each n.*

PROOF. The restriction of T to S has a closed graph and so, by Corollary 2 of [12], is sequentially continuous. The result now follows from Proposition 8. Q.E.D.

Corollary 10 *If T is an unbounded linear mapping between normed spaces, then the domain of T is infinite–dimensional.*

PROOF. Let V be a finite–dimensional, and therefore located, subspace of the domain of T. Then T is bounded, and hence sequentially continuous, on V; so, by Proposition 8, there exists x such that Tx is defined and $\rho\,(x, V) > 0$. Q.E.D.

In view of the last corollary, it is tempting to conjecture that if T is a linear mapping from a subset of a finite–dimensional Banach space X into a finite–dimensional Banach space Y, then the domain of T is finite–dimensional and hence T is bounded. The following example shows that this is not the case.

Let $a \in \mathbf{R}$, and define a linear subset S of $\mathbf{R}$ by

$$S = \{0\} \cup \{x \in \mathbf{R} : a \neq 0\}\,.$$

Now define a linear mapping $T : S \to \mathbf{R}$ by

$$Tx = \begin{cases} 0 & \text{if } x = 0 \\ a^{-1}x & \text{if } a \neq 0. \end{cases}$$

T is strongly extensional: for if $Tx \neq 0$, then $x \notin \{0\}$, so $a \neq 0$ and therefore $x = a\,(Tx) \neq 0$. T is not unbounded: for if it were unbounded, we would have $Tx \neq 0$ for some x, so T would be bounded by a^{-1}, a contradiction. But T is not even sequentially continuous, let alone bounded. To see this, construct an increasing binary sequence (λ_n) such that

$$\begin{aligned} \lambda_n = 0 &\;\Rightarrow\; |a| < 1/n, \\ \lambda_n = 1 &\;\Rightarrow\; |a| > 1/\,(n+1)\,. \end{aligned}$$

If $\lambda_n = 0$, set $x_n = 0$; if $\lambda_n = 1 - \lambda_{n-1}$, set $x_n = a$ and $x_k = 0$ for all $k \geq n+1$. Then (x_n) is a Cauchy sequence in S, and $x_n \to 0$. Supposing that $Tx_n \to 0$, choose N such that $|Tx_n| < 1$ for all $n \geq N$. If $\lambda_N = 1$, then $a \neq 0$. If $\lambda_N = 0$ and $\lambda_m = 1 - \lambda_{m-1}$ for some $m > N$, then $x_m = a$ and $Tx_m = a^{-1}a = 1$, which contradicts our choice of N. So if $\lambda_N = 0$, then $\lambda_n = 0$ for all n, and therefore $a = 0$. We see from all this that the statement

> *If T is a linear mapping defined, and not unbounded, on a linear subset of $\mathbf{R}$, then T is sequentially continuous*

implies LPO.

This example shows that the following result is the best of its type.

Proposition 11 *Let X be a finite–dimensional Banach space, D a linear subset of X, and T a linear mapping of D into a normed space Y. Then T is not unbounded.*

PROOF. First consider the case where $X = \mathbf{C}u$ is 1–dimensional. Supposing that T is unbounded, choose a sequence (α_n) converging to 0 in $\mathbf{C}$ such that $|\alpha_n| \, \|Tu\| = \|T(\alpha_n u)\| > 1$ for each n. Then $|\alpha_n| > \|Tu\|^{-1} > 0$ for each n, which is absurd. Thus the theorem holds if $\dim(X) = 1$.

Now assume that the theorem holds for all spaces X of dimension $n-1$, and consider a space X of dimension n. Suppose that T is unbounded, and let u be any nonzero vector in X. Write $X = X' \bigoplus \mathbf{C}u$, where $\dim(X') = n - 1$. Choose a sequence (x_n) in X' and a sequence (α_n) in $\mathbf{C}$ such that $x_n + \alpha_n u \to 0$ and $\|T(x_n + \alpha_n u)\| > 1$ for each n. Then (see pages 34–37 of [8]) $x_n \to 0$ in X' and $\alpha_n \to 0$ in $\mathbf{C}$. If Tu is defined, then $Tx_n = T(x_n + \alpha_n u) - \alpha_n Tu$ is defined, and, as $\lim_{n\to\infty} \alpha_n Tu = 0$, we have $\|Tx_n\| > 1$ for all sufficiently large n. This contradicts our induction hypothesis. Hence $u \notin D$ for every nonzero vector $u \in X$. It follows that the only possible element of D is 0, which is absurd since T is unbounded. This completes the induction. Q.E.D.

Returning to the example immediately preceding Proposition 11, we note that $(n^{-1}\lambda_n)$ is a Cauchy sequence in S and therefore converges to a limit ξ in $\mathbf{R}$. If $\xi \in S$, then either $\xi = 0$, in which case $\lambda_n = 0$ for all n and therefore $a = 0$; or else $\xi \in \{x \in \mathbf{R} : a \neq 0\}$ and so $a \neq 0$. Thus if S is complete, then T is bounded and hence sequentially continuous. More generally, we have

Corollary 12 *If, under the hypotheses of* Proposition 11, *the domain of T is closed in X, then T is sequentially continuous.*

PROOF. Let (x_n) be a sequence in D that converges to 0, and let $\varepsilon > 0$. Since D is complete, we see from Lemma 2 of [11] that either $\|Tx_n\| > \varepsilon/2$ for infinitely many n, or else $\|Tx_n\| < \varepsilon$ for all sufficiently large n. Since the first alternative implies that T is unbounded, it is ruled out by Proposition 11. Q.E.D.

3. BOUNDED *VS* NOT UNBOUNDED

We define the **complement** of a subset S of a normed space X to be

$$\backsim S = \{x \in X : \forall y \in S \ (x \neq y)\}.$$

In this final section we explore briefly the distinction between the statements "T is undefined at x" and "x is in the complement of the domain of T".

Let (e_n) be an orthonormal basis of unit vectors in an infinite-dimensional separable Hilbert space H, and let a be a real number such that $\neg(a = 0)$. Construct an unbounded linear mapping T on the linear span of $\{e_n : n \geq 1\}$ by setting $Te_n = n^2 e_n$ for each n. Let $x = a\sum_{n=1}^{\infty} n^{-1} e_n \in H$. If Tx is defined, then $\neg(a \neq 0)$ and so $a = 0$, a contradiction; hence Tx is undefined. But $T0 = 0$, and if $x \neq 0$, then $a \neq 0$. Hence the proposition

> *If T is an unbounded densely defined linear mapping on a Hilbert space H, and Tx is undefined, then x is in the complement of the domain of T*

entails Markov's Principle.

In general, we have the following amusing little result.

Proposition 13 *Let T be an unbounded linear mapping of a normed space X into a normed space Y, and x a point of X such that Tx is undefined. For each y in the domain of T there exists $z \in \widehat{X}$ such that if Tz is defined, then $x \neq y$.*

PROOF. Choose a sequence (x_n) in the domain of T such that for each n, $\|x_n\| < 1/n$ and $\|Tx_n\| > n$. Define an increasing binary sequence (λ_n) such that

$$\begin{aligned} \lambda_n = 0 \ &\Rightarrow \ \|x - y\| < 1/n, \\ \lambda_n = 1 \ &\Rightarrow \ \|x - y\| > 1/(n+1). \end{aligned}$$

If $\lambda_1 = 1$ we may take $z = y$; so we may assume that $\lambda_1 = 0$. If $\lambda_n = 0$, set $z_n = y + x_n$; if $\lambda_n = 1$, set $z_n = z_{n-1}$. Then (z_n) is a Cauchy sequence in X, and so converges to a limit $z \in \widehat{X}$. Suppose that Tz is defined, and choose a positive integer $N > \|T(z-y)\|$. If $\lambda_N = 0$ and $\lambda_{m+1} = 1 - \lambda_m$ for some $m \geq N$, then $z = y + x_m$ and therefore

$$\|T(z-y)\| = \|Tx_m\| > m \geq N,$$

a contradiction. So if $\lambda_N = 0$, then $\lambda_n = 0$ for all n, and therefore $x = y$, which is impossible. We conclude that $\lambda_N = 1$. Q.E.D.

References

[1] Michael Beeson, *Foundations of Constructive Mathematics*, Springer-Verlag, Heidelberg, 1985.

[2] Errett Bishop, *Foundations of Constructive Analysis*, McGraw-Hill, New York, 1967.

[3] Errett Bishop and Douglas Bridges, *Constructive Analysis*, Grundlehren math. Wiss. **279**, Springer-Verlag, Heidelberg, 1985.

[4] Douglas Bridges, *Computability: A Mathematical Sketchbook*, Graduate Texts in Mathematics **146**, Springer-Verlag, Heidelberg, 1994.

[5] Douglas Bridges, "Constructive mathematics and unbounded operators—a reply to Hellman", J. Philos. Logic **24**, 549–561, 1995.

[6] Douglas Bridges and Hajime Ishihara, "Linear mappings are fairly well-behaved", Arch. Math. **54**, 558–562, 1990.

[7] Douglas Bridges, William Julian, and Ray Mines, "A constructive treatment of open and unopen mapping theorems", Zeit. math. Logik und Grundlagen der Math. **35**, 29–43, 1989.

[8] Douglas Bridges and Fred Richman, *Varieties of Constructive Mathematics*, London Math. Soc. Lecture Notes **97**, Cambridge University Press, 1987.

[9] Douglas Bridges, Fred Richman, and Peter Schuster, "Adjoints, absolute values, and polar decompositions", J. Operator Theory **44**, 243–254, 2000.

[10] Geoffrey Hellman, "Constructive mathematics and quantum mechanics: unbounded operators and the spectral theorem", J. Phil. Logic **22,** 221–248, 1993.

[11] Hajime Ishihara, "Continuity and nondiscontinuity in constructive mathematics", J. Symbolic Logic **56**(4), 1349–1354, 1991.

[12] Hajime Ishihara, "A constructive proof of Banach's inverse mapping theorem", New Zealand J. Math. **23**(1), 71–75, 1994.

[13] R.V. Kadison and J.R. Ringrose, *Fundamentals of the Theory of Operator Algebras* (2 Vols), Academic Press, New York, 1983.

[14] Fred Richman, "Intuitionism as generalization", Philosophia Math. **5**, 124–128, 1990.

[15] Fred Richman, "Interview with a constructive mathematician", Modern Logic **6**, 247–271, 1996.

[16] Halsey L. Royden, "Aspects of constructive analysis", in *Errett Bishop: Reflections on Him and His Research*, Contemporary Mathematics **39**, Amer. Math. Soc., Providence RI, 1985.

[17] Walter Rudin, *Functional Analysis* (2nd edn), McGraw-Hill, New York, 1991.

[18] A.S. Troelstra and D. van Dalen, *Constructivism in Mathematics: An Introduction* (two volumes). North-Holland, Amsterdam, 1988.

THE POINTS OF (LOCALLY) COMPACT REGULAR FORMAL TOPOLOGIES

Giovanni Curi*
Dipartimento di Matematica Pura ed Applicata
Via Belzoni 7 - 35131 Padova, Italy
gcuri@math.unipd.it

Keywords: Formal Topology, Type Theory, Locale Theory, Compactness, Regularity

Mathematical Subject Classification: 54A05(06D20, 03F55, 03B15)

Abstract: In a paper appeared in 1990, C.J. Mulvey established a constructive characterization of completely prime filters on a compact regular locale L; although proved by intuitionistic logic, the result relies on a notion of maximality which contains an impredicative second-order quantification. In this note we present an alternative concept of maximality, entirely phrased in first-order terms, and give a predicative characterization of the points of a compact regular formal topology (equivalently, we give a characterization of the points of a compact regular locale which can be entirely carried out within Intuitionistic Type Theory). This result is then generalized to locally compact regular formal topologies (resp. locally compact regular locale).

Introduction

Formal Topology[1] was conceived with the aim of developing point-free topology (Locale Theory) in a constructive and predicative foundational setting, such as Martin-Löf's Intuitionistic Type Theory. Quite recently, the topological notion of regularity has been predicatively formulated in

*Partially supported by the project "Metodi Costruttivi in Topologia, Algebra e Analisi dei Programmi" of the Ministero dell' Università e della Ricerca Scientifica e Tecnologica.
[1]Formal Topology was introduced in [15]; a more recent presentation is contained in [16], [17].

P. Schuster, U. Berger and H. Osswald (eds.),
Reuniting the Antipodes – Constructive and Nonstandard Views of the Continuum, 39–54.

this framework, and the class of compact regular formal topologies has shown to have nice and promising properties, particularly from a constructive standpoint (cf. e.g. [5], [3] and [14]). In this note we establish a constructive characterization of the points of a compact regular formal topology, in which formal points are shown to coincide with particular subsets of basic neighbourhoods, the *maximal regular* ones. The specific feature of this characterization is that regular subsets will be considered to be maximal according to an entirely first-order criterion of maximality.

This result can then be seen to improve a previous characterization appeared in the context of Locale Theory: in [9] indeed Chris Mulvey introduces a particular formulation of the notion of maximality for regular filters which allows to prove intuitionistically that the completely prime filters on a compact regular locale coincide with the maximal regular filters (cf. [9]). In such a notion of maximality, however, an impredicative second-order quantification appears which makes the result incompatible with foundational settings for constructive mathematics such as Martin-Löf's Intuitionistic Type Theory and Aczel's Constructive Set Theory. A natural relation then exists between Formal Topology and Locale Theory (cf. [15]) that allows to give the following reading of our result: a characterization of completely prime filters on a compact regular locale by means of maximal regular filters can be obtained intuitionistically *and* predicatively, and such a characterization can be entirely carried out within Intuitionistic Type Theory.

Few modifications allow to generalize this result to *locally* compact regular formal topologies (and thus to locally compact regular locales). Then, in particular, examples of topologies for which these characterizations are valid are (that giving rise to) the Continuum, Cantor space and the spaces $\mathcal{L}(\mathcal{A})$ of linear functionals of norm ≤ 1 from a semi-normed space A to the reals[2].

1. PRELIMINARIES

We recall the basic definitions of Formal Topology ([15]). The reader is referred to [15], [16] and [17] for a detailed account (the presentation we are to adopt appears in [16]). We use a special notation for subsets,

[2]Endowed with the *weak** topology, cf. [4].

introduced and motivated in [18], which allows to work within Intuitionistic Type Theory (henceforth simply Type Theory, cf. [8], [13]) with essentially the usual mathematical formalism: for the present purpose it will suffice to know that a *subset* U of a set S is a unary predicate (dependent type) on S, $U(x)(x \in S)$, and that a *set-indexed family of subsets* is a binary predicate $U(x,i)(x \in S, i \in I)$ on the sets S and I, where for each $\bar{i}$, $U(x,\bar{i})(x \in S)$ is the subset of index $\bar{i}$ (for simplicity, we will also use the traditional notations $\{a \in S : U(a)\}$, to indicate the subset U, and $U_i (i \in I)$ for a family of subsets). Finally, we will write $a \epsilon U$ to mean $a \in S$ *and* $U(a)$ *true* (in the expression $a \epsilon U$ the symbol 'ϵ' is used, instead of '$\in$', to recall that we are considering a subset, i.e. a propositional function, and not a set; cf. [18]).

1.1 A (formal) topology is a triple $\mathcal{S} \equiv (S, \lhd, \mathsf{Pos})$ where S is a set, called the *base*, $\lhd$ is a relation between elements and subsets of S which satisfies the following conditions:

$$\textit{(reflexivity)} \qquad \frac{a \epsilon U}{a \lhd U}$$

$$\textit{(transitivity)} \qquad \frac{a \lhd U \qquad U \lhd V}{a \lhd V}$$

$$\textit{(}\downarrow\textit{-right)} \qquad \frac{a \lhd U \qquad a \lhd V}{a \lhd U \downarrow V}$$

where

$$\begin{array}{rcl} U \lhd V & \equiv & (\forall u \epsilon U) u \lhd V \\ U \downarrow V & \equiv & \{d : S | \ (\exists u \epsilon U)\ (d \lhd \{u\})\ \&\ (\exists v \epsilon V)\ (d \lhd \{v\})\} \end{array}$$

and Pos is a subset of S which satisfies

$$\textit{(monotonicity)} \qquad \frac{\mathsf{Pos}(a) \quad a \lhd U}{(\exists b \epsilon U)\mathsf{Pos}(b)}$$

$$\textit{(positivity)} \qquad \frac{\mathsf{Pos}(a) \rightarrow a \lhd U}{a \lhd U}.$$

We will write $\mathsf{Pos}(U)$ for $(\exists a \epsilon U)\mathsf{Pos}(a)$. The relation $\lhd$ is called *cover* and Pos *positivity predicate* (we pronounce $a \lhd U$ as 'U covers a', and

$\mathsf{Pos}(a)$ as 'a is positive'). For simplicity, when a basic neighbourhood a is covered by a singleton subset $\{b\}$ we will often write $a \lhd b$ instead of $a \lhd \{b\}$.

One can think intuitively of the elements of S as of indexes for the basic neighbourhoods of a topological space; the cover relation can then be seen as a formal description of the inclusion between basic neighborhoods and subsets of S, and the predicate Pos as a positive way to express that a certain neighbourhood is non-empty. Then, for instance, 'monotonicity' has the following intuitive reading: if a non-empty neighbourhood is covered by a family of neighbourhoods (indexed by U) then there must be at least one member of the family which is non-empty.

An equivalent formulation of positivity (cf. [15]) which we will use in the following is

$$\frac{a \lhd U}{a \lhd U^+},$$

where $U^+ \equiv \{b \,\epsilon\, U : \mathsf{Pos}(b)\}$ (that is, only non-empty neighbourhoods contribute to the cover).

Finally, we recall that given two subsets U, V of S, $U =_S V$ means exactly that $U \lhd V \;\&\; V \lhd U$, and that for $U \subseteq S$, the *(pseudo-)complement* U^* of U is given by $U^* \equiv \{b : \neg\mathsf{Pos}(b \downarrow U)\}$.

1.2 In a formal topology $\mathcal{S}$ a *formal point* is a subset $\alpha \subseteq S$ such that

i. $(\exists a)(a \,\epsilon\, \alpha)$

ii. $(a \,\epsilon\, \alpha \;\&\; b \,\epsilon\, \alpha) \rightarrow (\exists c)(c \,\epsilon\, a \downarrow b \;\&\; c \,\epsilon\, \alpha)$

iii. $\dfrac{a \,\epsilon\, \alpha \quad a \lhd U}{(\exists b)(b \epsilon U \;\&\; b \,\epsilon\, \alpha)}$

iv. $a \,\epsilon\, \alpha \rightarrow \mathsf{Pos}(a)$.

We will denote by $\mathsf{Pt}(\mathcal{S})$ the collection of formal points. (Condition *iv.* is actually known to be derivable from *iii.* and positivity and could thus be skipped[3]).

1.3 The relation with Locale Theory can be sketched as follows (a detailed discussion of this subject is contained in [15], [17]): defining, for

[3]A proof is recalled in [12]. A generalized definition of Formal Topology can however be considered in which the positivity rule is not required, cf. [16].

$U \subseteq S$, $\mathcal{S}U$ to be the subset $\{a \in S : a \lhd U\}$, we say that U is *saturated* if $U = \mathcal{S}U$; denoting then by $Sat(\mathcal{S})$ the collection of saturated subset of $\mathcal{S}$, $Sat(\mathcal{S})$ endowed with the operations

$$\mathcal{S}U \wedge \mathcal{S}V \equiv \mathcal{S}U \cap \mathcal{S}V = \mathcal{S}(U \downarrow V)$$

and

$$\bigvee_{i \in I} \mathcal{S}U_i \equiv \mathcal{S}(\bigcup_{i \in I} U_i)$$

forms a *frame* (or locale, or complete Heyting algebra).

From a non-constructive point of view the converse is also valid (that is, any frame can be obtained as the frame of saturated subsets of a formal topology $\mathcal{S}$). Finally, the points of a formal topology $\mathcal{S}$ are easily shown to correspond to completely prime filters on $Sat(\mathcal{S})$.

1.4 A formal topology $\mathcal{S} \equiv (S, \lhd, \mathsf{Pos})$ is said to be *compact* if whenever $S \lhd U$ there exists a finite[4] subset $U_0 \subseteq U$ such that $S \lhd U_0$.

The notion of regularity have been recently introduced in Formal Topology as the predicative counterpart of that given in the context of locales (see for instance [6]): for a, b in S, we say that b is *well-covered* by a iff $S \lhd a \cup b^*$; defining $wc(a)$ to be the subset of neighbourhoods b which are well-covered by a, $wc(a) \equiv \{b : S \lhd a \cup b^*\}$, a formal topology $\mathcal{S}$ is then said to be *regular* if for all a in S, $a \lhd wc(a)$[5]. A topology $\mathcal{S}$ will be said to be *compact regular* if it is compact and regular.

The following lemmas will be used in the following, often without an explicit mention; they obtain in any formal topology $\mathcal{S}$.

Lemma 1.5. *Let V, W, Z be subsets of S, and $U_i (i \in I)$ be a family of subsets of S. We have*

i) $V \cup (W \downarrow Z) =_{\mathcal{S}} (V \cup W) \downarrow (V \cup Z)$,

ii) $(\bigcup_i U_i) \downarrow V = \bigcup_i (U_i \downarrow V)$.

[4]Note that here and in the following a set, or a subset, is considered to be 'finite' if its elements can be listed; cf. the notions of finite and sub-finite in [2].

[5]This definition appeared in [14]; in case of compactness, it is equivalent to the one introduced in [5], [3].

Proof. i) Let $a \epsilon V \cup (W \downarrow Z)$, i.e. $(a \epsilon V \vee a \epsilon (W \downarrow Z))$. If $a \epsilon V$, then trivially $a \epsilon (V \cup W) \downarrow (V \cup Z)$; if $a \epsilon (W \downarrow Z)$ then $(\exists b \epsilon W)(\exists c \epsilon Z)(a \lhd b \ \& \ a \lhd c)$, whence $(\exists b \epsilon W \cup V)(\exists c \epsilon Z \cup V)(a \lhd b \ \& \ a \lhd c)$, i.e. $a \epsilon (V \cup W) \downarrow (V \cup Z)$. Conversely, let $a \epsilon (V \cup W) \downarrow (V \cup Z)$; by definition, there is $b \epsilon (V \cup W)$ and $c \epsilon (V \cup Z)$ such that $a \lhd b$ and $a \lhd c$. From $b \epsilon (V \cup W)$ and $c \epsilon (V \cup Z)$ we have that $b \epsilon V$ or $b \epsilon W$, and $c \epsilon V$ or $c \epsilon Z$. It is then straightforward to see that a is covered by $V \cup (W \downarrow Z)$.

ii) By definition, $(\bigcup_i U_i) \downarrow V \equiv \{a : (\exists b)(\exists c)(b \epsilon \bigcup_i U_i \ \& \ c \epsilon V \ \& \ a \lhd b \ \& \ a \lhd c)\} = \{a : (\exists b)(\exists c)((\exists i \in I) b \epsilon U_i \ \& \ c \epsilon V \ \& \ a \lhd b \ \& \ a \lhd c)\} = \{a : (\exists i \in I)(\exists b)(\exists c)(b \epsilon U_i \ \& \ c \epsilon V \ \& \ a \lhd b \ \& \ a \lhd c))\} = \bigcup_i (U_i \downarrow V)$.

Lemma 1.6. *Let b and V be respectively an element and a subset of S; if $S \lhd b^* \cup V$, then $b \lhd V$.*

Proof. From $S \lhd b^* \cup V$ we have $b \lhd b^* \cup V$; since $b \lhd b$, we have (by $\downarrow -right$) that $b \lhd (b^* \cup V) \downarrow b$. Then $b \lhd (b^* \downarrow b) \cup (V \downarrow b)$, and hence (by positivity) $b \lhd (V \downarrow b)^+ \lhd V$.

Lemma 1.7. *Let $U_i (i \in I)$ be a family of subsets of S. Then $(\bigcup_i U_i)^* = \bigcap_i (U_i^*)$; in particular, for $U, V \subseteq S$, $(U \cup V)^* = U^* \cap V^* = U^* \downarrow V^*$.*

Proof. We have that: $a \epsilon (\bigcup_i U_i)^* \iff \neg \mathsf{Pos}(a \downarrow \bigcup_i U_i) \iff \neg \mathsf{Pos}(\bigcup_i (a \downarrow U_i)) \iff \neg \mathsf{Pos}(\{b : (\exists i) b \epsilon (a \downarrow U_i)\}) \iff \neg(\exists c)(c \epsilon \{b : (\exists i) b \epsilon (a \downarrow U_i)\} \ \& \ \mathsf{Pos}(c)) \iff \neg(\exists c)((\exists i) c \epsilon (a \downarrow U_i) \ \& \ \mathsf{Pos}(c)) \iff (\forall i)\neg(\exists c)(c \epsilon (a \downarrow U_i) \ \& \ \mathsf{Pos}(c)) \iff (\forall i)\neg \mathsf{Pos}(a \downarrow U_i) \iff a \epsilon \{d : (\forall i \in I) d \epsilon U_i^*\} \iff a \epsilon \bigcap_i (U_i^*)$. Moreover, for $U, V \subseteq S$, $(U^* \cap V^*) = (U^* \downarrow V^*)$: indeed clearly $(U^* \cap V^*) \subseteq (U^* \downarrow V^*)$; let then $a \epsilon (U^* \downarrow V^*)$, there are then $b \epsilon U^*$ and $c \epsilon V^*$ such that $a \lhd b$ and $a \lhd c$. Thus (since $b \epsilon U^* \equiv \neg \mathsf{Pos}(b \downarrow U)$), we have that $\neg \mathsf{Pos}(a \downarrow U)$, and similarly $\neg \mathsf{Pos}(a \downarrow V)$; hence $a \epsilon (U^* \cap V^*)$.

2. MAXIMAL REGULAR POINTS

In [9] a characterization of completely prime filters on a compact regular locale L is obtained by means of intuitionistic logic: completely prime filters are proved to coincide with maximal regular filters, where a regular filter P is said to be *maximal* if and only if it is proper and *for all regular filter F*, if P is contained in F, then

$$a \in F \rightarrow (a \in P \vee 0 \in F),$$

for all $a \in L$. The impredicative second-order quantification on filters appearing in this definition cannot be accepted in the context of a constructive and predicative foundational setting: in particular such a characterization could not be formalized within Intuitionistic Type Theory ([8], [13]), nor assuming Peter Aczel's Constructive Set Theory ([1]).

In [9] it is also pointed out that the strength to prove the characterization intuitionistically is provided by the first-order quantification on elements appearing in the definition of maximal regular filter; we are then going to prove that an entirely first-order notion of maximality can be formulated which suffices to prove such a characterization in a completely constructive way. The result is worked out in the context of Formal Topology, thus yielding a proof which can automatically be formalized within Martin-Löf's Type Theory; recalling however the natural relation between Formal Topology and Locale Theory (cf. [15]), it will be easily recognized that the same arguments could as well be phrased in the context of Locale Theory.

2.1 Let $\mathcal{S} \equiv (S, \lhd, \mathsf{Pos})$ be a formal topology. We define a subset α of S to be *regular* if it satisfies

1. $(\exists a)(a \,\epsilon\, \alpha)$
2. $(a \,\epsilon\, \alpha \;\&\; b \,\epsilon\, \alpha) \leftrightarrow (\exists c \in S)(c \,\epsilon\, a \downarrow b \;\&\; c \,\epsilon\, \alpha)$
3. $a \,\epsilon\, \alpha \rightarrow \mathsf{Pos}(a)$
4. $a \,\epsilon\, \alpha \rightarrow (\exists b \in S)(b \,\epsilon\, wc(a) \;\&\; b \,\epsilon\, \alpha)$

We say that α is a *maximal regular subset* of $\mathcal{S}$[6] if moreover, for all a, b in S,

5. $b \,\epsilon\, wc(a) \rightarrow ((\exists c \in S)(c \,\epsilon\, \alpha \;\&\; \neg Pos(b \downarrow c)) \vee a \,\epsilon\, \alpha)$

[6]The idea for defining maximality for regular subsets in this way was inspired by the notion of *maximal approximation*, as formulated in [7].

(that is, if a neighbourhood b is well-covered by a neighbourhood a, either a is a neighbourhood of α, or there exists a neighbourhood c of α disjoint from b).

Observe that maximal regular subsets are maximal (between regular subsets) in the usual sense: let $\mathcal{S}$ be any formal topology and let $\alpha \subseteq S$ be a maximal regular subset. If $\beta \subseteq S$ is a regular subset and $\alpha \subseteq \beta$ we have $\alpha = \beta$: indeed, let $a \in \beta$; by regularity of β there is $b \in \beta$ such that $b \in wc(a)$. Then (by maximality) either there is $c \in \alpha$ such that $\neg Pos(b \downarrow c)$, or $a \in \alpha$; but $\alpha \subseteq \beta$ implies that $\mathsf{Pos}(b \downarrow c)$ for all $c \in \alpha$, hence $a \in \alpha$.

On a compact locale a regular filter is prime if and only if it is completely prime (cf. e.g. [6], pg. 135). The following lemma expresses this property in our context.

Lemma 2.2. *Let $\mathcal{S}$ be a compact formal topology and let α be a regular subset of $\mathcal{S}$. If U is any subset of S, a is an element of α and $a \lhd U$, then there is a finite subset V_0, $V_0 \subseteq U$, and a basic neighbourhood b in α, such that $b \lhd V_0$.*

Proof. Let U be any subset of S, $a \in \alpha$ and $a \lhd U$. By regularity of α there is $b \in wc(a)$, $b \in \alpha$. Then $S \lhd b^* \cup a$, whence also $S \lhd b^* \cup U$. By compactness, there is a finite U_0, $U_0 \subseteq b^* \cup U$ such that $S \lhd U_0$. Since $U_0 \subseteq b^* \cup U$ means that $U_0(a)$ *true* implies $b^*(a)$ *true* $\vee$ $U(a)$ *true*, we have (by the disjunction property) $U_0 = V_0 \cup W_0$, with V_0 finite subset of U and W_0 finite subset of b^*. Thus, by lemma 1.6, $b \lhd V_0$.

By a simple modification of the corresponding argument in [9] we have:

Lemma 2.3. *Let $\mathcal{S}$ be a compact regular formal topology and let α be a regular subset of $\mathcal{S}$. If $a \in \alpha$ and $a \lhd \{a_1, ..., a_t\}$, there exist $b \in \alpha$ and finite subsets $V_1, ..., V_t$ such that $V_1 \subseteq wc(a_1), ..., V_t \subseteq wc(a_t)$ and $b \lhd V_1 \cup ... \cup V_t$.*

Proof. Suppose that $a \epsilon \alpha$ and $a \lhd \{a_1, a_2, ..., a_t\}$. By regularity of α, we can find b in α such that $b \epsilon wc(a)$, i.e. such that $S \lhd b^* \cup a$. By $a \lhd \{a_1, a_2, ..., a_t\}$ we then have $S \lhd b^* \cup \{a_1, a_2, ..., a_t\}$, which, by regularity of $\mathcal{S}$, gives $S \lhd b^* \cup wc(a_1) \cup ... \cup wc(a_t)$. Compactness then allows to extract a finite subset U_0 of $b^* \cup wc(a_1) \cup wc(a_2) \cup ... \cup wc(a_t)$ such that $S \lhd U_0$. Thus $S \lhd b^* \cup V_1 \cup V_2 \cup ... \cup V_t$, where, for $i = 1, ..., t$, V_i is a finite subset of $wc(a_i)$; by lemma 1.6, we then obtain $b \lhd V_1 \cup ... \cup V_t$, as desired.

In a regular topology points form a sub-collection of the collection of maximal regular subsets:

Proposition 2.4. *Let $\mathcal{S}$ be a regular formal topology. If α is a formal point of $\mathcal{S}$, α is also a maximal regular subset of $\mathcal{S}$.*

Proof. Let $\alpha \in \mathsf{Pt}(\mathcal{S})$ be a formal point. Then α clearly satisfies *1.*, *2.* and *3.* of 2.1 (to prove that $(\exists c \in S)(c \epsilon a \downarrow b \ \&\ c \epsilon \alpha)$ implies $a \epsilon \alpha$ and $b \epsilon \alpha$, simply observe that $c \epsilon a \downarrow b$ means $c \lhd a$ and $c \lhd b$, then use condition *iii.* of 1.2). Since $\mathcal{S}$ is regular, we have $a \lhd wc(a)$ for all a, hence (by *iii.* of 1.2) we have that also *4.* is satisfied. To prove maximality, let a, b such that $b \epsilon wc(a)$, that is, such that $S \lhd b^* \cup a$. Let d be a neighbourhood of α (which exists by *i.* of 1.2); by $S \lhd b^* \cup a$, we have $d \lhd b^* \cup a$, and thus (by *iii.* of 1.2) we obtain $(\exists c \epsilon (b^* \cup a))(c \epsilon \alpha)$. This precisely means that there is $c \epsilon \alpha$ such that $\neg\mathsf{Pos}(c \downarrow b)$, or $a \epsilon \alpha$, as required.

We can now prove our main result: if a topology is compact and regular, we have that also the converse of proposition 2.4 is true:

Theorem 2.5. *For any compact regular formal topology $\mathcal{S}$, the formal points of $\mathcal{S}$ are precisely the maximal regular subsets of $\mathcal{S}$.*

Proof. By the preceding proposition, formal points of $\mathcal{S}$ are maximal regular subsets of $\mathcal{S}$. To prove the converse, let $\alpha \subseteq S$ be maximal regular. Obviously *i.*, *ii.* and *iv.* of 1.2 are satisfied. We have then to prove that from $a \epsilon \alpha$ and $a \lhd U$ we can deduce $(\exists b \epsilon U)(b \epsilon \alpha)$. By lemma 2.2, it is enough to prove the claim for finite U. Let then

$U = \{a_1, ..., a_t\}$. By lemma 2.3, we can find t finite subsets $V_1, ..., V_t$ of S and a basic neighbourhood b of α such that, for $i = 1, ..., t$, $V_i \subseteq wc(a_i)$ and $b \lhd V_1 \cup ... \cup V_t$. Let

$$V_1 = \{v_1^1, ..., v_{k_1}^1\}, ..., V_t = \{v_1^t, ..., v_{k_t}^t\}.$$

To fix ideas, let us consider V_1: by maximality of α we have that

$$(a_1 \epsilon \alpha \vee (\exists d_1 \epsilon \alpha)\neg \mathsf{Pos}(v_1^1 \downarrow d_1)) \& ... \& (a_1 \epsilon \alpha \vee (\exists d_{k_1} \epsilon \alpha)\neg \mathsf{Pos}(v_{k_1}^1 \downarrow d_{k_1})),$$

which yields

$$a_1 \epsilon \alpha \vee ((\exists d_1 \epsilon \alpha)\neg \mathsf{Pos}(v_1^1 \downarrow d_1) \& ... \& (\exists d_{k_1} \epsilon \alpha)\neg \mathsf{Pos}(v_{k_1}^1 \downarrow d_{k_1})).$$

If $d_i \epsilon \alpha$, for $i = 1, ..., k_1$, then, by *2.* of 2.1, there is $e^1 \epsilon \alpha$, with $e^1 \epsilon d_1 \downarrow d_2 \downarrow ... \downarrow d_{k_1}$. It follows that $a_1 \epsilon \alpha \vee (\exists e^1 \epsilon \alpha)\neg \mathsf{Pos}(e^1 \downarrow V_1)$. Similarly we obtain $a_j \epsilon \alpha \vee (\exists e^j \epsilon \alpha)\neg \mathsf{Pos}(e^j \downarrow V_j)$, for $j = 2, ..., t$. We can thus deduce that

$$a_1 \epsilon \alpha \vee ... \vee a_t \epsilon \alpha \vee ((\exists e^1 \epsilon \alpha)\neg \mathsf{Pos}(e^1 \downarrow V_1) \& ... \& (\exists e^t \epsilon \alpha)\neg \mathsf{Pos}(e^t \downarrow V_t)).$$

Suppose now that $(\exists e^1 \epsilon \alpha)\neg \mathsf{Pos}(e^1 \downarrow V_1) \& ... \& (\exists e^t \epsilon \alpha)\neg \mathsf{Pos}(e^t \downarrow V_t)$. Thus, since b and e^j belong to α, by *2.* of 2.1. we have that $(\exists f \epsilon \alpha)$ $(f \epsilon (b \downarrow e^1 \downarrow ... \downarrow e^t))$, whence, by *3.* of 2.1, $\mathsf{Pos}(b \downarrow e^1 \downarrow ... \downarrow e^t)$. From $b \lhd V_1 \cup ... \cup V_t$ we obtain $(b \downarrow e^1 \downarrow ... \downarrow e^t) \lhd V_1 \cup ... \cup V_t$. Then, using monotonicity, $\mathsf{Pos}((b \downarrow e^1 \downarrow ... \downarrow e^t) \downarrow (V_1 \cup ... \cup V_t))$. However, for $i = 1, ..., t$, also $\neg\mathsf{Pos}((b \downarrow e^1 \downarrow ... \downarrow e^t) \downarrow V_i)$ (since $\neg\mathsf{Pos}(e^i \downarrow V_i)$), whence $\neg\mathsf{Pos}((b \downarrow e^1 \downarrow ... \downarrow e^t) \downarrow (V_1 \cup ... \cup V_t))$, which gives a contradiction. Thus $a_1 \epsilon \alpha \vee ... \vee a_t \epsilon \alpha$, as desired.

Theorem 2.5 thus authorizes to speak of maximal regular *points* of a compact regular formal topology. A relevant example of a compact regular formal topology is that giving rise to the space of linear functionals of norm ≤ 1 from a semi-normed space A to the reals[7]: in [4] the topology is proved to be compact; following the ideas in [11] regularity can also be proved.

[7] The space $\mathcal{L}(\mathcal{A})$ was presented as a locale in [10], [11]; it has been introduced via a formal topology in [4]: linear functionals are obtained as formal points of a formal topology whose base is intuitively given by a base of the *weak* topology*.

3. A GENERALIZATION: LOCALLY COMPACT REGULAR FORMAL TOPOLOGIES

In this section we show that the characterization we have obtained can be extended with a few modifications to the wider class of locally compact regular formal topologies; an instance of such topologies is the (topology giving rise to the) continuum (or, more in general, the n-dimensional Euclidean space)[8].

3.1 We essentially follow [12] by defining a formal topology $\mathcal{S}$ to be *locally compact* if there is an indexed family $i(a)(a \in S)$ of subsets of S such that for all $a \in S$ and $U \subseteq S$,

$$a \lhd U \iff (\forall b \,\epsilon\, i(a))(b \lhd U_b),$$

where U_b is a finite subset (depending on b) of U. In [12] it is shown that if a topology $\mathcal{S}$ satisfies such a condition, we also have $i(a) =_{\mathcal{S}} a$.

We say that a topology is *locally compact regular* if it is locally compact and regular.

The following lemma states a relation between the subsets $wc(a)$ and $i(a)$ in a locally compact regular topology.

Lemma 3.2. *If $\mathcal{S}$ is a locally compact regular topology, $i(a) \subseteq wc(a)$ for all a in S. Hence, a locally compact formal topology is regular if and only if $i(a) \subseteq wc(a)$ for all a.*

Proof. Let $b \,\epsilon\, i(a)$, then, since $a \lhd wc(a)$ there is a finite subset $U_b \equiv \{c_1, c_2 ..., c_k\}$ of $wc(a)$ such that $b \lhd \{c_1, c_2 ..., c_k\}$; to prove that $b \,\epsilon\, wc(a)$, observe that $S \lhd (a \cup c_i^*)$, for $i = 1, ..., k$, implies that $S \lhd (a \cup c_1^*) \downarrow (a \cup c_2^*) \downarrow ... \downarrow (a \cup c_k^*)$; by lemma 1.5 we thus have $S \lhd a \cup (c_1^* \downarrow c_2^* \downarrow ... \downarrow c_k^*)$, and, by lemma 1.7, we obtain $S \lhd a \cup (c_1 \cup ... \cup c_k)^*$. It is then enough to show that $(c_1 \cup ... \cup c_k)^* \subseteq b^*$: let $d \,\epsilon\, (c_1 \cup ... \cup c_k)^*$, i.e. $\neg \mathsf{Pos}(d \downarrow (c_1 \cup ... \cup c_k))$; if we had $\mathsf{Pos}(d \downarrow b)$, we would also have $\mathsf{Pos}(d \downarrow (c_1 \cup ... \cup c_k))$ (since

[8] See e.g. [4] for the definition of the topology $\mathcal{R}$ used to introduce the continuum.

$b \lhd \{c_1, c_2..., c_k\}$), contradiction. Hence $\neg\mathsf{Pos}(d \downarrow b)$, i.e. $d \epsilon b^*$; thus $(c_1 \cup ... \cup c_k)^* \subseteq b^*$, as required.

Compact regular and locally compact regular topologies are in the expected convenient relation:

Proposition 3.3. *If $\mathcal{S}$ is compact regular, $\mathcal{S}$ is locally compact regular with $i(a) \equiv wc(a)$, for all $(a \in S)$.*

Proof. $\mathcal{S}$ compact regular means that $S \lhd U \rightarrow S \lhd U_0$ for U_0 finite subset of U. Define $i(a)(a \in S)$ to be the family $wc(a)(a \in S)$: if $a \lhd U$ and $b \epsilon wc(a)$, we have $S \lhd b^* \cup a \lhd b^* \cup U$. By compactness and lemma 1.6, there is then a finite U_0, $U_0 \subseteq U$ such that $b \lhd U_0$; the other implication (that is $(\forall b \epsilon wc(a))(b \lhd U_b) \rightarrow a \lhd U$, with U_b finite subset of U depending on b) is obtained recalling that (by regularity) for all a, $a \lhd wc(a)$.

3.4 Let $\mathcal{S}$ be a locally compact formal topology. We define $\alpha \subseteq S$ to be a *continuous subset* of $\mathcal{S}$[9] if it satisfies conditions *1.*, *2.* ,*3.* of 2.1 and, instead of *4.*,

4'. $a \epsilon \alpha \rightarrow (\exists b)(b \epsilon i(a)\ \&\ b \epsilon \alpha)$

We say that α is a *maximal continuous subset* of $\mathcal{S}$ if moreover, for all a, b in S

5'. $b \epsilon i(a) \rightarrow ((\exists c \in S)(c \epsilon \alpha\ \&\ \neg Pos(b \downarrow c)) \vee a \epsilon \alpha)$

As one can expect, if $\mathcal{S}$ is a locally compact formal topology, α and β are respectively a maximal continuous subset and a continuous subset of $\mathcal{S}$, we have that $\alpha \subseteq \beta$ implies $\alpha = \beta$ (argue as in the preceding section, simply replacing $wc(a)$ with $i(a)$).

We now show that results analogous to lemmas 2.2 and 2.3 obtain for locally compact formal topologies (observe that regularity is not required in lemma 3.6).

[9]The relation, established in [12], between (what we here call) locally compact formal topologies and distributive continuous lattices has motivated this terminology.

Lemma 3.5. *Let $\mathcal{S}$ be a locally compact topology, and let α be a continuous subset of S. If $a \epsilon \alpha$, U is any subset of S and $a \lhd U$, then a neighbourhood b of α can be found such that $b \lhd U_b$, where U_b is a finite subset of U.*

Proof. From $a \epsilon \alpha$ and *4'.*, we obtain an element b such that $b \epsilon \alpha$ and $b \epsilon i(a)$; by local compactness we have that $b \lhd U_b$, where U_b is a finite subset of U.

Lemma 3.6. *Let $\mathcal{S}$ be a locally compact formal topology, and let α be a continuous subset of S. If $a \epsilon \alpha$ and $a \lhd \{a_1, ..., a_t\}$, finite subsets $V_1, ..., V_t$ and a neighbourhood b of α can be found such that, for $k = 1, ..., t$, $V_k \subseteq i(a_k)$ and $b \lhd V_1 \cup ... \cup V_t$.*

Proof. Since α is continuous, there is b in α, $b \epsilon i(a)$; since $a_k \lhd i(a_k)$ for $k = 1, ..., t$, by local compactness we have $b \lhd U_0$, with U_0 finite subset of $i(a_1) \cup ... \cup i(a_t)$. Thus $b \lhd V_1 \cup ... \cup V_t$, with, for $k = 1, ..., t$, V_k are finite subsets of $i(a_k)$.

Lemmas 3.5 and 3.6 allow to prove that:

Proposition 3.7. *In a locally compact formal topology $\mathcal{S}$ a maximal continuous subset of $\mathcal{S}$ is a formal point of $\mathcal{S}$.*

Proof. Let α be a maximal continuous subset. The only non-obvious fact to prove is that, from $a \epsilon \alpha$ and $a \lhd U$, we can deduce $(\exists b \epsilon U)(b \epsilon \alpha)$. By lemma 3.5 we may consider only finite U. Let then $U = \{a_1, ..., a_t\}$, $a \epsilon \alpha$ and $a \lhd \{a_1, ..., a_t\}$, we want to prove that $a_1 \epsilon \alpha \vee ... \vee a_t \epsilon \alpha$. By lemma 3.6, we can find finite subsets $V_1, ..., V_t$ and $b \epsilon \alpha$ such that, for $k = 1, ..., t$, $V_k \subseteq i(a_k)$, and $b \lhd V_1 \cup ... \cup V_t$. The proof then proceeds in complete analogy with that of theorem 2.5.

Now a characterization of the points of locally compact regular formal topologies corresponding to the one proved in the previous section for compact regular topologies can be obtained:

Theorem 3.8. *For any locally compact regular formal topology $\mathcal{S}$, the formal points of $\mathcal{S}$ are precisely the maximal continuous subsets of $\mathcal{S}$.*

Proof. By the preceding proposition we only need to show that if $\alpha \in \mathsf{Pt}(\mathcal{S})$ is a formal point, then it is a maximal continuous subset of $\mathcal{S}$. We have already showed (proposition 2.4) that a formal point α satisfies *1., 2., 3.*. Since $\mathcal{S}$ is locally compact, we have $a \lhd i(a)$ for all a, and thus by *iii*) of 1.2 we have that also *4'*. is satisfied. To prove *5'*., let a, b such that $b \in i(a)$; by lemma 3.2, $b \in wc(a)$, i.e. $S \lhd b^* \cup a$. Then we may argue exactly as in the proof of proposition 2.4 to conclude that there is $c \in \alpha$ such that $\neg\mathsf{Pos}(c \downarrow b)$, or $a \in \alpha$.

Observe that the characterization we have proved in this section is a proper generalization of that obtained for compact regular topologies: indeed, by proposition 3.3, we could have derived the characterization provided in 2.5 by theorem 3.8.

Acknowledgments

The author is grateful to Giovanni Sambin and Peter Schuster, whose comments and suggestions have led to several improvements.

References

[1] Aczel, Peter, "The type-theoretic interpretation of constructive set theory". In: S. Miettinen, S. Väänänen, eds. *Proceedings of Symposia in Mathematical Logic*, Oulu 1974, and Helsinki, 1975. Report no.2 (University of Helsinki, Department of Philosophy, 1977) pp. 1-32.

[2] Bishop, Errett, and Bridges, Douglas, *Constructive Analysis* (Springer-Verlag), Berlin, 1985.

[3] Cederquist, Jan, and Coquand, Thierry, "Entailment relations and distributive lattices". Logic Colloquium '98 (Prague), 127–139, *Lecture Notes Logic*, 13, Assoc. Symbol. Logic, Urbana, IL, 2000.

[4] Cederquist, Jan, Coquand, Thierry, and Negri, Sara, "The Hahn-Banach theorem in Type Theory". In: *Twenty-Five years of Constructive Type Theory*, Proceedings of a Congress held in Venice, October 1995, G. Sambin and J. Smith eds. (Oxford University Press), 1998, pp. 57-72.

[5] Coquand, Thierry, "Formal Compact Hausdorff Spaces", unpublished.

[6] Johnstone, Peter T., *Stone Spaces*, Cambridge University Press, 1982.

[7] Martin-Löf, Per, *Notes on Constructive Mathematics* (Almqvist and Wiksell) Stockholm, 1970.

[8] Martin-Löf, Per, *Intuitionistic Type Theory*. Studies in Proof Theory. Lecture Notes, 1. (Bibliopolis), Napoli, 1984.

[9] Mulvey, Chris J., "The maximality of filters". Special issue in honor of B. Banaschewski. *Journal of Pure and Applied Algebra* 68 (1990), no. 1-2, 253–258.

[10] Mulvey, Chris J., and Pelletier, Joan W., "The dual locale of a seminormed space". *Cahiers de Topologie et Geometrie Differentielle*, vol. 23 (1987), n.1, pp.73-92.

[11] Mulvey, Chris J., and Pelletier, Joan W., "A globalization of the Hahn-Banach theorem". *Advances in Mathematics* 89 (1991), pp. 1-60.

[12] Negri, Sara, "Continuous lattices in formal topology, 1997". In *Lecture Notes in Computer Science*, E. Gimenez and C. Paulin (eds), selected papers of the Types meeting, Aussois, 1996, Lecture Notes in Computer Science 1512, pp. 333-353, Springer, 1998.

[13] Nordström, Bengt, Petersson, Kent, and Smith, Jan, *Programming in Martin-Löf's Type Theory* (Clarendon Press), Oxford, 1990.

[14] Sambin, Giovanni, A monograph on Formal Topology, unpublished manuscript.

[15] Sambin, Giovanni, "Intuitionistic formal spaces - a first communication". In *Mathematical Logic and its Applications*, D. Skordev, ed. Plenum, New York - London 1987, 187-204

[16] Sambin, Giovanni, "Formal topology and domains". Proceedings of the Workshop on Domains IV, Informatik-Bericht Nr. 99-01, Universität-GH Siegen, 1999.

[17] Sambin, Giovanni, and Gebellato, Silvia, "A preview of the basic picture: a new perspective on formal topology". In *Types for Proofs and Programs. International Workshop, TYPES '98, Kloster Irsee, Germany, March 27-31, 1998, selected papers.* Lecture Notes in Computer Science, 1657, Springer, eds. T. Altenkirch and W. Naraschewski and B. Reus, 1999, pp. 194-207.

[18] Sambin, Giovanni, and Valentini, Silvio, "Building up a toolbox for Martin-Löf's type theory: subset theory". In: *Twenty-Five years of Constructive Type Theory.* Proceedings of a Congress held in Venice, October 1995, G. Sambin and J. Smith eds., Oxford University Press, 1998, pp. 221-244.

EMBEDDING A LINEAR SUBSET OF $\mathcal{B}(H)$ IN THE DUAL OF ITS PREDUAL

Luminiţa Vîţă (Dediu)
Department of Mathematics and Statistics, University of Canterbury
Private Bag 4800, Christchurch, New Zealand
Luminita.DEDIU@ugal.ro

Keywords: weak-operator topology, ultraweakly continuous, predual, trace class

Abstract: The embedding of a linear set of bounded operators on a separable Hilbert space as a dense subset of the dual of its predual is explored constructively.

In this paper we continue the study of spaces of operators on a Hilbert space within constructive mathematics, as part of a programme for the systematic constructive development of the theory of operator algebras [6, 7].

The constructive framework within which we operate was erected by the late Errett Bishop [2], under the requirement that "existence" must be strictly interpreted as "computability" relative to some notion of algorithm. By not specifying formally what he meant by an algorithm, other than insisting that it must be executable by a finite number of human beings or computers in a finite time, Bishop enabled his work to have a variety of interpretations; in particular, all theorems of Bishop's constructive mathematics hold within classical (that is, traditional) mathematics and Brouwer's intuitionistic mathematics, and under a recursive interpretation.

Note, incidentally, that Bishop's algorithmic interpretation of existence makes no demands about the complexity of the algorithms used; at present, constructive mathematics addresses questions of computability in principle, rather than computability in practice. (See, however, [13].)

The main practical difference between constructive and classical mathematics is one of logic. The algorithmic interpretation of existence forces us to re–examine the interpretation of each logical connective and quantifier, and leads, it seems inevitably, to the use of intuitionistic logic, as

P. Schuster, U. Berger and H. Osswald (eds.),
Reuniting the Antipodes – Constructive and Nonstandard Views of the Continuum, 55–61.

originally codified by Heyting [11]. We also have to modify the underlying non-logical principles—for example, the axioms for the real line **R** [5] or the axioms of Zermelo–Fraenkel set theory [14]—so as to ensure that one cannot derive from them classical principles, such as the law of excluded middle, that are independent of intuitionistic logic. Once these precautions have been taken, we are, to all (non-philosophical) intents and purposes, working with intuitionistic Zermelo–Fraenkel set theory using intuitionistic logic. (For additional background material on constructive mathematics see [1, 2, 3, 5, 9, 15].)

Let H be a Hilbert space (not necessarily separable), $\mathcal{B}(H)$ the space of bounded linear operators on H, and $\mathcal{B}_1(H)$ the unit ball of $\mathcal{B}(H)$. Recall that the **weak–operator** topology τ_w on $\mathcal{B}(H)$ is the weakest topology with respect to which the mapping $T \mapsto \langle Tx, y\rangle$ is continuous for all $x, y \in H$. This topology is determined by the seminorms $T \mapsto |\langle Tx, y\rangle|$, where x, y run through any dense subset of $\mathcal{B}_1(H)$. Classically, $\mathcal{B}_1(H)$ is τ_w–compact; but constructively the most we can say, in general, is that it is τ_w–totally bounded [6].

Let $\mathcal{R}$ be a linear subset of $\mathcal{B}(H)$, let $\mathcal{R}1 = \mathcal{R} \cap \mathcal{B}_1(H)$, and let $R_\sharp$ denote the linear space of all linear functionals on R that are **ultraweakly continuous**—that is, τ_w–uniformly continuous on $\mathcal{R}_1$. If $\mathcal{R}_1$ is τ_w–totally bounded, then

$$\|f\| = \sup\{|f(T)| : T \in \mathcal{R}_1\}$$

defines a norm on $\mathcal{R}_\sharp$; taken with this norm, $\mathcal{R}_\sharp$ is called the **predual** of $\mathcal{R}$.

For convenience, we denote by $f_{x,y}$ the ultraweakly continuous functional $T \mapsto \langle Tx, y\rangle$ on $\mathcal{B}(H)$. We also recall that the weak* topology on the dual X^* of a locally convex space is the topology defined by the seminorms $f \mapsto |f(x)|$ $(x \in X)$; see [9] and [10].

Theorem 1 *Let $\mathcal{R}$ be a linear subset of $\mathcal{B}(H)$ such that $\mathcal{R}_1$ is totally bounded in the weak–operator topology τ_w, and define a mapping ϕ of $\mathcal{R}$ into the dual space $\mathcal{R}_\sharp^*$ of $\mathcal{R}_\sharp$ by*

$$\phi(T)(f) = f(T) \quad (T \in \mathcal{R}).$$

Then ϕ is one–one and linear, and is uniformly continuous on $\mathcal{R}_1$. Moreover, $\phi(\mathcal{R}_1)$ is weak–dense in the unit ball of $\mathcal{R}_\sharp^*$, and the restriction of ϕ^{-1} to $\phi(\mathcal{R}_1)$ is uniformly continuous with respect to the weak*–topology on $\mathcal{R}_\sharp^*$ and the weak–operator topology on $\mathcal{R}_1$.*

PROOF. Since ϕ is clearly linear, to prove that it is one–one we need only show that its kernel is $\{0\}$. But if $\phi(T) = 0$, then we have $\langle Tx, y\rangle = \phi(T)\,(f_{x,y}) = 0$ for all $x, y \in H$; whence $T = 0$.

For each $f \in \mathcal{R}_\sharp$ the mapping $T \mapsto \phi(T)(f)$ equals f and so is τ_w–uniformly continuous on $\mathcal{R}_1$. It follows immediately that ϕ is uniformly continuous as a mapping of $(\mathcal{R}_1, \tau_w)$ into $\mathcal{R}_\sharp^*$ (with the weak*–topology). Hence $K = \phi(\mathcal{R}_1)$ is weak*–totally bounded, and therefore located, in $\mathcal{R}_\sharp^*$ (see [7, 10]). Let u be an element of $\mathcal{R}_\sharp^*$, let $\{f_1, \dots, f_N\}$ be a finitely enumerable subset of $\mathcal{R}_\sharp$, and let ε be a positive number. To prove that $\phi(\mathcal{R}_1)$ is weak*–dense in the unit ball $(\mathcal{R}_\sharp^*)_1$ of $\mathcal{R}_\sharp^*$, it is enough to show that K intersects the neighbourhood

$$V = \left\{v \in \left(\mathcal{R}_\sharp^*\right)_1 : |(u-v)\,(f_k)| < 3\varepsilon \;\; (1 \le k \le N)\right\}$$

of u in $(\mathcal{R}_\sharp^*)_1$. To this end, choose a finite–dimensional subspace $\mathcal{G}$ of $\mathcal{R}_\sharp$ such that

$$\inf\{\|f_k - g\| : g \in \mathcal{G}\} < \varepsilon \quad (1 \le k \le N)$$

([3], page 308, Lemma (2.5)); for each k $(1 \le k \le N)$, then choose $g_k \in \mathcal{G}$ such that $\|f_k - g_k\| < \varepsilon$. The dual space $\mathcal{G}^*$ of $\mathcal{G}$ is a finite–dimensional Banach space with respect to the usual norm defined by

$$\|v\|' = \sup\{|v(g)| : g \in \mathcal{G},\, \|g\| \le 1\}\,.$$

Since $K \subset (\mathcal{R}_\sharp^*)_1$, and $(\mathcal{R}_\sharp^*)_1$ is a subset of the unit ball of $(\mathcal{G}^*)_1$, we can regard u and, for each $T \in \mathcal{R}_1$, the functional $\phi(T)$ as elements of $(\mathcal{G}^*)_1$. Now suppose that

$$\inf\left\{\|u - \phi(T)\|' : T \in \mathcal{R}_1\right\} \ne 0. \tag{5.1}$$

Note that K, being weak*–totally bounded, is located in $\mathcal{G}_1$. By Corollary (4.4) on page 341 of [3], there exists a linear functional ψ, with norm 1, on $\mathcal{G}^*$ such that

$$|\psi(u)| > \sup\{|\psi(v)| : v \in K\}\,.$$

Since $\mathcal{G}$ is finite–dimensional, ψ is weak*–uniformly continuous on $(\mathcal{G}^*)_1$; so, by Corollary (6.9) on page 357 of [3], there exists $g \in \mathcal{G}$ such that $\psi(v) = v(g)$ for all $v \in \mathcal{G}^*$. In particular,

$$\begin{aligned}|u(g)| &> \sup\{|v(g)| : v \in K\}\\ &= \sup\{|\phi(T)(g)| : T \in \mathcal{R}_1\}\\ &= \sup\{g(T) : T \in \mathcal{R}_1\}\\ &= \|g\|\,,\end{aligned}$$

which is absurd since $u \in (\mathcal{R}_\sharp^*)_1$. We conclude that (1) is false, and hence that

$$\|u - \phi(T_0)\|' < \frac{\varepsilon}{M+1}$$

for some $T_0 \in \mathcal{R}_1$, where

$$M = \max_{1 \leq k \leq N} \|g_k\| .$$

For each k $(1 \leq k \leq N)$ we now have

$$\begin{aligned} |(u - \phi(T_0))\,(f_k)| &\leq |(u - \phi(T_0))\,(f_k - g_k)| + |(u - \phi(T_0))\,(g_k)| \\ &\leq 2\,\|f_k - g_k\| + \|u - \phi(T_0)\|'\,\|g_k\| \\ &< 2\varepsilon + \frac{\varepsilon}{M+1}M \\ &< 3\varepsilon; \end{aligned}$$

in other words, $\phi(T_0) \in V$. Since $\phi(T_0) \in K$, this completes the proof that $\phi(\mathcal{R}_1)$ is dense in $(\mathcal{R}_\sharp^*)_1$.

Finally, the uniform continuity of the inverse mapping on K follows from the identity

$$|\langle Tx, y\rangle| = |\phi(T)\,(f_{x,y}))| \quad (x, y \in H;\ T \in \mathcal{R}_1),$$

with reference to the definitions of the weak*– and weak–operator topologies, and to Proposition 1.2.8 on page 19 of [12]. Q.E.D.

We proved in [7] (see also [8]) that, under the hypotheses of Theorem 1, the ultraweakly continuous linear functionals on $\mathcal{R}$ extend to ultraweakly continuous linear functionals on $\mathcal{B}(H)$ and are precisely those functionals f_A mapping T to Trace(TA), with A a trace–class operator on H. The norm of f_A on $\mathcal{R}$ is

$$\|f_A\|_{\mathcal{R}} = \sup\{|\mathrm{Trace}(TA)| : T \in \mathcal{R}_1\},$$

which in the case $\mathcal{R} = \mathcal{B}(H)$ equals the trace–class norm

$$\|A\|_2 = \mathrm{Trace}\,(A)$$

of A (see [4]).

Taken with Theorem 1, these observations lead to

Corollary 2 *Let $\mathcal{R}$ be a linear subset of $\mathcal{B}(H)$ such that $\mathcal{R}_1$ is totally bounded in the weak–operator topology τ_w. Let $\mathcal{T}(H)$ denote the set of trace–class operators on H, taken with the norm*

$$\|A\|_{\mathcal{R}} = \sup\{|\mathrm{Trace}(TA)| : T \in \mathcal{R}_1\},$$

Then

$$\Phi(T)(A) = \mathrm{Trace}(TA) \quad (T \in \mathcal{R},\ A \in \mathcal{T}(H))$$

defines a one–one linear mapping Φ of $\mathcal{R}$ into the dual space $\mathcal{T}(H)^*$ with the following properties.

(i) *$\Phi(\mathcal{R}_1)$ is dense in the unit ball $\mathcal{T}(H)^*_1$ of $\mathcal{T}(H)^*$.*

(ii) *Φ is uniformly continuous on $\mathcal{R}_1$.*

(iii) *the restriction of Φ^{-1} to $\Phi(\mathcal{R}_1)$ is uniformly continuous relative to the weak**–topology on $\Phi(\mathcal{R}_1)$ and the weak–operator topology on $\mathcal{R}_1$.*

Corollary 3 *Under the hypotheses of* Theorem 1 *and* Corollary 2, *the following conditions are equivalent.*

(i) *$\mathcal{R}_1$ is weak–operator complete.*

(ii) *ϕ maps $\mathcal{R}_1$ onto the unit ball of $\mathcal{R}^*_\sharp$.*

(iii) *Φ maps $\mathcal{R}_1$ onto the unit ball of $\mathcal{T}(H)^*$ relative to the norm $\|\cdot\|_{\mathcal{R}}$.*

PROOF. This is a special case of the following general lemma about metric spaces, whose straightforward proof we omit. Q.E.D.

Lemma 4 *Let X be a metric space, Y a complete metric space, and ϕ a one–one uniformly continuous mapping of X onto a dense subset of Y such that ϕ^{-1} is uniformly continuous on $\phi(X)$. Then X is complete if and only if $\phi(X) = Y$.*

Classically, any von Neumann algebra—that is, weak–operator closed *–subalgebra of $\mathcal{B}(H)$—can be identified, via the mapping ϕ, with the dual of its predual $\mathcal{R}_\sharp$ ([12], Vol. 2, page 482). If this were provable constructively, then we could use Theorem 1 to prove that $\mathcal{B}_1(H)$ is τ_w–complete, which, as mentioned above, we cannot do within constructive mathematics. Thus Theorem 1 appears to be the best general constructive result of its type.

References

[1] Michael J. Beeson, *Foundations of Constructive Mathematics*, Springer–Verlag, Heidelberg, 1985.

[2] Errett Bishop, *Foundations of Constructive Analysis*, McGraw–Hill, New York, 1967.

[3] Errett Bishop and Douglas Bridges, *Constructive Analysis*, Grundlehren der math. Wissenschaften **279**, Springer–Verlag, Heidelberg, 1985.

[4] Douglas Bridges, "A constructive look at positive linear functionals on $L(H)$", Pacific J. Math. **95**(1), 11–25, 1981.

[5] Douglas Bridges, "A constructive look at the real number line", in *Synthèse: Real Numbers, Generalizations of the Reals and Theories of Continua* (P. Ehrlich, ed.), pp. 29–92, Kluwer Academic Publishers, Amsterdam, 1994.

[6] Douglas Bridges and Luminiţa Dediu, "Weak–operator continuity and the existence of adjoints", Math. Logic Quart. **45**, 203–206, 1999.

[7] Douglas Bridges and Luminiţa Dediu, "Constructing extensions of ultraweakly continuous linear functionals", J. Functional Analysis, in press.

[8] D.S. Bridges and N.F. Dudley Ward, "Constructing ultraweakly continuous functionals on $\mathcal{B}(H)$", Proc. Amer. Math. Soc. **126**(11), 3347–3353, 1998.

[9] Douglas Bridges and Fred Richman, *Varieties of Constructive Mathematics*, London Math. Soc. Lecture Notes **97**, Cambridge University Press, 1987.

[10] Luminiţa Dediu and Douglas Bridges, "Constructive notes on uniform and locally convex spaces" in *Proceedings of 12th International Symposium, FCT '99 Iaşi, Romania*, Springer Lecture Notes in Computer Science **1684**, 195–203, 1999.

[11] A. Heyting, "Die formalen Regeln der intuitionistischen Logik", Sitzungsber. Preuß. Akad. Wiss. Berlin, 42–56, 1930.

[12] R.V. Kadison and J.R. Ringrose, *Fundamentals of the Theory of Operator Algebras* (Vol. I), Academic Press, New York, 1983.

[13] Ker-i Ko, *Complexity Theory of Real Functions*, Birkhaüser, Boston, 1991.

[14] John Myhill, "Some properties of intuitionistic Zermelo–Fraenkel set theory", in: *Cambridge Summer School in Mathematical Logic* (A. Mathias, H. Rogers, eds.), Lecture Notes in Mathematics **337**, Springer–Verlag, Berlin, 1973, pp. 206–231.

[15] A.S. Troelstra and D. van Dalen, *Constructivism in Mathematics: An Introduction* (two volumes), North Holland, Amsterdam, 1988.

NONSTANDARD ANALYSIS BY MEANS OF IDEAL VALUES OF SEQUENCES

Mauro Di Nasso
Università di Pisa, Dipartimento di Matematica Applicata "U. Dini"
Via Bonanno Pisano 25/B, 56126 Pisa, Italy
dinasso@dma.unipi.it

Keywords: Nonstandard analysis, ultrapowers

Mathematical Subject Classification: Primary: 26E35, 03E70, 03H05. Secondary: 03C20, 03E35.

Abstract: A notion of *ideal* value of **N**-sequences is axiomatized through elementary properties. The resulting theory ZFC[$\boldsymbol{\alpha}$] provides nonstandard analysis with a general foundational framework.

Introduction

In this paper, the axiomatic system ZFC[$\boldsymbol{\alpha}$] is presented. It is a generalization in a set theoretic context of an approach to nonstandard analysis recently given by V. Benci in [Be]. In that paper, nonstandard analysis is presented by postulating, besides a nonstandard embedding $* : V(X) \to V(X)$ from a superstructure into itself, the existence of a *pre-extension* for each function defined on the set **N** of natural numbers. Axioms are then given that correlate pre-extensions with the enlarging map $*$. Here, we formalize the idea of pre-extension by means of a notion of *ideal value.* Precisely, we consider a new *ideal* natural number $\boldsymbol{\alpha}$, we postulate a (unique) ideal value $\varphi(\boldsymbol{\alpha})$ for each **N**-sequences φ, and axiomatize the properties of such ideal values.

The idea of adding a new logical constant to behave as an "infinite" natural number is not new. Precisely, it goes back to the pioneering work by C. Schmieden and D. Laugwitz [SL], appeared before Robinson's presentation of nonstandard analysis. A general way of formally giving a "nonstandard extension" of a given first order theory by means of a new constant symbol, can be found in Laugwitz' work [La].

P. Schuster, U. Berger and H. Osswald (eds.),
Reuniting the Antipodes – Constructive and Nonstandard Views of the Continuum, 63–73.

In this paper, an approach is given in the full generality of set theory. As a consequence of our axioms, a nonstandard extension $^*A$ is determined for each set A, and all basic principles of nonstandard analysis, including *transfer* and countable *saturation*, are proved. All principles of mathematics as formalized by Zermelo-Fraenkel set theory with choice are assumed, with the only exception of regularity. Thus, working in ZFC[$\boldsymbol{\alpha}$], one can manipulate sets as in ordinary mathematics with no restrictions. Also, nonstandard arguments can be iterated because the enlarging map $*$ is defined for all sets. For instance, one can consider the *hyper*-hyperreal numbers $^*(^*\mathbf{R})$ or the nonstandard extension $^*\mathcal{L}$ of a Loeb measure space $\mathcal{L}$ etc. Another feature of our axiomatic system ZFC[$\boldsymbol{\alpha}$] is that it consists of "usual" mathematics plus a few simple properties for $\boldsymbol{\alpha}$ expressed in truly elementary terms. In principle, this makes it possible to present ZFC[$\boldsymbol{\alpha}$] even at the freshman level. For basic notions from set theory and model theory, we refer to [JE] and [CK], respectively.

1. THE AXIOMS

Let $\boldsymbol{\alpha}$ denote an "ideal" constant. ZFC[$\boldsymbol{\alpha}$] is the theory determined by the following seven groups of axioms.

1. *ZFC⁻*

All principles of mathematics, as axiomatized by Zermelo-Fraenkel set theory with choice but without regularity, are assumed. The replacement and separation schemata are also considered for formulas where the symbol $\boldsymbol{\alpha}$ occurs.

2. $\boldsymbol{\alpha} \notin \mathbf{N}$.

3. *Every sequence $\varphi : \mathbf{N} \to A$ has a unique ideal value $\varphi(\boldsymbol{\alpha})$.*

We remark that in general $\varphi(\boldsymbol{\alpha}) \notin A$. For simplicity, we will use the same symbol to denote a sequence $\varphi : \mathbf{N} \to A$ and its *pre-extension* $\varphi : \mathbf{N} \cup \{\boldsymbol{\alpha}\} \to A \cup \{\varphi(\boldsymbol{\alpha})\}$. The remaining axioms give some natural properties of the ideal values. In the following, by sequence we will always mean a function whose domain is the set $\mathbf{N}$ of natural numbers.

4. *The ideal value of the constant map $c_\emptyset : n \mapsto \emptyset$ is $c_\emptyset(\boldsymbol{\alpha}) = \emptyset$, and the pre-extension of the identity map $1_\mathbf{N} : n \mapsto n$ on $\mathbf{N}$ is the identity map on $\mathbf{N} \cup \{\boldsymbol{\alpha}\}$, i.e. $1_\mathbf{N}(\boldsymbol{\alpha}) = \boldsymbol{\alpha}$.*

5. *For every sequence φ of nonempty sets:*

$$\varphi(\boldsymbol{\alpha}) = \{\psi(\boldsymbol{\alpha}) : \psi \textit{ is a sequence with } \psi(n) \in \varphi(n) \textit{ for all } n \in \mathbf{N}\}$$

In other words, the elements of the ideal value $\varphi(\boldsymbol{\alpha})$ are precisely the ideal values of sequences that are "point-wise" elements of φ. The next axiom states that unions and intersections are preserved at the ideal point $\boldsymbol{\alpha}$.

6. *If $\vartheta(n) = \varphi(n) \cup \psi(n)$ for all $n \in \mathbf{N}$ then $\vartheta(\boldsymbol{\alpha}) = \varphi(\boldsymbol{\alpha}) \cup \psi(\boldsymbol{\alpha})$, and if $\vartheta(n) = \varphi(n) \cap \psi(n)$ for all $n \in \mathbf{N}$ then $\vartheta(\boldsymbol{\alpha}) = \varphi(\boldsymbol{\alpha}) \cap \psi(\boldsymbol{\alpha})$.*

The last axiom gives a natural *coherence* property of the ideal values with respect to compositions.

7. *If φ, ψ are sequences taking values in a set E, and if f is any function defined on E, then $\varphi(\boldsymbol{\alpha}) = \psi(\boldsymbol{\alpha}) \Rightarrow (f \circ \varphi)(\boldsymbol{\alpha}) = (f \circ \psi)(\boldsymbol{\alpha})$*

We remark that the above presentation of ZFC[$\boldsymbol{\alpha}$] is somewhat "informal". In particular, axiom 3 cannot be formalized in the language $\mathcal{L} = \{\in, \boldsymbol{\alpha}\}$ where $\boldsymbol{\alpha}$ is a constant symbol. A rigorous formalization can be given instead by using a binary relation symbol Ω, which is postulated to be a function whose domain is the class of all $\mathbf{N}$-sequences. In fact, by denoting $\Omega(\varphi) = \varphi(\boldsymbol{\alpha})$, all axioms of ZFC[$\boldsymbol{\alpha}$] can be reformulated as first order sentences of the language $\mathcal{L} = \{\in, \Omega\}$ (see [BD] for details). Here, we decided to give the "informal" presentation because we believe it is simpler and closer to intuition.

Quite surprisingly, we will see that the above axioms are all that is needed in order to get a countably saturated nonstandard embedding $*$ defined for all sets. We remark that in the given presentation, not a single technical notion from mathematical logic is used. In fact, axiom 1 can be skipped by implicitly assuming all "usual" principles of mathematics, and the remaining axioms are all expressed in truly basic terms, such as unions, intersections, or compositions of functions. Let us start developing our theory. The following proposition itemizes a number of basic set operations which are preserved at the ideal values.

Proposition 1. *Let φ, ψ, ϑ be any sequences. Then*

(i) $\vartheta(n) = \{\varphi(n), \psi(n)\}$ *for all* $n \Rightarrow \vartheta(\boldsymbol{\alpha}) = \{\varphi(\boldsymbol{\alpha}), \psi(\boldsymbol{\alpha})\}$.

(ii) $\vartheta(n) = \langle\varphi(n), \psi(n)\rangle$ *for all* $n \Rightarrow \vartheta(\boldsymbol{\alpha}) = \langle\varphi(\boldsymbol{\alpha}), \psi(\boldsymbol{\alpha})\rangle$.

(iii) $\vartheta(n) = \varphi(n) \setminus \psi(n)$ *for all* $n \Rightarrow \vartheta(\boldsymbol{\alpha}) = \varphi(\boldsymbol{\alpha}) \setminus \psi(\boldsymbol{\alpha})$.

(iv) $\vartheta(n) = \varphi(n) \times \psi(n) \neq \emptyset$ *for all* $n \Rightarrow \vartheta(\boldsymbol{\alpha}) = \varphi(\boldsymbol{\alpha}) \times \psi(\boldsymbol{\alpha})$.

Proof. The proofs are straightforward applications of the axioms, thus here we only give two of them as examples. Let us consider pairs. By axiom 5, if $f(n) = \{\varphi(n)\}$ for all n then $f(\boldsymbol{\alpha}) = \{\varphi(\boldsymbol{\alpha})\}$. Thus by axiom 6, $\vartheta(n) = \{\varphi(n), \psi(n)\} = \{\varphi(n)\} \cup \{\psi(n)\}$ for all n, implies $\vartheta(\boldsymbol{\alpha}) = \{\varphi(\boldsymbol{\alpha})\} \cup \{\psi(\boldsymbol{\alpha})\} = \{\varphi(\boldsymbol{\alpha}), \psi(\boldsymbol{\alpha})\}$. As for the setminus operation, notice that $\vartheta(n) = \varphi(n) \setminus \psi(n) \Leftrightarrow \vartheta(n) \cap \psi(n) = c_{\emptyset}(n)$ and $\vartheta(n) \cup (\varphi(n) \cap \psi(n)) = \varphi(n)$. By applying axioms 4 and 6, we get $\vartheta(\boldsymbol{\alpha}) \cap \psi(\boldsymbol{\alpha}) = \emptyset$ and $\vartheta(\boldsymbol{\alpha}) \cup (\varphi(\boldsymbol{\alpha}) \cap \psi(\boldsymbol{\alpha})) = \varphi(\boldsymbol{\alpha})$, hence $\vartheta(\boldsymbol{\alpha}) = \varphi(\boldsymbol{\alpha}) \setminus \psi(\boldsymbol{\alpha})$. ⊣

If a sequence is constantly equal to some natural number, then it takes the same value at the ideal point as well.

Proposition 2. *Let n be any given natural number and let c_n denote the sequence constantly equal to n. Then $c_n(\boldsymbol{\alpha}) = n$.*

Proof. Recall that in Zermelo-Fraenkel set theory, natural numbers are coded as von Neumann naturals, i.e. 0 is the emptyset and, by induction, $n + 1$ is the set $n \cup \{n\}$. Thus the case $n = 0$ is given by axiom 4. Now proceed by induction and consider c_{n+1}. For all $m \in \mathbf{N}$, $c_{n+1}(m) = n + 1 = n \cup \{n\} = c_n(m) \cup \{c_n(m)\}$. By the inductive hypothesis, $c_{n+1}(\boldsymbol{\alpha}) = c_n(\boldsymbol{\alpha}) \cup \{c_n(\boldsymbol{\alpha})\} = n \cup \{n\} = n + 1$. ⊣

The next proposition shows that sequences that are eventually equal have the same ideal values, while sequences that are eventually different have different ideal values.

Proposition 3. *Let φ and ψ be two sequences.*

(i) If $\varphi(n) = \psi(n)$ for all but finitely many n, then $\varphi(\boldsymbol{\alpha}) = \psi(\boldsymbol{\alpha})$.

(ii) If $\varphi(n) \neq \psi(n)$ for all but finitely many n, then $\varphi(\boldsymbol{\alpha}) \neq \psi(\boldsymbol{\alpha})$.

Proof. *(i)* Let $\{n \in \mathbf{N} : \varphi(n) \neq \psi(n)\} = \{n_1, \ldots, n_k\}$ be finite. Fix $n_0 \in \mathbf{N}$ with $\varphi(n_0) = \psi(n_0)$ and define the sequence ζ as follows. $\zeta(n) = n_0$ if $\varphi(n) = \psi(n)$, and $\zeta(n) = n$ otherwise. We claim that $\zeta(\boldsymbol{\alpha}) = n_0$. Notice that for all $n \in \mathbf{N}$, $\{\zeta(n)\} \setminus \{n\} \subseteq \{n_0\}$, hence $\{\zeta(\boldsymbol{\alpha})\} \setminus \{\boldsymbol{\alpha}\} \subseteq \{n_0\}$. Moreover, $\zeta(\boldsymbol{\alpha}) \neq \boldsymbol{\alpha}$ because $\zeta(n) \in \{n_0, n_1, \ldots, n_k,\}$ for all $n \in \mathbf{N} \Rightarrow \zeta(\boldsymbol{\alpha}) \in \{n_0, n_1, \ldots, n_k,\}$, while $\boldsymbol{\alpha} \notin \mathbf{N}$. Now, take $A = \text{range}(\varphi)$ and define the sequence ξ by setting $\xi(n_0) = \emptyset$ and $\xi(n) = A$ for $n \neq n_0$. Then $\{\varphi(n)\} \setminus \{\psi(n)\} \subseteq \xi(\zeta(n))$ for all $n \in \mathbf{N}$ implies $\{\varphi(\boldsymbol{\alpha})\} \setminus \{\psi(\boldsymbol{\alpha})\} \subseteq (\xi \circ \zeta)(\boldsymbol{\alpha}) =$ (by axiom 7) $(\xi \circ c_{n_0})(\boldsymbol{\alpha}) = c_{\emptyset}(\boldsymbol{\alpha}) = \emptyset$, hence $\varphi(\boldsymbol{\alpha}) = \psi(\boldsymbol{\alpha})$. *(ii)* Since $\{\varphi(n)\} \setminus \{\psi(n)\} = \{\varphi(n)\}$ for all but finitely many n, by *(i)* it follows that $\{\varphi(\boldsymbol{\alpha})\} \setminus \{\psi(\boldsymbol{\alpha})\} = \{\varphi(\boldsymbol{\alpha})\}$, hence $\varphi(\boldsymbol{\alpha}) \neq \psi(\boldsymbol{\alpha})$. $\dashv$

2. THE NONSTANDARD MAPPING

We are now ready to define the nonstandard mapping $*$.

Definition. For any set A, let ${}^*A$ be the ideal value of the sequence $c_A : n \mapsto A$ constantly equal to A, that is ${}^*A = c_A(\boldsymbol{\alpha})$.

The map $*$ is a 1-1 $\in$-homomorphism from the universal class of all sets into itself. Moreover, $*$ preserves all basic operations of sets (except powerset). [1]

Proposition 4. *For every sets A, B, the following hold.*

(i) ${}^*\{A, B\} = \{{}^*A, {}^*B\}$.

(ii) $A \subseteq B \Leftrightarrow {}^*A \subseteq {}^*B$.

(iii) $A \in B \Leftrightarrow {}^*A \in {}^*B$.

(iv) $A = B \Leftrightarrow {}^*A = {}^*B$.

(v) ${}^*(\bigcup A) = \bigcup({}^*A)$. [2]

(vi) ${}^*(A \cup B) = {}^*A \cup {}^*B$.

(vii) ${}^*(A \cap B) = {}^*A \cap {}^*B$.

(viii) ${}^*(A \setminus B) = {}^*A \setminus {}^*B$.

(ix) ${}^*\langle A, B\rangle = \langle {}^*A, {}^*B\rangle$.

(x) ${}^*(A \times B) = {}^*A \times {}^*B$.

(xi) ${}^*\{\langle x, x\rangle : x \in A\} = \{\langle \xi, \xi\rangle : \xi \in {}^*A\}$.

(xii) ${}^*\{\langle x, y\rangle : x \in y \in A\} = \{\langle \xi, \eta\rangle : \xi \in \eta \in {}^*A\}$.

(xiii) ${}^*\{\langle x, y\rangle : \langle y, x\rangle \in A\} = \{\langle \xi, \eta\rangle : \langle \eta, \xi\rangle \in {}^*A\}$.

(xiv) ${}^*\{\langle x, y, z\rangle : \langle x, z, y\rangle \in A\} = \{\langle \xi, \eta, \zeta\rangle : \langle \xi, \zeta, \eta\rangle \in {}^*A\}$.

[1] As a consequence, the *transfer* principle could be directly proved from proposition 4 (see for instance [D2]). In this paper, the transfer principle is proved in theorem 10 by other means, through the representation theorem 8.

[2] Recall that, by definition, $\bigcup X = \{y : y \in x \text{ for some } x \in X\}$.

(xv) $^*\{x : \exists y\, \langle x, y\rangle \in A\} = \{\xi : \exists \eta\, \langle \xi, \eta\rangle \in {}^*A\}$, *i.e.* $^*dom(A) = dom(^*A)$.

(xvi) $^*\{y : \exists x\, \langle x, y\rangle \in A\} = \{\eta : \exists \xi\, \langle \xi, \eta\rangle \in {}^*A\}$, *i.e.* $^*range(A) = range(^*A)$.

Proof. Properties *(i)-(x)* straightforwardly follow from the axioms and from proposition 1. The remaining properties are proved similarly one to the other. For instance, let us consider *(xv)*. If ξ is a sequence with $\xi(n) \in \mathrm{dom}(A)$ for all $n \in \mathbf{N}$, then there is a sequence η with $\langle \xi(n), \eta(n)\rangle \in A$ for all $n \in \mathbf{N}$. Hence $\langle \xi(\boldsymbol{\alpha}), \eta(\boldsymbol{\alpha})\rangle \in {}^*A$, and $\xi(\boldsymbol{\alpha}) \in \mathrm{dom}(^*A)$. Vice versa, let $p \in \mathrm{dom}(^*A)$, i.e. $\langle p, q\rangle \in {}^*A$ for some q. Then $\langle p, q\rangle = \zeta(\boldsymbol{\alpha})$ for some sequence $\zeta : \mathbf{N} \to A$. Since $^*A \subseteq {}^*D \times {}^*R$ where $D = \mathrm{dom}(A)$ and $R = \mathrm{range}(A)$, there are sequences $\xi : \mathbf{N} \to D$ and $\eta : \mathbf{N} \to R$ with $p = \xi(\boldsymbol{\alpha})$ and $q = \eta(\boldsymbol{\alpha})$. Notice that $\langle p, q\rangle = \nu(\boldsymbol{\alpha})$ where ν is the sequence $n \mapsto \langle \xi(n), \eta(n)\rangle$. Now, let $\pi : D \times R \to D$ be the projection on the first coordinate. $\xi(n) = (\pi \circ \nu)(n)$ for all $n \in \mathbf{N}$ implies $p = \xi(\boldsymbol{\alpha}) = (\pi \circ \nu)(\boldsymbol{\alpha})$. But $\nu(\boldsymbol{\alpha}) = \langle p, q\rangle = \zeta(\boldsymbol{\alpha})$, thus by axiom 7, $p = (\pi \circ \nu)(\boldsymbol{\alpha}) = (\pi \circ \zeta)(\boldsymbol{\alpha})$ where $\pi \circ \zeta : \mathbf{N} \to D$. We conclude that $p \in {}^*D = {}^*\mathrm{dom}(A)$. $\dashv$

For every A, clearly $\{^*a : a \in A\} \subseteq {}^*A$. The following result gives a characterization of the fixed points of $*$.

Proposition 5. $\{^*a : a \in A\} = {}^*A$ *if and only if* A *is finite.*

Proof. If $A = \{a_1, \ldots, a_k\}$ is finite, then $^*A = \{^*a_1, \ldots, {}^*a_k\}$ by *(i)* in the previous proposition. Vice versa, if A is infinite, then there is a 1-1 mapping $f : \mathbf{N} \to A$. Take a left inverse $g : A \to \mathbf{N}$, i.e. a mapping such that $g(f(n)) = n$ for all $n \in \mathbf{N}$. We claim that $f(\boldsymbol{\alpha}) \neq {}^*a$ for every $a \in A$. Assume by contradiction that $f(\boldsymbol{\alpha}) = {}^*a = c_a(\boldsymbol{\alpha})$ for some $a \in A$. Then, $\boldsymbol{\alpha} = 1_{\mathbf{N}}(\boldsymbol{\alpha}) = (g \circ f)(\boldsymbol{\alpha}) = (g \circ c_a)(\boldsymbol{\alpha}) = c_{g(a)}(\boldsymbol{\alpha}) = {}^*g(a)$. This is impossible, because (by proposition 2) $^*g(a) = g(a) \in \mathbf{N}$ while $\boldsymbol{\alpha} \notin \mathbf{N}$. $\dashv$

For each subset $A \subseteq \mathbf{N}$, let $\chi_A : \mathbf{N} \to \{0, 1\}$ be its characteristic function. That is, $\chi_A(n) = 1$ if $n \in A$ and $\chi_A(n) = 0$ otherwise. Notice that $\chi(\boldsymbol{\alpha}) \in \{0, 1\}$.

Theorem 6. $\mathcal{U} = \{A \subseteq \mathbf{N} : \chi_A(\boldsymbol{\alpha}) = 1\}$ *is an ultrafilter over* $\mathbf{N}$ *that contains no finite sets.*

Proof. In the following, recall that 0 and 1 are identified with the emptyset $\emptyset$ and with the singleton $\{\emptyset\}$, respectively. If A be a finite subset of $\mathbf{N}$, then $\chi_A(n) = c_\emptyset(n)$ for all but a finite number of n, and by proposition 3, $\chi_A(\boldsymbol{\alpha}) = c_\emptyset(\boldsymbol{\alpha}) = 0$, i.e. $A \notin \mathcal{U}$. Now, let us assume $A, B \in \mathcal{U}$, i.e. $\chi_A(\boldsymbol{\alpha}) = \chi_B(\boldsymbol{\alpha}) = 1$. $\chi_{A\cap B}(n) = \chi_A(n) \cap \chi_B(n)$ for all $n \Rightarrow \chi_{A\cap B}(\boldsymbol{\alpha}) = \chi_A(\boldsymbol{\alpha}) \cap \chi_B(\boldsymbol{\alpha}) = 1$, i.e. $A \cap B \in \mathcal{U}$. Similarly, if $A \supseteq B \in \mathcal{U}$, then $\chi_A(n) \supseteq \chi_B(n)$ for all $n \Rightarrow \chi_A(\boldsymbol{\alpha}) \supseteq \chi_B(\boldsymbol{\alpha}) = 1$, and so $\chi_A(\boldsymbol{\alpha}) = 1$. We are left to show the ultrafilter property. If we denote by $A^c = \mathbf{N} \setminus A$ then for each subset $A \subseteq \mathbf{N}$, $\chi_A(n) \cup \chi_{A^c}(n) = 1$ for all $n \Rightarrow \chi_A(\boldsymbol{\alpha}) \cup \chi_{A^c}(\boldsymbol{\alpha}) = 1$ and so $A \notin \mathcal{U}$ implies $A^c \in \mathcal{U}$. $\dashv$

Proposition 7. *Let φ, ψ be two sequences. Then*

(i) $\varphi(\boldsymbol{\alpha}) = \psi(\boldsymbol{\alpha}) \Leftrightarrow \{n \in \mathbf{N} : \varphi(n) = \psi(n)\} \in \mathcal{U}$.

(ii) $\varphi(\boldsymbol{\alpha}) \in \psi(\boldsymbol{\alpha}) \Leftrightarrow \{n \in \mathbf{N} : \varphi(n) \in \psi(n)\} \in \mathcal{U}$.

Proof. *(i)* Denote by $\Lambda = \{n \in \mathbf{N} : \varphi(n) = \psi(n)\}$. Assume first $\Lambda \in \mathcal{U}$. Let $A = \text{range}(\varphi) \cup \text{range}(\varphi)$ and let ξ be the sequence such that $\xi(n) = A$ if $n \in \Lambda$ and $\xi(n) = 0$ otherwise. Notice that $\xi = f \circ \chi_\Lambda$ where $f : \{0, 1\} \to \{0, A\}$ is the function $0 \mapsto 0$ and $1 \mapsto A$. $\chi_\Lambda(\boldsymbol{\alpha}) = 1 = c_1(\boldsymbol{\alpha}) \Rightarrow \xi(\boldsymbol{\alpha}) = (f \circ \chi_\Lambda)(\boldsymbol{\alpha}) = (f \circ c_1)(\boldsymbol{\alpha}) = c_A(\boldsymbol{\alpha}) = {}^*A$. For all $n \in \mathbf{N}$, $[\{\varphi(n)\} \setminus \{\psi(n)\}] \cap \xi(n) = 0$, thus $[\{\varphi(\boldsymbol{\alpha})\} \setminus \{\psi(\boldsymbol{\alpha})\}] \cap \xi(\boldsymbol{\alpha}) = 0$. But $\varphi(\boldsymbol{\alpha}) \in {}^*A = \xi(\boldsymbol{\alpha}) \Rightarrow \{\varphi(\boldsymbol{\alpha})\} \setminus \{\psi(\boldsymbol{\alpha})\} = 0 \Rightarrow \varphi(\boldsymbol{\alpha}) = \psi(\boldsymbol{\alpha})$. Vice versa, assume $\varphi(\boldsymbol{\alpha}) = \psi(\boldsymbol{\alpha})$. For all $n \in \mathbf{N}$, $[\{\varphi(n)\} \setminus \{\psi(n)\}] \cup \chi_\Lambda(n) \neq 0$, so $[\{\varphi(\boldsymbol{\alpha})\} \setminus \{\psi(\boldsymbol{\alpha})\}] \cup \chi_\Lambda(\boldsymbol{\alpha}) \neq 0$. Since $\{\varphi(\boldsymbol{\alpha})\} \setminus \{\psi(\boldsymbol{\alpha})\} = 0$ by hypothesis, we conclude $\chi_\Lambda(\boldsymbol{\alpha}) \neq 0$, i.e. $\Lambda \in \mathcal{U}$. *(ii)*. Notice that $x \in y \Leftrightarrow \{x\} \cap y = \{x\}$ and apply *(i)* to the sequences $f(n) = \{\varphi(n)\} \cap \psi(n)$ and $g(n) = \{\varphi(n)\}$. $\dashv$

$\mathcal{U}$ yields a nonatomic finitely additive two-valued measure $\mu : \mathcal{P}(\mathbf{N}) \to \{0, 1\}$ defined for all subsets of $\mathbf{N}$. Thus, for instance, the property *(i)* in the above proposition could be reformulated as follows: "Two sequences have the same ideal value if and only if they are equal almost everywhere (with respect to μ)". For any set A, let $A^{\mathbf{N}}/_{\mathcal{U}}$ be the *ultrapower* of A modulo the ultrafilter $\mathcal{U}$, and let $d_A : A \to A^{\mathbf{N}}/_{\mathcal{U}}$ be the diagonal embedding of A into its ultrapower. For every sequence $\varphi : \mathbf{N} \to A$, define $\pi_A(\varphi(\boldsymbol{\alpha})) = [\varphi] \in A^{\mathbf{N}}/_{\mathcal{U}}$, the equivalence class of φ modulo $\mathcal{U}$.

Representation Theorem 8. *For every set A, $\pi_A : {}^*A \to A^{\mathbf{N}}/_{\mathcal{U}}$ is an $\in$-isomorphism that makes the following diagram commutative:*

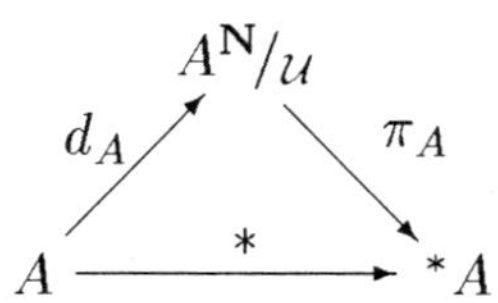

In particular, for all $\in$-formulas $\sigma(x_1, \ldots, x_n)$ and elements $a_1, \ldots$ $\ldots, a_n \in A$:

$$\langle A, \in \rangle \models \sigma(a_1, \ldots, a_n) \Leftrightarrow \langle {}^*A, \in \rangle \models \sigma({}^*a_1, \ldots, {}^*a_n)$$

Proof. By the previous proposition, $\varphi(\boldsymbol{\alpha}) = \psi(\boldsymbol{\alpha}) \Leftrightarrow \{n \in \mathbf{N} : \varphi(n) = \psi(n)\} \in \mathcal{U} \Leftrightarrow [\varphi] = [\psi]$. Thus π_A is 1-1, and also it is trivially onto. Besides, $\varphi(\boldsymbol{\alpha}) \in \psi(\boldsymbol{\alpha}) \Leftrightarrow \{n \in \mathbf{N} : \varphi(n) \in \psi(n)\} \in \mathcal{U} \Leftrightarrow A^{\mathbf{N}}/_{\mathcal{U}} \models [\varphi] \in [\psi]$ and so π_A is an $\in$-homomorphism. The commutativity of the diagram is immediate. In fact, for every $a \in A$, $\pi_A({}^*a) = \pi(c_a(\boldsymbol{\alpha})) = [c_a] = d_A(a)$, where $c_a : n \mapsto a$ is the sequence constantly equal to a. Finally, the equivalence for $\in$-formulas σ is given by Łos theorem on ultrapowers (see [CK] §4.1). $\dashv$

3. TRANSFER PRINCIPLE AND SATURATION PROPERTY

We are now ready to prove the two fundamental principles of nonstandard analysis, namely the *transfer* principle and the *saturation* property. As in the superstructure approach, by definition we say that a set x is *internal* if $x \in {}^*A$ for some A. Equivalently, x is internal if $x = \varphi(\boldsymbol{\alpha})$ is the ideal value of some sequence φ. Notice that the class $\mathbf{I}$ of internal elements is transitive, i.e. elements of internal sets are internal.

Theorem 9. *Countable Saturation.*
Let $\mathcal{F}$ be a countable family of internal sets. If every finite subfamily of $\mathcal{F}$ has nonempty intersection, then $\mathcal{F}$ has nonempty intersection.

Proof. Let $\mathcal{F} = \{F_n : n \in \mathbf{N}\}$ be a family of internal sets. Take A with $F_n \in {}^*A$ for all n. For each $n \in \mathbf{N}$, pick $a_n \in \bigcap_{i=0}^{n} F_i$ and consider the function $\varphi : \mathbf{N} \to A$ where $n \mapsto a_n$. For each $k \in \mathbf{N}$, $\{n \in \mathbf{N} : \varphi(n) \in F_k\} = \{n \in \mathbf{N} : n \geq k\}$ is a cofinite set, hence an element of $\mathcal{U}$. By the Representation theorem, $A^{\mathbf{N}}/_{\mathcal{U}} \models [\varphi] \in d_A(F_k) \Leftrightarrow \langle {}^*A, \in \rangle \models \pi_A^{-1}([\varphi]) \in F_k$. Since this holds for each $k \in \mathbf{N}$, we have found an element $a = \pi_A^{-1}([\varphi]) \in \bigcap \mathcal{F}$. $\dashv$

Recall that a formula σ is *bounded* if every quantifier occurs in the bounded form $\forall x \in \mathbf{I}$ (i.e. $\forall x\, x \in \mathbf{I} \to \ldots$) or $\exists x \in \mathbf{I}$ (i.e. $\exists x\, x \in \mathbf{I} \wedge \cdots$). By $\sigma^{\mathbf{I}}$ we denote the $\mathbf{I}$-relativization of σ, i.e. the formula obtained from σ by replacing each quantifier $\forall x$ and $\exists x$ with its restricted form $\forall x \in \mathbf{I}$ and $\exists x \in \mathbf{I}$, respectively.

Theorem 10. *Transfer Principle.*
For every bounded $\in$-formula $\sigma(x_1, \ldots, x_n)$ and elements $a_1, \ldots, a_n$:

$$\sigma(a_1, \ldots, a_n) \Leftrightarrow \sigma^{\mathbf{I}}({}^*a_1, \ldots, {}^*a_n) \Leftrightarrow \sigma({}^*a_1, \ldots, {}^*a_n)$$

Thus $* : \mathbf{V} \to \mathbf{V}$ *is a bounded elementary embedding of the universe into itself.*

Proof. Take a transitive set A containing $a_1, \ldots, a_n$. It is easily proved that ${}^*A$ is a transitive set as well, containing ${}^*a_1, \ldots, {}^*a_n$. Since bounded formulas are absolute between transitive classes (see for instance [JE] §14), by the representation theorem we get:
$\sigma(a_1, \ldots, a_n) \Leftrightarrow \langle A, \in\rangle \models \sigma(a_1, \ldots, a_n) \Leftrightarrow \langle {}^*A, \in\rangle \models \sigma({}^*a_1, \ldots, a_n) \Leftrightarrow \sigma^{\mathbf{I}}({}^*a_1, \ldots, {}^*a_n) \Leftrightarrow \sigma({}^*a_1, \ldots, {}^*a_n)$. ⊣

Finally, the next theorem gives a foundational justification to ZFC[$\boldsymbol{\alpha}$].

Theorem 11.
ZFC[$\boldsymbol{\alpha}$] is equiconsistent with ZFC.

Proof. If M is a model of ZFC[$\boldsymbol{\alpha}$], then its wellfounded part $\{m \in M : M \models$"m is wellfounded"$\}$ is a model of ZFC. Vice versa, let us assume the consistency of ZFC. We need to find a model of ZFC[$\boldsymbol{\alpha}$]. In the following, we will work in ZFBC, an equiconsistent variant of Zermelo-Fraenkel set theory where a (definable) bijection $\mathbf{C} : \mathrm{ON} \to \mathbf{V}$ between the class of ordinals and the universal class of all sets is available (thanks to a new binary relation C considered in the language), and where regularity is replaced by Boffa's *superuniversality* axiom. Superuniversality is a strong anti-foundation axiom which implies the existence of a transitive collapse for every (not necessarily wellfounded) extensional structure. [3] The construction below closely follows arguments in the proof of the Extension Principle of [BH]. Thus we give here only a sketch and refer the reader to that paper for details. [4] The universal class $\mathbf{V}$ is an increasing union of transitive sets indexed by ordinals. For instance, $\mathbf{V} = \bigcup\{W_\alpha : \alpha \in \mathrm{ON}\}$ where W_α is the transitive closure of the set $\{\mathbf{C}(\beta) : \beta < \alpha\}$. Now fix a nontrivial ultrafilter D over $\mathbf{N}$. As a consequence of superuniversality, one gets transitive collapses $\pi_\alpha : W_\alpha{}^{\mathbf{N}}/_D \to T_\alpha$ such that for all $\alpha \leq \beta$, π_α is the restriction of π_β to $W_\alpha{}^{\mathbf{N}}/_D$:

---

[3] ZFBC was first proposed as a foundational framework for the practice of nonstandard methods by D. Ballard and K. Hrbáček in [BH] (we refer to that paper for a precise formulation). That theory was also adopted by the author for the construction of superstructure-like nonstandard universes in [D1]. A proof of the consistency of superuniversality can be found in [Bo].

[4] See also [BD], where a stronger theory than ZFC[α] is given a faithful interpretation in ZFC.

$$\begin{array}{ccc} {W_\alpha}^{\mathbf{N}}/_D & \xrightarrow{\pi_\alpha} & \langle T_\alpha, \in \rangle \\ \downarrow & & \downarrow \\ {W_\beta}^{\mathbf{N}}/_D & \xrightarrow{\pi_\beta} & \langle T_\beta, \in \rangle \end{array}$$

The class $\pi \doteq \bigcup\{\pi_\alpha : \alpha \in \mathrm{ON}\}$ is a transitive collapse for the D-ultrapower of the universe $\mathbf{V}^{\mathbf{N}}/_D = \{[\varphi] : \varphi$ is a function with $\mathrm{dom}(\varphi) = \mathbf{N}\}$. We remark that in order to make each D-equivalence class $[\varphi]$ a set (and not a proper class) some caution is needed. In ZFC, this problem is overcome by using Scott's trick. In our nonwellfounded context (where a rank is not available) one can define $[\varphi]$ as the least function in the well-ordering of the universe (given by $\mathbf{C}$) which is D-equivalent to φ. Doing so, the collection $\{[\varphi] : \varphi$ is a function with $\mathrm{dom}(\varphi) = \mathbf{N}\}$ is actually the extension of a formula in ZFBC. Now, let $\pi([1_{\mathbf{N}}])$ be the interpretation of $\boldsymbol{\alpha}$, where $1_{\mathbf{N}}$ denotes the identity map on $\mathbf{N}$. For all sequences φ, let us interpret $\varphi(\boldsymbol{\alpha}) = \pi([\varphi])$. Notice that the nonstandard embedding $*$ corresponds to the bounded elementary embedding $(\bigcup_{\alpha \in ON} \pi_\alpha \circ d_\alpha) : \mathbf{V} \to \mathbf{V}$, where $d_\alpha : W_\alpha \to {W_\alpha}^{\mathbf{N}}/_D$ are the diagonal embeddings. All axioms of ZFC$[\boldsymbol{\alpha}]$ directly follow from properties of the ultrapower construction. $\dashv$

References

[BE] V. BENCI, *A Construction of a Nonstandard Universe*, in "Advances in Dynamical Systems and Quantum Physics", S. Albeverio et al., eds., World Scientific, 1995, pp. 207-237.

[BD] V. BENCI AND M. DI NASSO, *Expanding ZFC⁻ for the Practice of Nonstandard Analysis*, in preparation.

[BH] D. BALLARD AND K. HRBÁČEK, *Standard Foundations for Nonstandard Analysis*, J. Symb. Logic, vol. 57 (1992), pp. 471-478.

[Bo] M. BOFFA, *Forcing et Négation de l'Axiome de Fondement*, Mém. Acad. Sc. Belg., tome XL, fasc. 7, 1972.

[CK] C.C. CHANG AND H.J. KEISLER, "Model Theory", 3rd edition, North-Holland, 1990.

[D1] M. DI NASSO, *Pseudo-Superstructures as Nonstandard Universes*, J. Symb. Logic, vol. 63 (1998), pp. 222-236.

[D2] M. DI NASSO, *Elementary Embeddings of the Universe in a Nonwellfounded Context*, in preparation.

[JE] T. JECH, "Set Theory", Academic Press, 1978.

[LA] D. LAUGWITZ, *Ω-Calculus as a Generalization of Field Extension - An Alternative Approach to Nonstandard Analysis*, in "Nonstandard Analysis - Recent Developments" (Proceedings Victoria Conference 1980), A. Dold and B. Eckmann, eds., Lecture Notes in Mathematics 983, Springer-Verlag, 1983, pp. 120-133.

[SL] C. SCHMIEDEN AND D. LAUGWITZ, *Eine Erweiterung der Infinitesimalrechnung*, Math. Zeitschr., vol. 69 (1958), pp. 1-39.

NILPOTENT INFINITESIMALS AND SYNTHETIC DIFFERENTIAL GEOMETRY IN CLASSICAL LOGIC

Paolo Giordano
Accademia di architettura, Università della Svizzera italiana
Largo Bernasconi, CH-6658 Mendrisio, Switzerland
pgiordano@arch.unisi.ch

Keywords: Synthetic differential geometry; Differential manifolds, foundations

Abstract: An extension of $\mathbb{R}$ (and ${}^*\mathbb{R}$) with nilpotent infinitesimals (e.g. $h \neq 0$ but $h^2 = 0$) is presented in order to obtain results similar to Kock-Lawvere's synthetic differential geometry [3], but in a classical and not intuitionistic context. The same extension can be used to add new infinitesimal points to spaces similar to Chen's ones [1]. In the category of extended spaces we can develop differential geometry not only of usual manifolds but also of infinite dimensional spaces, without coordinates and with a strong geometric intuition, that is in a way that we will call "synthetic".

Introduction

In the original works of G.W. Leibniz, S. Lie or E. Cartan, we find the use of two kinds of infinitesimals: invertible, formalized in NSA, and nilpotent. In 1967 F.W. Lawvere [5] proposed a set of axioms for nilpotent infinitesimals as a base for a synthetic differential geometry (SDG), that is a theory that could develop differential geometry with infinitesimals tools without coordinates systems, but talking directly of geometric objects and their properties. Since the seventies SDG has had a great development, see [3] [4], with applications in the usual manifolds and also in infinite dimensional spaces, or in geometric spaces with singular points. One of the central points of Lawvere's idea is to consider a ring R satisfying the following axiom (called Kock-Lawvere's axiom): let $x \in R$ and $f : R \longrightarrow R$ **any** function from R in itself, then there exists one and only one $m \in R$ with the property

$$\forall h \in D : \quad f(x+h) = f(x) + h \cdot m,$$

P. Schuster, U. Berger and H. Osswald (eds.),
Reuniting the Antipodes – Constructive and Nonstandard Views of the Continuum, 75–92.

where $D = \{h \in R \mid h^2 = 0\}$ is the set of *first order infinitesimals*. In other words in SDG any function restricted to the first order neighborhood of x becomes a straight line. Obviously in SDG we set $f'(x) := m$. The problem of this axiom is that it is incompatible with classical logic: if we set

$$f(x) := \begin{cases} 0 & \text{if } x = 0 \\ 1 & \text{if } x \neq 0 \end{cases}$$

then it follows that $D = R = \{0\}$ (see [4]). SDG admits only models in intuitionistic logic, where we can suppose to have smooth functions only. In this work we want to modify the previous axiom so as to obtain a theory with final results similar to SDG's ones, but in a classical context and without great logical problems or difficult models.

1. THE RING OF EXTENDED REALS

A natural idea to avoid the previous incompatibility with classical logic, is to restrict the universal quantification for functions $f : R \longrightarrow R$ to a particular subset $\mathcal{F} \subset R^R$:

$$\forall f \in \mathcal{F}\ \exists!\, m \in R : \ \forall h \in D : \quad f(x+h) = f(x) + h \cdot m. \tag{7.1}$$

If we try to construct a model for (7.1), or a similar axiom as we will see, a natural idea is to think new numbers as equivalence classes $[h]$ of usual functions $h : \mathbb{R} \longrightarrow \mathbb{R}$. In such a way we can both include the real field using the classes generated by constant functions, and to hope that the class generated by $h(t) = t$ would be an infinitesimal number of first order. To realize how to define this equivalence relation we consider for example $x = 0$, and we see (7.1) in the following sense:

$$f(h(t)) \sim f(x) + h(t) \cdot f'(0). \tag{7.2}$$

If we think $h(t)$ "sufficiently similar to t", we can define $\sim$ so that (7.2) is equivalent to

$$\lim_{t \to 0} \frac{f(h(t)) - f(x) - h(t) \cdot f'(0)}{t} = 0,$$

that is

$$x \sim y \quad :\Longleftrightarrow \quad \lim_{t \to 0} \frac{x(t) - y(t)}{t} = 0, \tag{7.3}$$

in this way (7.2) is very near to the definition of differentiability for f in 0. It is important to note that, for De l'Hôpital's theorem

$$\mathcal{C}^1(\mathbb{R};\mathbb{R})/\sim \quad \simeq \quad \mathbb{R}[\varepsilon]/(\varepsilon^2 = 0)$$

so we obtain nothing new. But if we define $x \sim y$ for continuous functions, then we obtain a non-trivial ring with nilpotent elements, like that generated by $h(t) = |t|^{1/k}$. Finally if we want invertible infinitesimals also, then we only have to consider sequences of continuous functions, so as to obtain ordinary sequences of reals (like in $^*\mathbb{R}$) if we take constant functions. These are the motivations for the following definition

Definition 1.1 *Let $\mathcal{C}_0$ be the set of all the functions $x : \mathbb{R} \longrightarrow \mathbb{R}$ continuous in $t = 0$, and let $x, y : \mathbb{N} \longrightarrow \mathcal{C}_0$ be sequences of such functions, then we say that $x \sim y$ if*

$$\exists N \in \mathbb{N} : \forall n \geq N : \quad \lim_{t \to 0} \frac{x_n(t) - y_n(t)}{t} = 0$$

The quotient $\mathcal{C}_0^{\mathbb{N}}/\sim$ will be indicated with $^\bullet\mathbb{R}$.

It is easy to define $^\bullet\mathbb{R}$ so that to include the hyperreals of NSA. It is sufficient to define

$$x \sim y \quad :\Longleftrightarrow \quad \left\{ n \in \mathbb{N} \mid \lim_{t \to 0} \frac{x_n(t) - y_n(t)}{t} = 0 \right\} \in \mathcal{U}_\infty,$$

where $\mathcal{U}_\infty$ is an ultrafilter which contains the filter of cofinite sets.
In order to study firstly the properties of new infinitesimal numbers, we will omit the extension of the following results through the use of an ultrafilter, with the hope of performing them in a future work.
Obviously $\sim$ is a congruence relation with respect to pointwise operations, so $^\bullet\mathbb{R}$ is a commutative ring which we call *ring of extended reals.* To simplify the notations we write $x = y$ instead of $x \sim y$, changing the meaning of "=" on the set $\mathcal{C}_0^{\mathbb{N}}$; analogous changes will be made with other equivalence relations. The immersion of $\mathbb{R}$ is $r \longmapsto \hat{r}$ defined by $\hat{r}_n(t) := r$. The quotient set $\mathbb{R}^{\mathbb{N}}/\sim$, where $x \sim y$ iff $x_n = y_n \ \forall n \geq N$ for some N, will be indicated with $^\diamond\mathbb{R}$; it is included in the extended reals with $\gamma \in {}^\diamond\mathbb{R} \longmapsto \tilde{\gamma}_n(t) := \gamma_n$. Conversely if $x \in {}^\bullet\mathbb{R}$ then are well defined the maps

$$^\diamond x : n \in \mathbb{N} \longmapsto x_n(0) \in \mathbb{R},$$

$$\exists \lim_{n \to +\infty} (^\diamond x)_n \in \mathbb{R} \quad \Longrightarrow \quad ^\circ x := \lim_{n \to +\infty} (^\diamond x)_n.$$

We call $^\diamond x$ the *discrete part* of x, and $^\circ x$ the *standard part* of x.

1.1. The ideal of first order infinitesimal numbers

If we want that $f(h_n(t)) \sim f(0)+h_n(t)\cdot f'(0)$ then from Taylor's formula we obtain

$$\lim_{t\to 0} \frac{f(h_n(t)) - f(x) - h_n(t)\cdot f'(0)}{t} = \lim_{t\to 0} \frac{h_n(t)}{t}\cdot \sigma(t), \tag{7.4}$$

with $\sigma(t) \to 0$ for $t \to 0$. This suggests that for this ring we have not to define D with the condition $h^2 = 0$ (take for example $h_n(t) = t^{2/3}$ in (7.4)), but rather with

$$\limsup_{t\to 0} \left| \frac{h_n(t)}{t} \right| < +\infty.$$

Generally we write $x \approx y$ for

$$\exists N \in \mathbb{N} : \forall n \geq N : \quad \limsup_{t\to 0} \left| \frac{x_n(t) - y_n(t)}{t} \right| < +\infty$$

and we say that x *close to* y. This is a well-defined congruence on ${}^\bullet\mathbb{R}$ that coincides with equality on $\mathbb{R}$ and ${}^\diamond\mathbb{R}$.

Definition 1.2

$$D := \{h \in {}^\bullet\mathbb{R} \,|\, h \approx 0\}$$

The elements of D will be called "first order infinitesimals".

So we have $x = y$ if, $\forall n \geq N$ for some N, $x_n(t) = y_n(t) + \mathrm{o}(t)$, and $x \approx y$ if $x_n(t) = y_n(t) + \mathrm{O}(t)$, always for $t \to 0$. For example if $r, s \in \mathbb{R}$, then $h_n(t) := r\,|t|$ if $t \geq 0$ and $h_n(t) := s\,|t|$ if $t \leq 0$ is a first order infinitesimal; another one is $h_n(t) := r\cdot t\cdot \sin(\frac{1}{t})$, and obviously $h_n(t) := t$. Conversely, if $\alpha \in (0,1)$, then $x_n(t) := |t|^\alpha$ is not an element of D.

Theorem 1.1 *D is an ideal of ${}^\bullet\mathbb{R}$, and*

$$\forall h \in D : \quad h^2 = 0$$

Proof: It follows from elementary properties of lim sup; for example the inequalities

$$\limsup_{t\to 0} \left| \frac{h_n(t) - k_n(t)}{t} \right| \leq \limsup_{t\to 0} \left| \frac{h_n(t)}{t} \right| + \limsup_{t\to 0} \left| \frac{k_n(t)}{t} \right| < +\infty$$

$$\limsup_{t\to 0} \left| \frac{x_n(t)\cdot h_n(t)}{t} \right| \leq |x_n(0)| \cdot \limsup_{t\to 0} \left| \frac{h_n(t)}{t} \right| < +\infty$$

prove that D is an ideal, and the following

$$0 \leq \liminf_{t\to 0} \left| \frac{h_n(t)^2}{t} \right| \leq \limsup_{t\to 0} \left| \frac{h_n(t)^2}{t} \right| \leq |h_n(0)| \cdot \limsup_{t\to 0} \left| \frac{h_n(t)}{t} \right| = 0$$

prove that every element of D has square equal to zero.

■

Another interesting ideal is $D_k := \{h \in {}^\bullet\mathbb{R} \mid h^k \in D\}$: in fact from the equality

$$\frac{h_n(t)^i \cdot u_n(t)^{k-i}}{t} = \left[\frac{h_n(t)^k}{t}\right]^{\frac{i}{k}} \cdot \left[\frac{u_n(t)^k}{t}\right]^{\frac{k-i}{k}}$$

one can prove that

$$h, u \in D_k \quad \Longrightarrow \quad h^i \cdot u^{k-i} \in D,$$

and from this the conclusion follows applying Newton's formula and Theorem 1.1. It is also easy to prove that in the extended reals we always have $h \cdot k = 0$ if $h^2 = k^2 = 0$. In the following section we prove a useful cancellation law, which is a well substitute of the fact that ${}^\bullet\mathbb{R}$ is not a field.

1.1.1 Cancellation law for the extended reals

Theorem 1.2 *Let $x \in {}^\bullet\mathbb{R}$ and $x \neq 0$, then*

$$x \cdot r = x \cdot s \quad \textit{and} \quad r, s \in \mathbb{R} \quad \Longrightarrow \quad r = s.$$

Proof: We can write the hypothesis $x \cdot r = x \cdot s$ as

$$\exists N \in \mathbb{N} : \forall n \geq N : \quad \lim_{t\to 0} \frac{x_n(t)}{t} \cdot (r - s) = 0 = |r - s| \cdot \limsup_{t\to 0} \left| \frac{x_n(t)}{t} \right|;$$

but the $\limsup_{t\to 0} \left| \frac{x_n(t)}{t} \right| \neq 0$ for at least one $n \in \mathbb{N}$ because $x \neq 0$, and so $r = s$.

■

This law is true also if $r, s \in {}^\diamond\mathbb{R}$ and x is independent from n, but it is not true if r, s are generic extended reals, in fact we have

$${}^\diamond x = {}^\diamond y \quad \Longrightarrow \quad \forall h \in D : \quad h \cdot x = h \cdot y.$$

1.2. Extension of functions

Before considering the infinitesimal formula to derive functions, we have to solve the problem to extend functions $f : \mathbb{R} \longrightarrow \mathbb{R}$ to a certain ${}^\bullet f : {}^\bullet\mathbb{R} \longrightarrow {}^\bullet\mathbb{R}$.

Definition 1.3 *Let A be an open set in $\mathbb{R}^k$, and $f \in \mathcal{C}^0(A;\mathbb{R})$; let $x_1, \ldots, x_k \in {}^\bullet\mathbb{R}$ such that $({}^\circ x_1, \ldots, {}^\circ x_k) \in {}^\circ A$, then we define*

$$ {}^\bullet f(x_1, \ldots, x_k) \ : \ n \in \mathbb{N} \longmapsto f \circ (x_1, \ldots, x_k) \in \mathcal{C}_0 $$

that is

$$ {}^\bullet f(x_1, \ldots, x_k)_n(t) := f(x_{1n}(t), \ldots, x_{kn}(t)). $$

This definition is well posed if f is locally lipschitzian, but a less strong condition on f is sufficient:

Definition 1.4 *Let A open in $\mathbb{R}^k$, B open in $\mathbb{R}^n$ and $f : A \longrightarrow B$, then we say that "f is pathwise lipschitzian" if for any $x, y : \mathbb{R} \longrightarrow A$ continuous in $t = 0$ and with $x(0) = y(0)$, we can find $K \in \mathbb{R}$ and $\delta > 0$ s.t.*

$$ \forall t \in (-\delta, \delta) : \quad \|f(x_t) - f(y_t)\| \leq K \cdot \|x(t) - y(t)\|. $$

We will generalize this definition to other spaces which we will see later and where the notion of locally lipschitzian function has no sense. The set of all pathwise lipschitzian functions from A to B is indicated with $\mathrm{Lip}(A;B)$. It is easy to prove that all these functions are continuous in A, and that their composition is pathwise lipschitzian again. Finally we have chosen x, y continuous in $t = 0$ to have in $\mathrm{Lip}(A;B)$ the usual locally lipschitzian functions. If $f \in \mathrm{Lip}(A;\mathbb{R})$, then ${}^\bullet f$ is well defined, in fact for $t \in (-\delta, \delta)$ we have

$$ \begin{aligned} 0 \ &\leq \limsup_{t \to 0} \left| \frac{f(x_{1n}(t), \ldots, x_{kn}(t)) - f(y_{1n}(t), \ldots, y_{kn}(t))}{t} \right| \leq \\ &\leq k \cdot \limsup_{t \to 0} \sqrt{\sum_i \left(\frac{x_{in}(t) - y_{in}(t)}{t} \right)^2} = 0. \end{aligned} $$

The function ${}^\bullet f$ is an extension of f, that is

$$ {}^\bullet f(r_1, \ldots, r_k) = f(r_1, \ldots, r_k) \quad \text{in} \quad {}^\bullet\mathbb{R} \quad \text{for} \quad r_1, \ldots, r_k \in \mathbb{R} $$

so we can use the symbol $f(r_1, \ldots, r_k)$ without confusion. In conclusion of this section we enunciate the following useful transfer theorem for equalities whose proof follows directly from the previous definitions:

Theorem 1.3 *Let A be open in $\mathbb{R}^k$, and $\tau, \sigma \in Lip(A; \mathbb{R})$. Moreover let*

$$^\bullet A := \{(x_1, \ldots, x_k) \in {}^\bullet\mathbb{R} \mid ({}^\circ x_1, \ldots, {}^\circ x_k) \in {}^\circ A\}$$

Then it results

$$\forall (x_1, \ldots, x_k) \in {}^\bullet A : \quad {}^\bullet\tau(x_1, \ldots, x_k) = {}^\bullet\sigma(x_1, \ldots, x_k)$$

iff

$$\forall (r_1, \ldots, r_k) \in A : \quad \tau(r_1, \ldots, r_k) = \sigma(r_1, \ldots, r_k)$$

1.3. The derivation formula

Now we will prove the analogous, in this context, of the Kock-Lawvere axiom. Because of its differences with this axiom, we will call it *derivation formula.* It is natural to come up to expectation that it will be equivalent to the usual differentiability of a function, in fact we have

Theorem 1.4 *Let A open in $\mathbb{R}$, $x \in A$ and $f \in Lip(A; \mathbb{R})$, then the following conditions are equivalent:*

1) f is differentiable in x

2) $\exists!\, m \in \mathbb{R} : \forall h \in D : \quad f(x+h) = f(x) + h \cdot m.$

In this case we have $m = f'(x)$, where $f'(x)$ is the usual derivative of f in x.

There are three big differences from this formula and the Kock-Lawvere axiom: the first one is that the derivation formula is true for differentiable pathwise lipschitzian functions, while in the axiom we have $\forall f$ (intuitionistically defined); the second one is that we can use the formula only for classical real numbers x, and not for a generic $x \in {}^\bullet\mathbb{R}$; the last one is that we have $m = f'(x) \in \mathbb{R}$, i.e. the slope is an usual real number. In other words we can say that this formula allows us to differentiate the usual differentiable functions using a language with infinitesimal numbers.

Proof:

1) $\Longrightarrow$ 2): Let $m := f'(x)$ and $h \in D$, i.e. $\limsup_{t \to 0} \left| \frac{h_n(t)}{t} \right| < +\infty \; \forall n \geq$

N. For hypothesis f is differentiable in x, so we can find a function $\sigma : \mathbb{R} \longrightarrow \mathbb{R}$ such that

$$\forall u \in \mathbb{R} : \quad f(x+u) = f(x) + u \cdot m + u \cdot \sigma(u)$$
$$\lim_{u \to 0} \sigma(u) = \sigma(0) = 0.$$

Therefore

$$\limsup_{t \to 0} \left| \frac{f(x + h_n(t)) - f(x) - h_n(t) \cdot m}{t} \right| = \limsup_{t \to 0} \left| \frac{h_n(t) \cdot \sigma(h_n(t))}{t} \right|$$
$$\leq \sigma(h_n(0)) \cdot \limsup_{t \to 0} \left| \frac{h_n(t)}{t} \right| = 0.$$

This proves the existence; for the uniqueness we simply use the cancellation law (section 1.1.1)
2) $\Longrightarrow$ 1): For this implication it is sufficient to apply the hypothesis 2) with $h_n(t) := t$.

■

If we apply this theorem to the $\mathcal{C}^1$ function $p(r) := \int_x^{x+r} f(t)\,\mathrm{d}t$, then we obtain the following

Corollary 1 *Let A be open in $\mathbb{R}$, $x \in A$ and $f \in Lip(A;\mathbb{R})$. Then*

$$\forall h \in D : \quad \int_x^{x+h} f(t)\, dt = h \cdot f(x).$$

Moreover $f(x) \in \mathbb{R}$ is uniquely determined by this formula.

With the ideal D_k of the kth order infinitesimal numbers it is possible to prove the Taylor formula

$$\forall h \in D_k : \quad f(x+h) = \sum_{i=0}^{k} \frac{h^i}{i!} \cdot f^{(i)}(x)$$

with the standard reals $f^{(i)}(x)$ uniquely determined by this formula.

1.4. Order relation

Definition 1.5 *For $x, y \in {}^\bullet\mathbb{R}$, we say that $x \geq y$ if we can found $z \in {}^\bullet\mathbb{R}$ and $N \in \mathbb{N}$ such that $z = 0$ in ${}^\bullet\mathbb{R}$ and*

$$\forall n \geq N \; \exists \delta > 0 : \forall t \in (-\delta, \delta) : \quad x_n(t) - y_n(t) \geq z_n(t).$$

Observe that δ depends by n, and the reasons of this are in the definition of infinitesimal number. The Definition 1.5 is well posed, and for example we have that the infinitesimal number $h_n(t) = |t|$ is positive but not negative. It is easy to prove that this relation is reflexive and transitive, so it remains to prove that it is also anti-symmetric. If $x \geq y$ and $y \geq x$ then we have

$$
\begin{aligned}
& x_n(t) - y_n(t) \geq z_{1n}(t) && \forall n \geq N \quad \forall t \in (-\delta_{1n}, \delta_{1n}) \\
& y_n(t) - x_n(t) \geq z_{2n}(t) && \forall n \geq N \quad \forall t \in (-\delta_{2n}, \delta_{2n}) \\
& \lim_{t \to 0} \frac{z_{1n}(t)}{t} = 0 = \lim_{t \to 0} \frac{-z_{2n}(t)}{t} && \forall n \geq N.
\end{aligned}
$$

For $\delta_n := \min\{\delta_{1n}, \delta_{2n}\}$ we have

$$\forall t \in (0, \delta_n): \quad \frac{z_{1n}(t)}{t} \leq \frac{x_n(t) - y_n(t)}{t} \leq -\frac{z_{2n}(t)}{t}$$

$$\forall t \in (-\delta_n, 0): \quad \frac{-z_{2n}(t)}{t} \leq \frac{x_n(t) - y_n(t)}{t} \leq \frac{z_{1n}(t)}{t},$$

and so $x = y$ for $t \to 0$. With this relation, ${}^\bullet\mathbb{R}$ becomes an ordered ring. We prove it only for the product: if we have $x_n(t) - y_n(t) \geq z_{1n}(t)$, $w_n(t) \geq z_{2n}(t)$ with $z_1 = z_2 = 0$, then we set $z_n(t)$ equal to the minimum between $-|w \cdot z_1|_n(t)$ and $-|(x - y) \cdot z_2|_n(t)$ and we can prove that $(x \cdot w)_n(t) - (y \cdot w)_n(t) \geq z_n(t)$, with $z = 0$ in ${}^\bullet\mathbb{R}$. We also observe that $\geq$ extends the order relation in $\mathbb{R}$. It is also possible to prove the cancellation law for inequality, that is if $h \in {}^\bullet\mathbb{R}$ is different from zero and $r, s \in \mathbb{R}$, then from $|h| \cdot r \leq |h| \cdot s$ we can deduce $r \leq s$.

2. REGULAR SPACES AND FUNCTIONS

In these section we will define the notion of regular space and regular function of type $\mathcal{C}^n$ (briefly "$\mathcal{C}^n$-space" and "$\mathcal{C}^n$-function"). They represent the most general kind of spaces and functions extendible with our infinitesimal and infinite points. Any C^n manifold is also a $\mathcal{C}^n$-space, and the category of all $\mathcal{C}^n$-space is cartesian closed, so it contains many infinite-dimensional spaces like the set of all the usual C^n functions between two manifolds. The ideas used in this section are similar to analogous ideas of [Che] and [Fr;Kr].

In the sequel we will consider a category $\mathcal{F}$ with a forgetful functor $|-| : \mathcal{F} \longrightarrow$ **Set** which acts as the identity on arrows. For simplicity we indicate with $\mathcal{N}$ the set of objects of $\mathcal{F}$ and we set $\mathcal{F}_{HK} := \mathcal{F}(H; K)$.

The last hypothesis on $\mathcal{F}$ is that $\mathcal{F}_{HK}$ contains all the constant functions $f : |H| \longrightarrow |K|$. For example we can consider $\mathcal{F} = \mathbf{Man}^n$, the category of all $\mathcal{C}^n$ manifolds, with the forgetful functor, or the category $\mathcal{F}$ with objects all the set $\mathbb{R}^h$ with $h \in \mathbb{N}$, and with hom-set the usual $\mathcal{C}^n(\mathbb{R}^h; \mathbb{R}^k)$. We can think $\mathcal{F}$ as a category of "type of figures", or as a category of "spaces with a well known notion of regular space and regular function": with the category $\bar{\mathcal{F}}$ we will extend this notion to a more general type of spaces.

Definition 2.1 *If X is a set, then we say that $(\mathcal{D}, X, \mathcal{C})$ is in $\bar{\mathcal{F}}$ if $\mathcal{D} = \{\mathcal{D}_H\}_{H \in \mathcal{N}}$ and $\mathcal{C} = \{\mathcal{C}_K\}_{K \in \mathcal{N}}$ are families with*

$$\mathcal{D}_H \subseteq \mathbf{Set}(|H|; X) \qquad \text{and} \qquad \mathcal{C}_K \subseteq \mathbf{Set}(X; |K|)$$

we indicate, e.g., with the notation $\mathcal{D}_H \cdot \mathcal{C}_K$ the set of all the composition $(f \cdot g)(x) = g(f(x))$ of functions $f \in \mathcal{D}_H$ and $g \in \mathcal{C}_K$. The families $\mathcal{D}$ and $\mathcal{C}$ have finally to satisfy the following conditions for any H, K:

$$\mathcal{D}_H \cdot \mathcal{C}_K \subseteq \mathcal{F}_{HK} \tag{7.5}$$

$$\mathcal{F}_{JH} \cdot \mathcal{D}_H \subseteq \mathcal{D}_J$$

$$\forall d : |H| \longrightarrow X \text{ constant} \quad \Longrightarrow \quad d \in \mathcal{D}_H$$

For the condition (7.5) we can think $\mathcal{D}_H$ as the set of all the regular functions defined on the well known object H and with values in the new space X. There is no analogous condition for $\mathcal{C}_K$, and in fact we think it like a set of particular functions from X to K; for example if X is a manifold, we can put in $\mathcal{C}_K$ only the applications obtainable as extension from a chart on all X; if X is a normed space, we can put in it the continuous linear function $c : X \longrightarrow \mathbb{R}^k$ only. We will see that one criterion to choose the elements of $\mathcal{C}_K$ will be to obtain a "good" extension ${}^\bullet X$ of X with infinitesimal points. If $\mathcal{X} = (\mathcal{D}, X, \mathcal{C})$ is a space in $\bar{\mathcal{F}}$ we will also use the notation $\mathcal{D}(H; \mathcal{X})$ for $\mathcal{D}_H$, to clarify the distinction with other spaces. The definition of regular function between two spaces in $\bar{\mathcal{F}}$ is identical to that of [1]:

Definition 2.2 *If $\mathcal{X}$ and $\mathcal{Y}$ are $\bar{\mathcal{F}}$-objects with support sets X, Y, then we say that "$f : \mathcal{X} \longrightarrow \mathcal{Y}$ is an arrow in $\bar{\mathcal{F}}$" or is "regular" if*

$$X \xrightarrow{\ f\ } Y$$

and

$$\forall H \in \mathcal{N} : \quad \mathcal{D}(H; \mathcal{X}) \cdot f \subseteq \mathcal{D}(H; \mathcal{Y}). \tag{7.6}$$

With these definitions $\bar{\mathcal{F}}$ becomes a category. Note that it is, in general, in the second Grothendieck universe because $\mathcal{D}$ and $\mathcal{C}$ are $\mathcal{N}$-indexed families, and $\mathcal{N}$ is the class of the objects of $\mathcal{F}$. The simplest $\bar{\mathcal{F}}$-objects are $\bar{K} = (\mathcal{F}_{(-)K}, |K|, \mathcal{F}_{K(-)})$ for $K \in \mathcal{F}$, and for it we have that if $c \in \mathcal{C}_K$ then $c : \mathcal{X} \longrightarrow \bar{K}$ is regular, and $d \in \mathcal{D}_H$ iff $d : \bar{H} \longrightarrow \mathcal{X}$ is regular. Obviously we have $\mathcal{F}(H;K) = \bar{\mathcal{F}}(\bar{H};\bar{K})$, and so $\overline{(-)} : \mathcal{F} \longrightarrow \bar{\mathcal{F}}$ is a full functor, faithful on the arrows. Let's indicate with $\mathcal{C}^n_{\text{fin}}$ the category $\bar{\mathcal{F}}$ with figures in $\mathcal{F}$ of type $\mathbb{R}^h$ and with $\mathcal{C}^n$ the category $\bar{\mathcal{F}}$ with $\mathcal{F} = \mathbf{Man}^n$, the category of all C^n manifolds with charts in Banach spaces; the objects of $\mathcal{C}^n$ will be called *$\mathcal{C}^n$-spaces*. In $\mathcal{C}^n$ the functor $\overline{(-)}$ is faithful also on objects. In $\mathcal{C}^n_{\text{fin}}$ we have the following result: for a manifolds M we can define $\bar{M} = (\mathcal{D}, M, \mathcal{C})$ as the $\mathcal{C}^n_{\text{fin}}$-space with

$$\mathcal{D}_h := \mathbf{Man}^n(\mathbb{R}^h; M) \qquad \text{and} \qquad \mathcal{C}_k := \mathbf{Man}^n(M; \mathbb{R}^k)$$

It is easy to prove the following

Theorem 2.1 *Let M and N be finite-dimensional manifolds, then we have $f : \bar{M} \longrightarrow \bar{N}$ in $\mathcal{C}^n_{\text{fin}}$ iff $f : M \longrightarrow N$ in $\mathbf{Man}^n$. Moreover if $\bar{M} = \bar{N}$ and N admits partitions of unity, then $M = N$.*

Following [2] it is possible to prove that the category $\bar{\mathcal{F}}$ is complete and co-complete. Only an observation can be made for the lifting of $p_i : M \longrightarrow M_i$, with $(M_i)_{i \in I}$ manifolds in $\mathcal{F} = \mathbf{Man}^n$. We can define

$$\mathcal{D}(H;M) := \{d : |H| \longrightarrow M \mid \forall i \in I : \quad d \cdot p_i \in \mathcal{D}(H;M_i)\}$$

just as in [2], but in $\mathcal{C}(M;K)$ we can put the functions $c : M \longrightarrow |K|$ expressible as

$$c = (p_{i_1} \cdot c_1, \ldots, p_{i_h} \cdot c_h) \cdot f$$

with $i_1, \ldots i_h \in I$, $c_j \in \mathcal{C}(M_{i_j}; K_j)$, and $f \in \mathbf{Man}^n(K_1 \times \ldots \times K_h; K)$. In this way we can have in $\mathcal{C}(M_1 \times M_2; \mathbb{R}^{k_1+k_2})$ the usual charts of the product manifold, and moreover that

$$\overline{M \times N} = \bar{M} \times \bar{N}. \tag{7.7}$$

We will use this property later on. Finally if we define

$$\mathcal{D}_H := \{d : |H| \longrightarrow \bar{\mathcal{F}}(X;Y) \mid \bar{H} \times X \xrightarrow{d^\vee} Y \quad \text{in} \quad \bar{\mathcal{F}}\}$$

(obviously in the sequel we put $d^\vee(h,x) := d(h)(x)$ and $\mu^\wedge(x)(y) := \mu(x,y)$) and

$$\mathcal{C}_K := \{c : \bar{\mathcal{F}}(X;Y) \longrightarrow K \mid \forall H \in \mathcal{F}\ \forall d \in \mathcal{D}_H : \quad d \cdot c \in \mathcal{F}(H;K)\}$$

then $\langle \mathcal{D}, \bar{\mathcal{F}}(X;Y), \mathcal{C}\rangle$ is an $\bar{\mathcal{F}}$-object. With this definition, just as in [1], we can prove that $\bar{\mathcal{F}}$ is cartesian closed.

Examples: Using the property (7.7) we can prove that the following applications are $\mathcal{C}^\infty$ regular

1)

$$\begin{aligned} \partial_i : \quad \mathcal{C}^\infty(\mathbb{R}^n;\mathbb{R}^k) &\longrightarrow \mathcal{C}^\infty(\mathbb{R}^n;\mathbb{R}^k) \\ u &\longmapsto \partial u/\partial x_i \end{aligned}$$

2) The functional

$$\begin{aligned} i : \quad \mathcal{C}^\infty(\mathbb{R}^2;\mathbb{R}) &\longrightarrow \mathcal{C}^\infty(\mathbb{R};\mathbb{R}) \\ u &\longmapsto \int_a^b u(-,s)\,\mathrm{d}s \end{aligned}$$

3) Using 2) we can generalize it with the functional

$$\mathcal{I}(u)(t) := \int_a^b F[u(t,s), \partial_2 u(t,s), s]\,\mathrm{d}s$$

$$\mathcal{I} : \mathcal{C}^\infty(\mathbb{R}^2;\mathbb{R}^k) \longrightarrow \mathcal{C}^\infty(\mathbb{R};\mathbb{R})$$

where

$$F : \mathbb{R}^k \times \mathbb{R}^k \times \mathbb{R} \longrightarrow \mathbb{R}$$

is smooth. This application is $\mathcal{C}^\infty$ regular too.

2.1. The extension of spaces and functions

We now face up the problem to extend with infinitesimal and infinite points the objects and the arrows of some subcategory of $\mathcal{C}^n$-spaces. We will always work with the category $\mathcal{C}^n$ (with figures generic manifolds, that is with $\mathcal{F} = \mathbf{Man}^n$) and we will search for some subcategory in which the notion of pathwise lipschitzian function could have a sense. The idea to generalize from ${}^\bullet\mathbb{R}$ to the extension ${}^\bullet X$ of a $\mathcal{C}^n$-space is to take the limit

$$\lim_{t\to 0} \frac{\varphi[x_n(t)] - \varphi[y_n(t)]}{t} = 0$$

where $\varphi \in \mathcal{C}(X;K)$ and $K \in \mathbf{Ban}$, the category of Banach spaces. The following definitions born from this idea. If X is a $\mathcal{C}^n$-space we indicate with $\mathcal{C}_0(X)$ the set of all the functions $x : \mathbb{R} \longrightarrow X$ for which $x \cdot c : \mathbb{R} \longrightarrow K$ is continuous in $t = 0$ for any $c \in \mathcal{C}(X;K)$. We call this functions "*continuous in* $t = 0$ *(rel* X*)*". Moreover we write $x \asymp y$ in X, and we say "x, y *are identified in* X", to indicate that $\forall K \in \mathbf{Ban}\ \forall \varphi \in \mathcal{C}(X; K) : \varphi(x) = \varphi(y)$. If $x \asymp y$ implies $x = y$, then we say that "X *is separated*", for example if the space is a manifold with partitions of unity.

Definition 2.3 *If* X, Y *are* $\mathcal{C}^n$*-spaces, then we say Lip*$(X;Y)$ *the set of all the "pathwise lipschitzian functions", that is of all the* $f : X \longrightarrow Y$ *such that if* $K \in \mathbf{Ban}, \psi \in \mathcal{C}(Y; K)$ *and* $x, y \in \mathcal{C}_0(X)$ *with* $x(0) \asymp y(0)$, *then we can find* $J \in \mathbf{Ban}, \varphi \in \mathcal{C}(X; J), \delta > 0$ *such that*

$$\forall t \in (-\delta, \delta) : \quad \| \psi[f(x_t)] - \psi[f(y_t)] \| \leq \| \varphi(x_t) - \varphi(y_t) \|$$

This definition generalizes that given in Definition 1.4. We haven't a theorem to assure that any $\mathcal{C}^n$-function is pathwise lipschitzian, and for this reason we restrict to a full subcategory with the following

Definition 2.4 *We say that a* $\mathcal{C}^n$*-space* X *is "lipschitzian" if all the regular functions* $g : X \longrightarrow \overline{K}$, *with* $K \in \mathbf{Ban}$, *are pathwise lipschitzian.* $\mathcal{C}^n\mathcal{L}$ *and* $\mathcal{C}^n\mathcal{L}s$ *are the categories of* $\mathcal{C}^n$*-lipschitzian spaces, and* $\mathcal{C}^n$*-lipschitzian separated spaces.*

Note that any manifold is trivially a $\mathcal{C}^n$-lipschitzian space. Now any $\mathcal{C}^n$-function $f : X \longrightarrow Y$ defined on a lipschitzian space X is pathwise lipschitzian. The categories $\mathcal{C}^n\mathcal{L}$ and $\mathcal{C}^n\mathcal{L}s$ are complete, co-complete and cartesian closed. Finally, if we take $\mathcal{C}^n$-spaces as objects and pathwise lipschitzian functions as arrows, we obtain a category $\mathcal{C}^n\vec{\mathcal{L}}$.

Definition 2.5 *If* X *is a* $\mathcal{C}^n$*-space, and* x, y *are sequences of* $\mathcal{C}_0(X)$, *then we say* $x \sim y$ *if we can find* $N \in \mathbb{N}$ *such that for any* $n \geq N, K \in$ $\mathbf{Ban}$ *and* $\varphi \in \mathcal{C}(X; K)$ *we have*

$$\lim_{t \to 0} \frac{\varphi(x_n t) - \varphi(y_n t)}{t} = 0$$

Finally we set ${}^\bullet X := \mathcal{C}_0(X)^{\mathbb{N}}/_\sim$.

Note that at the moment on the set ${}^\bullet X$ there is no structure. If $f \in \mathrm{Lip}(X;Y)$, then the extension ${}^\bullet f(x)_n := f \circ x_n$ is well defined. To embed the space X in ${}^\bullet X$ we can consider $(\hat{x})_n(t) := x$; this is a real injection only if X is separated. Moreover we have ${}^\bullet f(\hat{x}) = \widehat{f(x)}$, so ${}^\bullet f$ is an extension of f only if X and Y are both separated.

2.2. The category ${}^\bullet\mathcal{C}^n$

Now our purpose is to search for a structure for the set ${}^\bullet X$ and its subsets, for example for ${}^\bullet\mathbb{R}$ and D, and with this structure to construct a new category ${}^\bullet\mathcal{C}^n$. For this aim we will follow an idea similar to that used for the passage from $\mathcal{F}$ to $\overline{\mathcal{F}}$, but with a modification. In fact some consideration must be done on the exponential object ${}^\bullet\mathbb{R}^D = {}^\bullet\mathcal{C}^n(D; {}^\bullet\mathbb{R})$. It seems natural to think that its support set must be $|{}^\bullet\mathbb{R}^D| = \{{}^\bullet f|_D \mid f \in \mathcal{C}^n(\mathbb{R};\mathbb{R})\}$. In this way, in fact, the derivation formula can be used for every function $f : D \longrightarrow {}^\bullet\mathbb{R}$ in ${}^\bullet\mathcal{C}^n$. In the same way every function $\tau : D \longrightarrow {}^\bullet M$ in ${}^\bullet\mathcal{C}^n$ will be a tangent vector of the manifold M if we define $|{}^\bullet M^D| = \{{}^\bullet f|_D \mid f \in \mathcal{C}^n(\mathbb{R}; M)\}$. But let's consider a constant function f in $|{}^\bullet\mathbb{R}^D|$, then $f = {}^\bullet g|_D$ for some $g \in \mathcal{C}^n(\mathbb{R};\mathbb{R})$, and so $f(0) = {}^\bullet g(0) = g(0) \in \mathbb{R}$ and f assumes necessarily a standard value: not all constant functions are in $|{}^\bullet\mathbb{R}^D|$. Likewise $\tau(0) \in M$ if $\tau \in |{}^\bullet M^D|$. This has a consequence for the property of cartesian closure of ${}^\bullet\mathcal{C}^n$: for the same reason there is no lifting $s^\wedge : D \longrightarrow {}^\bullet\mathbb{R}^D$ for the application

$$\begin{aligned} s: \quad & D \times D \longrightarrow {}^\bullet\mathbb{R} \\ & (h,k) \longmapsto h+k \end{aligned}$$

because $s^\wedge(h)(0) = h \notin \mathbb{R}$. Hence we will give only a sufficient condition for the lifting property. Let's indicate with ${}^\bullet\mathbf{Man}^n$ the category with objects all the pairs (A, H) with $A \subseteq {}^\bullet H$, for simplicity they will be indicated with $AH := (A, H)$, and arrows defined by

$$AH \xrightarrow{\ g\ } BK \quad :\Longleftrightarrow \quad \exists f \in \mathcal{C}^n(H;K): \quad g = {}^\bullet f|_A : A \longrightarrow B$$

Definition 2.6 *If X is a set, then $(\mathcal{D}, X, \mathcal{C})$ is an object of ${}^\bullet\mathcal{C}^n$ if for all $AH, BK, CJ \in {}^\bullet\mathbf{Man}^n$ we have*

$$\mathcal{D}_{AH} \subseteq \mathbf{Set}(A;X) \qquad \text{and} \qquad \mathcal{C}_{BK} \subseteq \mathbf{Set}(X;B)$$

$$\mathcal{D}_{AH} \cdot \mathcal{C}_{BK} \subseteq {}^\bullet\mathbf{Man}^n(AH;BK)$$

$${}^\bullet\mathbf{Man}^n(CJ, AH) \cdot \mathcal{D}_{AH} \subseteq \mathcal{D}_{CJ}.$$

The definition of arrow in ${}^\bullet\mathcal{C}^n$ is the same as in (7.6).

Note that we are considering subset $A \subseteq {}^\bullet H$ because, for the usual reason, if $\delta \in \mathcal{C}^n(H;\mathbb{R})$ and ${}^\bullet\delta : {}^\bullet H \longrightarrow D$, then δ is constant to 0, so ${}^\bullet\mathbf{Man}^n({}^\bullet HH; D\mathbb{R})$ is trivial; on the contrary $1_D \in {}^\bullet\mathbf{Man}^n(D\mathbb{R}; D\mathbb{R})$. The category ${}^\bullet\mathcal{C}^n$ is complete and co-complete, and again we choose to define the lifting as in section 2. Now the initial problem becomes to complete on the objects the definition of the functor ${}^\bullet(-) : \mathcal{C}^n\vec{\mathcal{L}} \longrightarrow {}^\bullet\mathcal{C}^n$. We solve it with the following

Definition 2.7 *If X is a $\mathcal{C}^n$-space and $S \subseteq {}^\bullet X$, we define an object of ${}^\bullet\mathcal{C}^n$ with support set S using the following structure*

$$d \in \mathcal{D}_{AH} \quad :\Longleftrightarrow \quad d : A \longrightarrow S$$

$$\exists\, \delta \in \mathcal{C}^n(\bar{H}; X) : \quad {}^\bullet\delta|_A = d$$

$$c \in \mathcal{C}_{BK} \quad :\Longleftrightarrow \quad c : S \longrightarrow B$$

$$\exists\, \gamma \in \mathcal{C}^n\vec{\mathcal{L}}(X; \bar{K}) : \quad {}^\bullet\gamma|_S = c.$$

It will be indicate with ${}^\bullet(SX)$; in particular ${}^\bullet X := {}^\bullet({}^\bullet XX)$.

Note that $\mathcal{C}^n(\bar{H}; X) = \mathcal{C}^n\vec{\mathcal{L}}(\bar{H}; X)$ because $\bar{H}$ is trivially lipschitzian if H is a manifold. For example we simply set $D := {}^\bullet(D\mathbb{R})$, and ${}^\bullet\mathbb{R} := {}^\bullet({}^\bullet\mathbb{R}\mathbb{R})$. With this definition ${}^\bullet(-)$ is actually a functor, full and faithfull on the category $\mathbf{Man}^n_1$ of manifolds which admits partitions of unity. The final situation of our categories is

$$\mathbf{Man}^n_1 \subseteq \mathbf{Man}^n \subseteq \mathcal{C}^n\mathcal{L}s \subseteq \mathcal{C}^n\mathcal{L} \subseteq \mathcal{C}^n\vec{\mathcal{L}} \subseteq \mathcal{C}^n, \quad \mathbf{Man}^n_1 \subseteq {}^\bullet\mathcal{C}^n$$

We now consider the definition of Y^X for $X, Y \in {}^\bullet\mathcal{C}^n$.

Definition 2.8 *If $X, Y \in {}^\bullet\mathcal{C}^n$, we put on ${}^\bullet\mathcal{C}^n(X;Y)$ the following structure: at first we define*

$$d \in \mathcal{D}^0_{CJ} \quad :\Longleftrightarrow \quad 1)\ \ C \subseteq J$$

$$2)\ \ C \xrightarrow{\ d\ } {}^\bullet\mathcal{C}^n(X;Y)$$

$$3)\ \ \overline{CJ} \times X \xrightarrow{\ d^\vee\ } Y \quad \text{in} \quad {}^\bullet\mathcal{C}^n$$

then we can set

$$\mathcal{D}_{AH} := \{f \cdot d \mid \exists CJ \in {}^\bullet\mathbf{Man}^n : \quad f \in {}^\bullet\mathbf{Man}^n(AH; CJ) \quad , \quad d \in \mathcal{D}^0_{CJ}\}$$

$$\mathcal{C}_{BK} := \{c \in B^{{}^\bullet\mathcal{C}^n(X;Y)} \mid \forall AH\ \forall d \in \mathcal{D}_{AH} : d \cdot c \in {}^\bullet\mathbf{Man}^n(AH; BK)\}.$$

Note that in $\mathcal{D}^0_{CJ}$ we consider $C \subseteq J$ not $C \subseteq {}^\bullet J$. This condition is important only to express the derivation formula by means of an ${}^\bullet\mathcal{C}^\infty$-isomorphism between ${}^\bullet\mathbb{R}^D$ and $\mathbb{R}^2$. In the category ${}^\bullet\mathcal{C}^n$ the "lowering property" of cartesian closure are still true, on the contrary the "lifting properties" are true for some spaces only:

Definition 2.9 *If X is a ${}^\bullet\mathcal{C}^n$-space, we say that "X is closed with respect to constants" if*

- $\forall AH \in {}^\bullet\mathbf{Man}^n \quad \forall f : A \longrightarrow X$ *constant* : $\quad f \in \mathcal{D}(AH; X)$;
- *any* $\delta \in \mathcal{D}(AH; X)$ *is of the form* $\delta = f \cdot d$ *where*

$$f \in {}^\bullet\mathbf{Man}^n(AH; CJ),\ d \in {}^\bullet\mathcal{C}^n(\overline{CJ}; X),\ C \subseteq J$$

  for some $CJ \in {}^\bullet\mathbf{Man}^n$.

Hence Y^X, and $\overline{AH}$ if $A \subseteq H$, are close with respect to constants and if X has this property then $X \times Y \xrightarrow{h} Z$ implies $X \xrightarrow{h^\wedge} Z^Y$.

2.3. Examples

Finally we present some examples of differential geometry without coordinates systems, and valid for objects of ${}^\bullet\mathcal{C}^n$. Almost all these results are re-formulations of analogous results of SDG.

2.3.1 Tangent vectors Exactly as for lines in ${}^\bullet\mathbb{R}$, any first order infinitesimal curve traced on an object $M \in {}^\bullet\mathcal{C}^n$ becomes a straight line, hence we define "tangent vector to M in m" any $t : D \longrightarrow M$ (always in ${}^\bullet\mathcal{C}^n$) such that $t(0) = m$. The object $M^D =: \mathrm{T}M$ is the tangent bundle of M; it has a natural projection

$$\pi \ : t \in M^D \longmapsto t(0) \in M$$

which is a ${}^\bullet\mathcal{C}^n$-arrow. Every tangent space $\mathrm{T}_m M := \pi^{-1}(\{m\}) = \{t \in M^D \mid t(0) = m\}$ can be thought as a subobject of $\mathrm{T}M$. Before adding tangent vectors we see that the differential of a function $f : M \longrightarrow N$ can be simply defined using the composition

$$\mathrm{d}f \ : \ t \in M^D \longrightarrow t \cdot f \in N^D.$$

For the properties of the category ${}^\bullet\mathcal{C}^n$ we always have that the applications

$$\mathrm{T}M \xrightarrow{\mathrm{d}f} \mathrm{T}N$$

$$\mathrm{T}_m M \xrightarrow{\mathrm{d}_m f} \mathrm{T}_{f(m)} N$$

are still ${}^\bullet\mathcal{C}^n$-regular. For a generic object M it is not usually possible to add tangent vectors (e.g. in a double point of a curve), thus we have to suppose that from k tangent vectors $t_i : D \longrightarrow M$, $i = 1, \ldots, k$, in the point $m \in M$ we can construct "the infinitesimal parallelogram" generated by the vectors, that is a function $p : D^k \longrightarrow M$ such that

$$\forall i = 1, \ldots, k : \quad p(0, \overset{i-1}{\ldots}, 0, h, 0, \ldots, 0) = t_i(h). \tag{7.8}$$

This is the idea of the following

Definition 2.10 *We say that the object M is "infinitesimally linear in m" if for any $k \in \mathbb{N}$ greater than 1 and for any $t_i \in T_m M$, $i = 1, \ldots, k$, there exists one and only one $p : D^k \longrightarrow M$ such that (7.8) is true for all $h \in D$.*

Hence we can define the sum of two vectors t_1, t_2 as the diagonal of the parallelogram p generated by t_1, t_2:

$$(t_1 + t_2)(h) := p(h, h) \qquad \forall h \in D$$

With this definition any $\mathrm{T}_m M$ becomes a real vector space, and the differential of f, which is always homogeneous for every object, becomes linear. Any manifolds is infinitesimally linear in any point.

2.3.2 Vector fields Let's define the vector fields of a separated $\mathcal{C}^n$-space $M = (\mathcal{D}, M_0, \mathcal{C})$, e.g. a manifold with partitions of unity, using its extension $\mathcal{M} := {}^\bullet M \in {}^\bullet\mathcal{C}^n$. First of all we can consider the space ${}^\bullet(M_0 M) \in {}^\bullet\mathcal{C}^n$ like the space M itself, without new points. Then we can define "vector field" an application $X : M \longrightarrow \mathcal{M}^D$ with the following properties

$$1) \quad \forall m \in M : \quad \pi[X(m)] = X(m)(0) = m$$

$$2) \quad \exists f \in \mathcal{C}^n[M; \mathcal{C}^n(\mathbb{R}; M)] : \forall m \in M : \quad X(m) = {}^\bullet[f(m)]|_D$$

Condition 2) says, in other words, that $X(m) = g(m, -) : D \longrightarrow \mathcal{M}$, for a $\mathcal{C}^n$-regular function $g : M \times \mathbb{R} \longrightarrow M$. In ${}^\bullet\mathcal{C}^n$ we always have $X^\vee : M \times D \longrightarrow \mathcal{M}$, but using the cartesian closure of $\mathcal{C}^n$ we can see X as an infinitesimal transformation of 1_M too. In fact starting from $f : M \longrightarrow \mathcal{C}^n(\mathbb{R}; M)$ we can consider its lowering

$$\mathbb{R} \times M \xrightarrow{\ i f^\vee\ } M$$

where $i : (m, r) \in M \times \mathbb{R} \longmapsto (r, m) \in \mathbb{R} \times M$. So we have

$$(i f^\vee)^\wedge : \mathbb{R} \longrightarrow M^M \quad \text{in} \quad \mathcal{C}^n$$

$$\bullet(if^\vee)^\wedge : \bullet\mathbb{R} \longrightarrow \bullet(M^M) \quad \text{in} \quad \bullet\mathcal{C}^n$$
$$\tau := \left[\bullet(if^\vee)^\wedge\right]|_D : D \longrightarrow \bullet(M^M) \quad \text{in} \quad \bullet\mathcal{C}^n$$

The space $\bullet(M^M)$ contains all the transformations of M in itself because M is separated. Moreover it is easy to prove that $\tau(0)(m) = X(m)(0)$ that is $\tau(0) = 1_M$, hence the vector field X can be thought as an infinitesimal transformation of the space M in itself.

References

[1] K.T. Chen, *On differentiable spaces*, Categories in continuum physics, Lecture Notes in Mathematics 1174, Springer-Verlag (1982), 38–42.

[2] A. Frölicher, A. Kriegl, *Linear spaces and differentation theory*, John Wiley & sons (1988).

[3] A. Kock, *Synthetic Differential Geometry*, London Math. Soc. Lect. Note Series 51, Cambrige Univ. Press (1981).

[4] R. Lavendhomme, *Lecons de géométrie différentielle synthétique naïve*, Ciaco (1987).

[5] F.W. Lawvere, *Categorical Dynamics*, Topos theoretic Methods in Geometry, Aarhus, Various Publications Series no. 30 (1979), 1–28.

ON HYPERFINITE APPROXIMATIONS OF THE FIELD *R*

Evgenii I. Gordon
Department of Mathematics, University of Illinois
273 Altgeld Hall, 1409 West Green Street, Urbana, IL 61801, USA
gordon3@.math.uiuc.edu

Olga A. Rezvova
Department of Mathematics, University of Nizhnii Novgorod
Gagarin Ave 23, Nizhnii Novgorod 603600, Russia

Keywords: Computer arithmetic, approximation, locally compact groups

Abstract: The notions of hyperfinite approximations of locally compact groups and of the field of reals are introduced. It is proved that the groups, which are approximated by hyperfinite groups are unimodular, but this condition is not sufficient – the group $SO(3)$ is not approximable by hyperfinite groups.

The algebraic systems, which approximate $\mathbf{R}$ are the hyperfinite versions of computer arithmetics implemented in existing computers. The nonapproximability of some matrix groups implies the impossibility to approximate $\mathbf{R}$ by a hyperfinite field in a such way that the operation of taking an inverse element in $\mathbf{R}$ is approximated by the similar operation in approximating hyperfinite field.

Introduction

A new definition of approximation of topological groups by finite ones was introduced and investigated by means of nonstandard analysis by the first author (cf. [Gor] and references there). According to this definition the finite groups G_n approximate a topological group G if their group operations approximate the group operation of G in some natural sense. In general, groups G_n are not subgroups of G (subsets only, see definition 1 below).

P. Schuster, U. Berger and H. Osswald (eds.),
Reuniting the Antipodes – Constructive and Nonstandard Views of the Continuum, 93–102.

The approximability of all locally compact abelian (LCA) groups was proved in [Gor]. This fact makes it possible to deduce the large part of harmonic analysis on LCA groups from the harmonic analysis on finite groups. The most part of this program was realized in [Gor], where also some new results concerning convergence of approximations of Fourier transform on LCA groups by finite Fouier transform were obtained on this way. On the base of these results the finite dimensional approximations of operators in the Hilbert spaces of functions on LCA groups were constructed and investigated in [AGK]. The similar results about approximations of operators may be obtained also for the case of approximable noncommutative groups, because as it was shown in [Gor] for compact groups and in [AGG] for discrete groups, the representations of approximable groups may be approximated by the representations of approximating finite groups.

In [VeGo] the approximability of a large class of nilpotent Lie groups was proved.

Unfortunately the class of approximable topological groups does not contain many groups, important for applications. For example, all approximable locally compact groups are unimodular, but this condition is not sufficient – simple compact Lie groups are not approximable as well [AGG]. Even not all discrete groups are approximable. The class of all approximable discrete groups was investigated in [VeGo]. Some applications of this class to ergodic theory were described there.

The fact that some matrix groups are not approximabale by finite ones implies some restrictions on approximations of the field $\mathbf{R}$ by finite algebraic systems. We prove in the Section 2 that it is impossible to approximate $\mathbf{R}$ by finite fields in such a way that the operation of taking an inverse element in $\mathbf{R}$ is approximated by the same operations in the finite fields. Notice that all known finite algebraic systems which are implemented in computers for simulation of the field $\mathbf{R}$ (cf. e.g. [Knu]) are not even rings – the associative and distributive laws fail there. Nevertheless the possibility to approximate $\mathbf{R}$ by finite algeraic systems implies the possibility to approximate all matrix groups and any other objects defined on the base of reals by appropriate finite objects. From this point of view the problem of construction of approximations of $\mathbf{R}$ with as good algebraic properties as possible, for example finite rings, seems to be of interest. The definitions and considerations of this paper make it possible to investigate this problem.

1. APPROXIMATIONS OF TOPOLOGICAL GROUPS

In this section we expose some results about approximations of locally compact groups by finite ones. The Definition 1 is contained in [Gor]. The Theorem 1 also may be found in [Gor], but we give here essentially simplier proof of this theorem. The main new result of this section is nonapproximability of $SO(3)$, which we deduce from the well-known Kazhdan's theorem about ε-representations [Kazh]. The idea to use Kazhdan's theorem in these problems was suggested by L.Glebsky.

In what follows we consider a locally compact separable group $\mathcal{G}$. To make the definition shorter we fix some metric d, which defines the topology on $\mathcal{G}$. Indeed the property of finite approximability, introduced here depends only on the topology of $\mathcal{G}$.

Definition 1 *We say that a locally compact separable group $\mathcal{G}$ is finitely approximable iff there exist a sequence of finite groups G_n and a sequence of injective maps $j_n : G_n \to \mathcal{G}$, which satisfy the following conditions:*

1. $\forall \varepsilon > 0 \ \forall K \in \mathcal{K} \ \exists n_0 \forall n_{>} n_0 \ \forall \xi \in K \ \exists g \in G_n d(j_n(g), \xi) < \varepsilon$, *where* $\mathcal{K}$ *is a family of all compact subsets of* $\mathcal{G}$*;*
2. $\forall \varepsilon > 0 \forall K \in \mathcal{K} \exists n_0 \forall n_{>} n_0 \forall g, h \in j_n^{-1}(K)$
   $d\left(j_n(g \cdot h^{\pm 1}), j_n(g) \cdot j_n(h)^{\pm 1}\right) < \varepsilon;$
3. $j_n(e_n) = e$, *where* e_n *and* e *are the units of groups* G_n *and* $\mathcal{G}$ *repectively.*

We say that $\langle G_n, j_n \rangle$ is an approximating sequence for $\mathcal{G}$

For compact groups this definition may be simplified.

Proposition 1 *A compact separable group $\mathcal{G}$ is approximable iff there exist a sequence of finite groups G_n and a sequence of injective maps $j_n : G_n \to \mathcal{G}$, which satisfy the following conditions:*

1. $\forall \varepsilon > 0 \exists n_0 \forall n_{>} n_0 \ \forall \xi \in \mathcal{G} \exists g \in G_n d(j_n(g), \xi) < \varepsilon,$
2. $\forall \varepsilon > 0 \exists n_0 \forall n_{>} n_0 \forall g, h \in G_n$
   $d\left(j_n(g \cdot h^{\pm 1}), j_n(g) \cdot j_n(h)^{\pm 1}\right) < \varepsilon;$
3. $j_n(e_n) = e.$

The reformulation of this definition on nonstandard language is following. We denote by $\mathrm{ns}({}^*\mathcal{G})$ the external subgroup of all nearstandard elements of a group ${}^*\mathcal{G}$.

Proposition 2 *1 A locally compact separable group $\mathcal{G}$ is finitely approximable iff there exist a hyperfinite group G and an internal map $j : G \to {}^*\mathcal{G}$ such that*

1. $\forall \xi \in ns({}^*\mathcal{G})\ \exists g \in G \quad j(g) \approx \xi;$
2. $\forall g_1, g_2 \in j^{-1}(ns({}^*\mathcal{G})) j(g_1 \cdot g_2^{\pm 1}) \approx j(g_1) \cdot j(g_2)^{\pm 1}.$
3. $j(e_G) = e.$

2 A compact separable group $\mathcal{G}$ is finitely approximable iff there exist a hyperfinite group G and an internal map $j : G \to {}^*\mathcal{G}$ such that

1. $\forall \xi \in \mathcal{G}\ \exists g \in G \quad j(g) \approx \xi;$
2. $\forall g_1, g_2 \in j^{-1}(\mathcal{G}) j(g_1 \cdot g_2^{\pm 1}) \approx j(g_1) \cdot j(g_2)^{\pm 1}.$
3. $j(e_G) = e.$

We omit the simple proofs of this propositions $\square$.

Theorem 1 *Any finitely approximable group $\mathcal{G}$ is unimodular.*

Proof. As usual we denote by $C_0(\mathcal{G})$ the space of all continuous functions with compact support on $\mathcal{G}$ and by $^\circ\alpha$ – the standard part of a real α. We use the nonstandard definition of finite approximability (Proposition 2).

Fix any relatively compact neighbourhood U of unit $e \in \mathcal{G}$. Let $A = j^{-1}({}^*U)$, $\|A\|$ denotes the hyperfinite cardinality of A, $\Delta = \|A\|^{-1}$, and define a positive functional I on $C_0(\mathcal{G})$ by the formula

$$I(f) = {}^\circ\left(\Delta \cdot \sum_{g \in G} {}^*f(j(g))\right),$$

for any standard function $f \in C_0(\mathcal{G})$ [1].

First of all we show that this functional is defined on the total space $C_0(\mathcal{G})$. Let $f \in C_0(\mathcal{G})$, $\sup f = C$ and $\operatorname{supp} f = K$. Consider an open neighbourhood of unit V such that $\bar{V} \subseteq U$, where $\bar{V}$ is a closure of V. By compactness of K we can take a standardly finite $H = \{h_1, \ldots, h_n\} \subseteq K$

[1] For the case of locally compact abelian groupsthis representation of the Haar integral was obtained in [Gor], Section 2.4.

such that $K \subseteq HV$. By the nonstandard definitionof finite approximability, there exist $g_1, \ldots, g_n \in G$ such that $j(g_i) \equiv h_i$, $i = 1, \ldots, n$. Notice that

$$^*K \subseteq \bigcap_{i=1}^{n} h_i{}^*V \subseteq \bigcap_{i=1}^{n} j(g_i)^*\bar{V} \subseteq \bigcap_{i=1}^{n} j(gi)^*U$$

In particular the hyperfinite cardinality $\|j^{-1}(^*K)\| \leq n \cdot \Delta^{-1}$. Then

$$|\Delta \sum_{g \in G} {}^*f(j(g)) \leq \Delta \cdot C \cdot \|j^{-1}(^*K)\| \leq C \cdot n$$

and thus $I(f)$ is finite.

Let $f \in C_0(\mathcal{G})$ be such that $0 \leq f \leq 1$ and $f|U = 1$. Then obviously $I(f) \geq 1$ and thus $I(f)$ is a non trivial functional.

Let $\xi \in \mathcal{G}$ and $f^\xi(\eta) = f(\xi\eta)$. By item 1. of Proposition 2 $\xi \approx j(h)$ for some $h \in G$. Using the considerations similar to those of the proof of finitness of $I(f)$ and item 2. of Proposition 2 we obtain that

$$I(f^\xi) = {}^\circ\left(\Delta \cdot \sum_{g \in G} {}^*f(j(gh))\right) = {}^\circ\left(\Delta \cdot \sum_{g \in G} {}^*f(j(g))\right) = I(f),$$

and thus I is left invariant. In the same way we can prove that it is right invariant $\Box$.

The well known example of a nonunimodular and thus not finitely approximable group is a group $\mathcal{G}_0$ of all matrices of the form

$$\begin{pmatrix} a & b \\ 0 & 1 \end{pmatrix},$$

where $a, b \in \mathbf{R}$ and $a > 0$ (cf. for example [HeRo]).

Not all unimodular groups are finitely approximable. We shall show now that the group $SO(3)$, which is compact and thus unimodular, is not finitely approximable. The proof is based on a well-known D.Kazhdan's theorem about ε-representations [Kazh]. We formulate here a slight modification of this theorem (cf. [AGG])

Let H be a metrizable group with an invariant metric ρ, G – an arbitrary group.

We say that a map $T : G \to H$ is an ε-homomorphism if

$$\forall g_1, g_2 \in G \; \rho\left(T(g_1 g_2), T(g_1)T(g_2)\right) < \varepsilon.$$

Theorem 2 *[Kazh] For any compact Lie group H there exists $\delta > 0$ such that for any amenable (in particular, finite) group G, for any $0 < \varepsilon < \delta$ and for any ε-homomorphism $\tau : G \to H$ there exists a homomorphism $\varphi : G \to H$ such that $\forall g \in G\ \rho(\varphi(g), \tau(g)) < 2\varepsilon$*

The proof of this theorem is almost exactly the same as the proof of Kazhdan's theorem - cf. [Kazh].

If in conditions of Theorem 2 we replace a compact Lie group by the group $U(H)$ of all unitary operators in a Hilbert space H and consider a metric induced by operator norm in this group we obtain exactly Kazhdan's theorem.

The following theorem is a simple corollary of Theorem 2.

Theorem 3 *The group $SO(3)$ is not finitely approximable.*

Proof. Let δ be the same as in theorem 2. If $\langle G_n, j_n\rangle$ is an approximating sequence for $SO(3)$, which satisfies the conditions of Proposition 1, then for any $0 < \varepsilon < \delta$ there exists such n_0 that for all $n > n_0$ $j_n : G_n \to SO(3)$ is an ε-homomorphism and $j_n(G_n)$ is an ε-grid in $SO(3)$ i.e. for any $h \in SO(3)$ there exists $g \in G_n$ such that $\rho(j_n(g), h) < \varepsilon$. By Theorem 2 there exists a homomorphism $i_n : G_n \to SO(3)$ such that $\forall g \in G_n \rho(j_n(g), i_n(g)) < 2\varepsilon$. Thus $i_n(G_n)$ is a subgroup of $SO(3)$ which is a 3ε-grid in $SO(3)$. Thus we have proved that for any $\varepsilon > 0$ there exists a finite subgroup $H \subseteq SO(3)$ such that H is an ε-grid in $SO(3)$. But this is impossible, because $SO(3)$ has only finitely many subgroups, which do not consist of rotations over a fixed axis (see for example[Kost]). Obviously rotations over a fixed axis can not form an ε-grid in $SO(3)$ □

2. HYPERFINITE COMPUTER ARITHMETICS

In this section we introduce the definition of approximation of the field **R** by finite algebraic systems. It is similar to Definition 1. We consider only nonstandard version of this definition. It is connected with an idealized computer, which has a hyperfinite memory size. In our opinion it is more natural to use the nonstandard language and, in particular, to consider the hyperfinite algebraic systems, approximating **R** rather then the sequences of finite algebraic systems in the problems of investigation of computer simulation of continuous objects.

The example of hyperfinite computer arithmetic, approximating **R**, which we consider here, is a nonstandard version of the computer arithmetic, described in the monograph [Knu]. This computer arithmetic is implemented in existing computers.

Definition 2 *Let $\mathcal{R} = \langle R, \{0\}\{\ominus, \oplus, \odot\}\rangle$ be a hyperfinite algebraic system with one constant 0, one unary operation $\ominus$ and two binary operations $\oplus, \odot$; $j : R \to {}^*\mathbf{R}$ – an injective map, which satisfy the following conditions:*

1. $\forall \xi \in ns({}^*\mathbf{R})\ \exists g \in R \quad j(g) \approx \xi$.
2. $\forall g_1, g_2 \in j^{-1}(ns({}^*\mathbf{R}))\ j(g_1 \odot g_2) \approx j(g_1) \odot j(g_2)$, $j(g_1 \oplus g_2) \approx j(g_1) \oplus j(g_2)$, $j(\ominus g) \approx -j(g), j(0) = 0$.
3. $\forall g \in R \ \ominus \ominus g = g$, $\ominus(g_1 \oplus g_2) = (\ominus g_1) \oplus (\ominus g_2)$, $g \oplus (\ominus g) = 0$.

Then we say that the pair $\langle \mathcal{R}, j\rangle$ is a hyperfinite computer arithmetic.

Neither commutativity nor associativity or distributivity of operation $\oplus$ and $\odot$ is assumed in this definition.

Let $\langle \mathcal{R}, j\rangle$ be a hyperfinite computer arithmetic. For $g_1, g_2 \in R$ we denote $g_1 \oplus (\ominus g_2)$ by $g_1 \ominus g_2$. Define a binary equivalence relation $\sim$ on R:

$$g_1 \sim g_2 \Longleftrightarrow j(g_1 \ominus g_2) \approx 0.$$

Let $R_b = j^{-1}(\mathrm{ns}({}^*\mathbf{R}))$. It is easy to see from Definition 2 that R_b is closed under the operations of $\mathcal{R}$ and thus $\langle R_b, \ominus, \oplus, \odot\rangle$ is an external subsystem of $\mathcal{R}$, which we denote by $\mathcal{R}_b$.

Theorem 4 *The binary relation $\sim$ is a congruence relation on $\mathcal{R}_b$. The quotient system $\mathcal{R}_b/\sim$ is isomorphic to the field **R**. An isomorphism $\varphi : \mathcal{R}_b/\sim \to \mathbf{R}$ may by defined by the formula*

$$\varphi(g^\#) = {}^\circ(j(g)),$$

where $g^\#$ is the equivalence class of an element $g \in R_b$.

The proof of this theorem follows directly from Definition 2. Let us show, for example, that $\sim$ is transitive (we do not assume that $\oplus$ is associative). Let $a, b, c \in R_b$, $a \sim b$, $b \sim c$. Then $j(a \ominus c) \approx j(a) - j(c) = j(a) - j(b) + j(b) - j(c) \approx j(a \ominus b) + j(b \ominus c) \approx 0$, thus $a \sim c$ □

Example. Recall that the normal (computer) form of a real α is its representation:

$$\pm 0.a_0 a_1 a_2 \ldots\ldots \cdot 10^p,$$

where $p \in \mathbf{Z}$, and $a_0a_1a_1....$ is a finite or infinite sequence of decimal digits ($0 \leq a_n \leq 9$) and $a_0 \neq 0$. The integer p is called the exponent of α and $a_0a_1a_1....$ – its mantissa.

Fix two hypernatural numbers $P, Q \in {}^*\mathbf{N} \setminus \mathbf{N}$ and let R be the internal hyperfinye set of all hyperreals α such that mantissa of α contains no more then Q digits and its exponent p satisfies the unequality $|p| < P$. We define the operations on R as follows. For $\alpha, \beta \in R$ let $\ominus\alpha = -\alpha$. To define $\alpha \oplus \beta$ we represent $\alpha + \beta$ in normal form $0.c_0c_1 \ldots c_Q..... \cdot 10^r$ (notice that mantissa of the sum may contain more then Q didgits). Now, if $|r| < P$, then $\alpha \oplus \beta = 10^r \cdot 0.c_0c_1 \ldots c_{Q-1}$.

If $r \geq P$ we say that an *exponent overfilling* occurs. In this case $\alpha \oplus \beta = 10^{P-1} \cdot 0.99 \ldots 9$ (Q digits).

If $r \leq -P$ we say that an *exponent vanishing* occurs. In this case $\alpha \oplus \beta = 0$.

The definition of $\alpha \odot \beta$ is similar.

We assume that $j : R \to {}^*\mathbf{R}$ is an identity injection. Then R_b consists of $\alpha \in R$ with standard exponent.

Theorem 5 *The pair $\langle \mathcal{R}, j \rangle$, where $\mathcal{R} = \langle R, \{0\}\{\ominus, \oplus, \odot\} \rangle$ is the algebraic system, defined in this example, is a nonstandard computer arithmetic.*

The proof of this theorem follows from the next lemma, which can be verified directly.

Lemma 1 *Let $\alpha, \beta \in R_b$*

1. *If in $\alpha + \beta$ the exponent vanishing occurs then $\alpha + \beta \approx 0$.*

2. *In $\alpha + \beta$ the exponent overfilling does not occur*

3. *If in $\alpha \odot \beta$ neither the exponent vanishing, nor the exponent overfilling occurs, then*

$$|\alpha \cdot \beta - \alpha \odot \beta| \leq \frac{|\alpha \cdot \beta|}{2 \cdot 10^{P-1}}.$$

It is demonstrated in [Knu] that operations $\oplus$ and $\odot$, defined here are commutative but they do not satisfy either the law of associativity or the law of distributivity. An interesting problem is if there exists a nonstandard computer arithmetic $\langle \mathcal{R}, j \rangle$, such that $\mathcal{R}$ is a hyperfinite ring. We are not able to answer this question here. The following theorem contains some restrictions on the properties of hyperfinite computer arithmetic.

We say that a hyperfinite computer arithmetic $\langle \mathcal{R}, j\rangle$ is *of a ring type* if $\mathcal{R} = \langle R, \{0,1\}\{\ominus, \oplus, \odot\}\rangle$ is an associative ring with a unit and it is of a *strong ring type* if an addition for any $\alpha \in R_b$ such that $j(\alpha) \not\approx 0$ there exists an inverse element $\alpha^{-1} \in R$, and $j(\alpha^{-1}) \approx ((j(\alpha)^{-1})$

Theorem 6 *There does not exist a hyperfinite computer arithmetic of a strong ring type.*

Proof. Let $\langle \mathcal{R}, j\rangle$, where $\mathcal{R} = \langle R, \{0,1\}\{\ominus, \oplus, \odot\}\rangle$ be a nonstandard computer arithmetic of a strong ring type, $G(R)$ is the internal group of all invertible elements of R. Consider the internal group G of of all invertible 2×2 matrixes of the form

$$\begin{pmatrix} \alpha & \beta \\ 0 & 1 \end{pmatrix},$$

where $\alpha \in G(R)$ and $b \in R$. Let $i : G \to {}^*\mathcal{G}_0$ (cf. p. 97) be such that

$$i\left(\begin{pmatrix} \alpha & \beta \\ 0 & 1 \end{pmatrix}\right) = \begin{pmatrix} j(\alpha) & j(\beta) \\ 0 & 1 \end{pmatrix}.$$

Now it is to see that $\langle G, i\rangle$ is a hyperfinite approximation of a group $\mathcal{G}_0$ in the sense of Proposition 2. This is a contradiction because $\mathcal{G}_0$ is not unimodular and hense not finitely approximable according to Theorem 1 □

The similar considerations show that if there exists a hyperfinite computer arithmetic of a ring tipe (not a strong one!) then $\mathcal{G}_0$ is approximable by finite semigroups.

References

[AGK] Albeverio S., Gordon E., Khrennikov A. (1998) Finite dimensional approximations of operators in the spaces of functions on locally compact abelian groups. *Reports from MASDA, Report 9841, Sept. 1998 MASDA, Växjö University, S - 351 95 Växjö. Sweden*

[AGG] Alekseev M., Glebsky L., Gordon E. (1999) *On approximations of groups, group actions and Hopf algebras.* Russian Academy of Science. Institute of Applied Physics. Preprint No. 491. (in Russian)

[Gor] Gordon E. (1997) *Nonstandard Methods in Commutative Harmonic Analysis* AMS, Providence, Rhode Island.

[HeRo] Hewitt E., Ross K. (1979) *Abstract Harmonic Analysis.* Second Ed. New York: Springer - Verlag.

[Kazh] Kazhdan D. (1982) On ε-representations. *Israel Journal of Math.* V. 43, No. 4, pp. 315 –323.

[Knu] Knut D. (1982) *Art of Comuter Programming.* V. II. Addison Wesley Pub. Co.

[Kost] Kostrikin A.I. (1982) *Introduction to algebra.* New York: Springer - Verlag.

[VeGo] Vershik A.M., Gordon E.I. (1998) Groups, that are locally embeddable in the class of finite groups *St.Petersbourg Math.J.* Vol. 9, No. 1, pp. 49 – 67.

VARIOUS CONTINUITY PROPERTIES IN CONSTRUCTIVE ANALYSIS

Hajime Ishihara
School of Information Science, Japan Advanced Institute of Science and Technology Tatsunokuchi, Ishikawa, Japan
ishihara@jaist.ac.jp

Ray Mines*
Department of Mathematical Sciences, New Mexico State University Las Cruces, New Mexico 88003, USA
ray@nmsu.edu

Keywords: Constructive, continuity, sequential continuity, sequential nondiscontinuity, weak bounded

Mathematical Subject Classification: Primary 03F65; Secondary 46S30

Abstract: This paper deals with continuity properties of functions $f: X \to Y$ between metric spaces X and Y within the framework of Bishop's constructive mathematics. We concentrate on the situation when X is not complete. We investigate the relations between the properties of nondiscontinuity, sequential continuity, mapping Cauchy sequences to totally bounded sequences, and a certain boundedness condition.

Introduction

In constructive mathematics, investigations into conditions that ensure the continuity of a function from one metric space to another go back at least to Brouwer, who proved that every function from the real numbers to a metric space must be continuous [3, 4].

In recursive constructive mathematics, Markov showed, in 1954, that every function $f\colon \mathbf{R} \to \mathbf{R}$ is nondiscontinuous[1]. Tsejtin [20], and

*The second author thanks the Japan Advanced Institute of Science and Technology and the JAIST Foundation for support of his stay in Japan from May 28 to August 2, 1998.

[1]The definition of this and other terms are given in Section 1.

P. Schuster, U. Berger and H. Osswald (eds.), Reuniting the Antipodes – Constructive and Nonstandard Views of the Continuum, 103–110.

Kreisel, Lacombe and Shoenfield [14] extended this to show that every function of a complete separable metric space into a separable metric space is continuous. Orevkov [16] proved that every sequentially continuous function from a separable metric space into a metric space is continuous. Corresponding results in Brouwer's intuitionistic mathematics were obtained by Troelstra [17].

Let $f: X \to Y$ be a function between metric spaces. When X is complete, Ishihara [9] showed that f is sequentially continuous if and only if it is both strongly extensional and nondiscontinuous. Moreover, the following statements are equivalent.

1 Every nondiscontinuous function of a complete metric space into a metric space is sequentially continuous.

2 Every function of a complete metric space into a metric space is strongly extensional.

3 Every pseudopositive real number is positive.

Ishihara [10] removed the assumption of the completeness of X, and showed that every sequentially continuous function from a separable metric space into a metric space is pointwise continuous if and only if every pseudobounded subset of $\mathbf{N}$ is bounded.

Recently, Bridges and Mines [6] proved that a linear function between linear spaces is sequentially continuous if and only if it preserves Cauchy sequences. Since a function that preserves Cauchy sequences is sequentially continuous [5] and since, in the case that X is complete, a sequentially continuous function preserves Cauchy sequences, their result suggests that it is worth studying functions that preserve Cauchy sequences in the case that X is not necessarily complete.

Bridges and Dediu [5] investigated functions that map Cauchy sequences to totally bounded sequences and showed that these functions are strongly extensional.

In this paper, we work within Bishop's constructive mathematics and investigate relations between various continuity properties (such as nondiscontinuous, sequentially continuous, mapping Cauchy sequences to totally bounded sequences, etc.) in the case that the domain is not necessarily complete. Bishop's constructive mathematics is the natural setting because it may be regarded as the constructive core of mathematics: both Brouwer's intuitionistic mathematics and recursive constructive mathematics of the Markov School are models of Bishop's constructive mathematics. For background material on Bishop's constructive mathematics see [1, 2, 7]; background on other constructive schools can be found in [8, 15, 18, 19]; and applications of sequential continuity in constructive analysis are given in [12, 13].

1. BASIC DEFINITIONS

For the readers convenience, we gather together the basic definitions that we will use in this paper.

Let $f\colon X \to Y$ by a function between two metric spaces. The various types of continuity that we will use in this paper are given in the following.
The function f is

1 *nondiscontinuous* if $x_n \to x$ and $\forall n(d(f(x_n), f(x)) \geq \delta)$ imply $\delta \leq 0$;

2 *sequentially continuous* if $x_n \to x$ implies $f(x_n) \to f(x)$;

3 *(pointwise) continuous* if for each x and $\epsilon > 0$ there exists $\delta > 0$ such that $d(x, y) < \delta$ implies $d(f(x), f(y)) < \epsilon$.

4 *strongly extensional* if $d(f(x), f(y)) > 0$ implies $d(x, y) > 0$.

We also need Markov's Principle (MP):

$$\forall x \in \mathbf{R}[\neg\neg(x > 0) \Rightarrow (x > 0)].$$

A real number a is *pseudopositive* if it satisfies $\forall x \in \mathbf{R}(\neg\neg(0 < x) \vee \neg\neg(x < a))$. We cannot show constructively that every pseudopositive real number is positive. This latter implication is weaker than MP, and is called Weak Markov Principle (WMP). See [9, 11] for an account of these two principles.

Ishihara [10] called a countable subset A of $\mathbf{N}$ *pseudobounded* if $\lim_{n\to\infty} a_n/n = 0$ for all sequences $\{a_n\}$ in A, and used BD-N to denote the statement "If A is a pseudobounded countable subset of $\mathbf{N}$, then A is bounded."

2. STRONGLY EXTENSIONAL AND SEQUENTIAL CONTINUITY

A mapping f between metric spaces is *hyperextensional* if for each a, b with $0 < a < b$ and each convergent sequence $\{x_n\}$ with a limit x, either

$d(f(x_n), f(x)) < b$ for all n or $d(f(x_n), f(x)) > a$ for some n. Clearly, a mapping which maps Cauchy sequences to totally bounded sequences is hyperextensional.

In the sequel, we show that hyperextensionality plays a role of strong extensionality in the case that X need not be complete. The following corresponds to Theorem 1 of [9].

Theorem 1 *Let f be a mapping between metric spaces X and Y. Then f is sequentially continuous if and only if f is nondiscontinuous and hyperextensional.*

Proof It is straightforward to see that if f is sequentially continuous, then it is nondiscontinuous and hyperextensional. Let $\{x_n\}$ be a sequence in X converging to a limit x, and let $\epsilon > 0$. Construct an increasing binary sequence $\{\lambda_n\}$ such that

$$\begin{aligned} \lambda_n = 0 \ &\Rightarrow \ d(f(x_k), f(x)) > \epsilon/2 \text{ for some } k \geq n, \\ \lambda_n = 1 \ &\Rightarrow \ d(f(x_k), f(x)) < \epsilon \text{ for all } k \geq n. \end{aligned}$$

We may assume that $\lambda_1 = 0$. Define a sequence $\{y_n\}$ in X as follows: if $\lambda_n = 1 - \lambda_{n-1}$, choose $k \geq n-1$ with $d(f(x_k), f(x)) > \epsilon/2$ and set $y_n := x_k$; otherwise set $y_n := x$. Then $\{y_n\}$ converges to the limit x, and hence either $d(f(y_n), f(x)) < \epsilon/2$ for all n or $d(f(y_n), f(x)) > \epsilon/4$ for some n. In the former case, if $\lambda_n = 1 - \lambda_{n-1}$, then $d(f(y_n), f(x)) > \epsilon/2$, a contradiction, and hence $\lambda_n = 0$ for all n. Therefore we can construct a subsequence $\{x_{k_n}\}$ such that $d(f(x_{k_n}), f(x)) > \epsilon/2$, and so $\epsilon/2 \leq 0$ by nondiscontinuity of f. This contradiction ensures that the latter must be the case. Thus there exists n such that $\epsilon/4 < d(f(y_n), f(x))$, and hence $\lambda_n = 1$. □

Lemma 2 *If f is a hyperextensional mapping between metric spaces, then it is strongly extensional.*

Proof Suppose that a mapping f between metric spaces X and Y is hyperextensional. Let x, y be points of X with $d(f(x), f(y)) > 0$. Construct an increasing binary sequence $\{\lambda_n\}$ such that

$$\begin{aligned} \lambda_n = 0 \ &\Rightarrow \ d(x, y) < 1/(n+1), \\ \lambda_n = 1 \ &\Rightarrow \ d(x, y) > 0. \end{aligned}$$

We may assume that $\lambda_1 = 0$. Define a sequence $\{z_n\}$ in X as follows: if $\lambda_n = 1 - \lambda_{n-1}$, set $z_n := y$; otherwise set $z_n := x$. Then $\{z_n\}$

converges to x, and hence either $d(f(z_n), f(x)) < d(f(x), f(y))$ for all n or $d(f(z_n), f(x)) > d(f(x), f(y))/2$ for some n. In the former case, if $\lambda_n = 1 - \lambda_{n-1}$ for some n, then $d(f(z_n), f(x)) = d(f(y), f(x)) < d(f(x), f(y))$, a contradiction, and hence $\lambda_n = 0$ for all n. Therefore $x = y$, and so $f(x) = f(y)$. This contradiction ensures that there exists n such that $d(f(x), f(y))/2 < d(f(x), f(z_n))$, and hence $\lambda_n = 1$. □

Proposition 3 *A mapping of a complete metric spaces into a metric space is hyperextensional if and only if it is strongly extensional.*

Proof By Lemma 2 and Lemma 1 of [9]. □

3. NONDISCONTINUITY AND SEQUENTIAL CONTINUITY

In this section we investigate the relation between nondiscontinuous and sequentially continuous functions whose domains are not necessarily complete. We introduce the concept of weakly-finite sequence and show that every nondiscontinuous function is sequentially continuous if and only if every weakly-finite sequence is finite.

A binary sequence $\{\lambda_n\}$ is **weakly finite** if it satisfies $\neg\forall m \exists n (\lambda_{m+n} = 1)$. We probably should say that $\{\lambda_n\}$ has weakly finite support, but there should be no confusion. The binary sequence $\{\lambda_n\}$ is **finite** if it has finite support, that is, if it satisfies $\exists m \forall n (\lambda_{m+n} = 0)$.

Following [18], we say that a subset A of the real numbers is **weakly bounded** if $\neg$(A is unbounded).

Proposition 4 *Every weakly-finite binary sequence is finite if and only if every countable weakly-bounded subset of* $\mathbb{N}$ *is bounded.*

Proof Assume that every weakly-finite binary sequence is finite, and assume that $S = \{a_n : n \in \mathbb{N}\}$ is weakly bounded. Define a binary sequence $\{\lambda_n\}$ by $\lambda_n = 1$ implies there exists $k \leq n$ such that $a_k > n$ and $\lambda_n = 0$ implies $a_k \leq n$ for all $k \leq n$. Since S is weakly bounded, it follows that $\{\lambda_n\}$ is weakly finite. Hence $\{\lambda_n\}$ is finite. This implies that S is bounded.

Assume that every countable weakly-bounded subset of $\mathbb{N}$ is bounded, and let $\{\lambda_n\}$ be a weakly-finite binary sequence. Set $a_n = \sum_{i=1}^{n} \lambda_i$. Then the set $\{a_n : n \in \mathbb{N}\}$ is weakly bounded, hence bounded. Consequently, the sequence $\{\lambda_n\}$ is finite. □

Theorem 5 *The following statements are equivalent.*

1 *Every mapping between metric spaces is hyperextensional.*

2 *Every strongly-extensional mapping between metric spaces is hyperextensional.*

3 *Every nondiscontinuous mapping between metric spaces is sequentially continuous.*

4 *Every strongly-extensional nondiscontinuous mapping between metric spaces is sequentially continuous.*

5 *Every linear mapping between normed spaces is hyperextensional.*

6 *Every strongly-extensional linear mapping between normed spaces is hyperextensional.*

7 *Every nondiscontinuous linear mapping between normed spaces is sequentially continuous.*

8 *Every strongly-extensional nondiscontinuous linear mapping between normed spaces is sequentially continuous.*

9 *Every weakly-finite binary sequence is finite.*

10 *LPO.*

Proof The proofs of 1) $\Rightarrow$ 2), 3) $\Rightarrow$ 4), 5) $\Rightarrow$ 6), 7) $\Rightarrow$ 8), 1) $\Rightarrow$ 5), and 4) $\Rightarrow$ 8) are clear.

By Theorem 3.1, we have 1) $\Rightarrow$ 3), 2) $\Rightarrow$ 4), 5) $\Rightarrow$ 7), and 6) $\Rightarrow$ 8).

The proof of 8) $\Rightarrow$ 9): Let E denote the finite span of the canonical basis $\{e_n : n > 0\}$ in l^1 and let $\|\cdot\|$ denote the l^1-norm. Let $\{\lambda_k\}$ be a weakly-finite binary sequence. Define the linear mapping $T : E \to E$ by $T(e_k) := k\lambda_k e_k$. It is clear that T is strongly extensional. We will show that T is nondiscontinuous. Let $\{a_k\}$ be a sequence of elements in E that converges to 0. For each k write $a_k = \sum_{i=1}^{\infty} a_{ki}e_i$. (Despite the appearance, this is a finite sum.) Suppose $T(a_k) \geq \delta > 0$ for all k. Let $m > 0$. Since the sequence $\{a_k\}$ converges to 0, there exists an integer N so that $k > N$ implies $\|a_k\| < \delta/2m$. Let $\Pi_m : E \to \bigoplus_{i=1}^{m} \mathbb{R}e_i$ be the projection onto the first m coordinates. Then for all $k > N$ we have $\|\Pi_m(a_k)\| \leq \|a_k\| < \delta/2m$. It follows that $\|T(\Pi_m(a_k))\| < \delta/2$. Therefore for all $k > N$ we have $\|\sum_{i=m+1}^{\infty} i\lambda_i a_{ki}e_i\| = \|T(a_k) - T(\Pi_m(a_k))\| > \delta/2$. This implies there is an $n > 0$ so that $\lambda_{m+n} = 1$. This shows that $\forall m \exists n(\lambda_{m+n} = 1)$. This contradicts the assumption that the sequence $\{\lambda_n\}$ is weakly finite; therefore, T is nondiscontinuous.

Since T is a strongly-extensional nondiscontinuous linear mapping, it is, by assumption, sequentially continuous. The sequence $\{(1/k)e_k\}$ clearly converges to 0 in E. Since $\|T((1/k)e_k)\| = \lambda_k$, it follows, by sequential continuity, that there is an N so that $k > N$ implies $\lambda_k = 0$. Thus the sequence $\{\lambda_n\}$ is finite.

The proof of 9) $\Rightarrow$ 10): Let $\{\lambda_n\}$ be a binary sequence. We may assume, without loss of generality, that $\{\lambda_n\}$ contains at most one 1. Then $\neg\forall m \exists n(\lambda_{m+n} = 1)$, and hence there exists m such that $\forall n(\lambda_{m+n} = 0)$. If $\lambda_k = 0$ for all $k \leq m$, then $\lambda_n = 0$ for all n; or else $\lambda_k = 1$ for some $k \leq m$, and hence $\lambda_n = 1$ for some n.

The proof of 10) $\Rightarrow$ 1) is clear. □

References

[1] E. Bishop, *Foundations of Constructive Analysis*, McGraw-Hill, New York, 1967.

[2] E. Bishop and D. Bridges, *Constructive Analysis*, Grundlehren der math. Wiss. 279, Springer-Verlag, Berlin, 1985.

[3] L.E.J. Brouwer, *Begründung der Funktionenlehre unabhängig vom logischen Satz vom ausgeschlossenen Dritten. Erster Teil, Stetigkeit, Messbarkeit, Derivierbarkeit,* Nederl. Akad. Wetensch. Verhandelingen 1[e] sectie 13 no. 2, 1–24, (1923).

[4] L.E.J. Brouwer, *Beweis dass jede volle Funktion gleichmässigstetig ist,* Nederl. Akad. Wetensch. Proc. 27, 189-193 (1924).

[5] D. Bridges and L. Dediu, *Weak continuity properties in constructive analysis*, preprint, 1998.

[6] D. Bridges and R. Mines, *Sequentially continuous linear mappings in constructive analysis*, J. Symbolic Logic 63 (1998), 579-583.

[7] D. Bridges and F. Richman, *Varieties of constructive mathematics*, London Math. Soc. Lecture Notes 97, Cambridge Univ. Press, 1987.

[8] A. Heyting, *Intuitionism - an Introduction* (3rd edn.), North-Holland, Amsterdam, 1971.

[9] H. Ishihara, *Continuity and nondiscontinuity in constructive mathematics*, J. Symbolic Logic 56 (1991), 1349-1354.

[10] H. Ishihara, *Continuity properties in constructive mathematics*, J. Symbolic Logic 57 (1992), 557-565.

[11] H. Ishihara, *Markov's principle, Church's thesis and Lindelöf's theorem*, Indag. Mathem. 4 (1993), 321-325.

[12] H. Ishihara, *A constructive version of Banach's inverse mapping theorem*, New Zealand J. Math. 23 (1994), 71-75.

[13] H. Ishihara, *Sequential continuity of linear mappings in constructive mathematics*, J. Universal Computer Science 3 (1997), 1250-1254.

[14] G. Kreisel, D. Lacombe and J. Shoenfield, *Fonctionelles récursivement définissables et fonctionelles récursives*, C. R. Acad. Sci. Paris, Ser. A-B 245 (1957), 399-402.

[15] B. A. Kushner, *Lectures on Constructive Mathematical Analysis*, American Mathematical Society, 1985.

[16] V. P. Orevkov, *Equivalence of two definitions of continuity* (Russian), Zap. Nauchn. Semin. Leningr. Otd. Mat. Inst. Steklov. 20, 145-159, 286 (1971); translation: J. Sov. Math. 1 92-99 (1973).

[17] A. S. Troelstra, *Intuitionistic continuity*, Nieuw Archief voor Wiskunde (3) 15, 2-6 (1967).

[18] A. S. Troelstra and D. van Dalen, *Constructivism in Mathematics*, Vol. 1, North-Holland, Amsterdam, 1988.

[19] A. S. Troelstra and D. van Dalen, *Constructivism inMathematics*, Vol. 2, North-Holland, Amsterdam, 1988.

[20] G. S. Tsejtin, *Algorithmic operators in constructive complete-separable metric spaces* (Russian), Dokl. Akad. Nauk SSSR 128 (1959),49-52.

LOEB MEASURES AND BOREL ALGEBRAS

H. Jerome Keisler

Department of Mathematics, University of Wisconsin-Madison
480 Lincoln Drive, Madison, Wisconsin 53706-1388, USA
keisler@math.wisc.edu

Yeneng Sun

Department of Mathematics, National University of Singapore
2 Science Drive 2, Singapore 117543, Republic of Singapore
matsuny@nus.edu.sg

Keywords: Loeb measures, Lebesgue meassure, Borel algebras, bijections

Abstract: It is shown that a measurable function from an atomless Loeb probability space $(\Omega, \mathcal{A}, P)$ to a Polish space is at least continuum-to-one valued almost everywhere. It follows that there is no injective mapping $h : [0,1] \to \Omega$ such that $h([a,b])$ is Loeb measurable for each $0 \leq a < b \leq 1$ and $P(h([0,1])) > 0$. Thus, when an atomless Loeb measurable algebra on an internal set of cardinality continuum is imposed on the unit interval $[0,1]$ through a bijection, it cannot contain the Borel algebra.

1. INTRODUCTION

The Loeb measure construction introduced in [L] is a method from nonstandard analysis which is a valuable tool for proving standard results in mathematics (see, for example, [AFHL], [A] and [K1]). Recently, a number of special measure-theoretic properties of Loeb spaces were discovered and formulated in conventional terms. One is the neocompact approach developed in [FK1], [FK2], [K2] and [K3] which extracted the essential features of a nonstandard existence proof using Loeb spaces and provided a general procedure for going from approximate to limit. The others include a new theory of correspondences in [S2] and [S3] and

P. Schuster, U. Berger and H. Osswald (eds.),
Reuniting the Antipodes – Constructive and Nonstandard Views of the Continuum, 111–117.

a systematic study of the fundamental probabilistic concept of independence in [S1], [S4] and [S5] using Loeb product spaces.

A natural question which arises when the Loeb construction is used to obtain standard results is: what is a Loeb space? In applications one usually only needs to work with hyperfinite Loeb probability spaces in a nonstandard universe which is a countably indexed ultrapower. In that case, a hyperfinite set has the external cardinality of the continuum (the cardinality of the continuum will be denoted by $\mathbf{c}$), so there is a bijection between the hyperfinite set and the unit interval $[0, 1]$. Using this bijection, one can impose the Loeb measure structure on $[0, 1]$. Thus, a Loeb measure can simply be viewed as a measure on the unit interval. In this setting, the special measure-theoretic properties of Loeb spaces can be restated as properties of *some measure* on the most familiar underlying space $[0, 1]$.

Once a Loeb measure structure is imposed on the unit interval $[0, 1]$ through a bijection, one may want to know its relationship with the Lebesgue measure structure. In particular, whether the imposed Loeb measurable algebra on $[0, 1]$ could contain the Borel algebra, or more strongly, whether the imposed Loeb measure structure on $[0, 1]$ could be an extention of the Lebesgue measure structure. An affirmative answer would alleviate the anxiety caused by unfamiliar notions in Loeb measure theory, such as internal sets, hyperfinite sets, or rich probability spaces. However, the results in this paper show that the answer to the above question is negative.

Note that if an imposed Loeb measurable algebra on $[0, 1]$ contains the Borel algebra, then the relevant bijection from the original Loeb space must be Loeb measurable. Motivated by this observation, we consider the behavior of a general Loeb measurable mapping f from an atomless Loeb probability space $(\Omega, \mathcal{A}, P)$ to a Polish space, and show that almost all points with respect to the image measure have at least continuum many preimages. Of course, this includes the particular case $X = [0, 1]$. As a consequence, for any injective mapping $h : [0, 1] \to \Omega$ in the other direction, if $h([a, b])$ is Loeb measurable for each $0 \leq a < b \leq 1$ then $P(h([0, 1])) = 0$. In the case that the mapping is a bijection, this shows that the answer to the previous question is negative.

For the convenience of the reader, the results are rigorously stated in Section 2 and the proofs are given in a separate Section 3. As usual, we work in an $\aleph_1$-saturated nonstandard universe (see [HL] for this basic definition).

2. THE RESULTS

The following proposition shows that a measurable mapping f from an atomless Loeb probability space to a Polish space X is almost everywhere at least continuum-to-one valued.

Proposition 1 *Let f be a Loeb measurable mapping from an atomless Loeb probability space $(\Omega, \mathcal{A}, P)$ to a Polish space X with distribution ν on X. Then, the inverse image set $f^{-1}(x)$ of x has external cardinality at least the continuum for ν-almost all $x \in X$.*

The result cannot be improved by replacing "ν-almost all $x \in X$" by "all $x \in X$". In fact, starting from the above mapping f and any ν-null set B in X, one can construct a Loeb measurable mapping g with distribution ν such that for any $x \in B$, $g^{-1}(x)$ is at most a singleton. This mapping g can be obtained by modifying the values of f on the P-null set $f^{-1}(B)$.

Proposition 1 covers the case of a mapping from a Loeb space to $[0,1]$. In the other direction, we now consider injective mappings from $[0,1]$ to an atomless Loeb space. The next proposition shows that if the image set of each subinterval is Loeb measurable under such a mapping h, then the image of h must be a P-null set.

Corollary 2 *Let $(\Omega, \mathcal{A}, P)$ be a Loeb probability space, and let $h : [0,1] \to \Omega$ be an injective mapping such that $h([a,b])$ is Loeb measurable whenever $0 \le a \le b \le 1$. Then $P(h([0,1])) = 0$.*

If our nonstandard universe is a countably indexed ultrapower, then any hyperfinite set has the same external cardinality as $[0,1]$. Thus, there are $2^{\mathfrak{c}}$ many possible bijections between a hyperfinite set and $[0,1]$ in this case. Given any probability space $(\Omega, \mathcal{A}, P)$, any bijection f from Ω to the unit interval $[0,1]$ imposes a probability space $([0,1], \mathcal{I}, \mu)$ on $[0,1]$ where

$$\mathcal{I} = f(\mathcal{A}) = \{f(A) : A \in \mathcal{A}\}, \qquad \mu(f(A)) = P(A).$$

Let us say that such a probability space $([0,1], \mathcal{I}, \mu)$ is **isomorphic** to $(\Omega, \mathcal{A}, P)$. Given the abundance of such bijections, the result in the following corollary, which says that the induced isomorphic measure structures on [0,1] can *never* contain the Borel structure, is somewhat surprising.

Corollary 3 *Let* $([0,1], \mathcal{I}, \mu)$ *be a probability space which is isomorphic to an atomless Loeb probability space* $(\Omega, \mathcal{A}, P)$. *Then* $\mathcal{I}$ *cannot contain the Borel* σ*-algebra* $\mathcal{B}$.

It is interesting to point out that given any atomless probability space $(\Omega, \mathcal{A}, P)$, if there is an isomorphic probability space on $[0,1]$ which contains the Borel algebra, then there is actually an isomorphic probability space on $[0,1]$ which is an extension of the Lebesgue measure structure. To see this, suppose that there is a probability space $([0,1], \mathcal{I}, \mu)$ isomorphic to $(\Omega, \mathcal{A}, P)$ through a bijection $f : \Omega \to [0,1]$ such that the σ-algebra $\mathcal{I} = f(\mathcal{A})$ contains the Borel algebra $\mathcal{B}$. Then, $([0,1], \mathcal{B}, \mu|_{\mathcal{B}})$ is an atomless Borel probability space. The classical isomorphism theorem on Borel sets and Borel measures (see [P]) implies the existence of a Borel isomorphism h between the measure $\mu|_{\mathcal{B}}$ and the Lebesgue measure λ on $[0,1]$. Let g be the composition of f and h. Then, the measure structure on $[0,1]$ imposed by g extends the Lebesgue measure structure.

3. THE PROOFS

We begin with a lemma.

Lemma 4 *Let* G *be an internally measurable mapping from an atomless, internal probability space* $(\Omega, \bar{A}, \bar{P})$ *to* $^*X$, *where* X *is a Polish space. Assume that* $G(\Omega)$ *is near standard in* $^*X$ *for every* $\omega \in \Omega$. *Let* $g(\omega) = {}^\circ G(\omega)$ *for all* $\omega \in \Omega$. *Let* μ *be the distribution on* X *induced by the mapping* g *on the Loeb probability space* $(\Omega, \mathcal{A}, P)$. *Then, for any* x *in the support of* μ, *the external cardinality of* $g^{-1}(x)$, *the inverse image of the singleton* $\{x\}$, *is at least the continuum.*

Proof Let d be a metric on X and O_n the open ball $B(x, 1/n)$ for every $n \geq 1$. Since x is in the support of μ, $\mu(O_n) = P(g^{-1}(O_n)) > 0$. Thus, we can find an internally measurable subset A_n of Ω such that $A_n \subseteq g^{-1}(O_n)$ with $\bar{P}(A_n) > \mu(O_n)/2$. Hence, for any natural number $n \geq 1$,

(1) the internal cardinality $|A_n|$ of A_n is greater than n,

(2) $d(x, G(A_n)) < 1/n$.

By $\aleph_1$-saturation, the sequence $A_n, n \in \mathbf{N}$ can be extended to internally measurable subsets of Ω, $A_n, n \in {}^*\mathbf{N}$. By overspill, one can find an infinite $\gamma \in {}^*\mathbf{N}$ such that for all $1 \leq n \leq \gamma$, Properties (1) and (2) hold. This means that $|A_\gamma| > \gamma$ and $d(x, G(A_\gamma)) < 1/\gamma$, which implies that

$A_\gamma \subseteq g^{-1}(x)$. Hence the external cardinality of $g^{-1}(x)$ is at least the continuum. □

Proof of Proposition 1

Let F be an internal lifting of f. Take an increasing sequence of internally measurable sets B_n in Ω such that $\bar{P}(B_n) > 1 - 1/2^n$ and ${}^\circ F(\omega) = f(\omega)$ for all $\omega \in B_n$.

Let $\bar{P}_n$ be the atomless, internal probability measure on B_n defined by normalizing $\bar{P}$ restricted to B_n and let P_n be the relevant Loeb measure. Let μ_n be the distribution on X induced by $f|_{B_n}$. Then, Lemma 1 implies that for all x in the support S_n of μ_n, the external cardinality of $(f|_{B_n})^{-1}(x)$ is at least the continuum.

Let $S = \cup_{n=1}^{\infty} S_n$. Then

$$\nu(S) \geq \nu(S_n) = P(f^{-1}(S_n)) \geq P_n((f_{B_n})^{-1}(S_n)) \cdot P(B_n)$$
$$= \mu_n(S_n) \cdot P(B_n) = P(B_n),$$

which implies that $\nu(S) = 1$. Hence, for all x in S, the external cardinality of $f^{-1}(x)$ is at least the continuum. □

Proof of Corollary 2

If there were such an h, and f was the extension of h^{-1} which sends the complement of $h([0,1])$ to 0, then f would be a Loeb measurable function with distribution ν such that the inverse image set $f^{-1}(x)$ is a singleton for any x in the set $(0,1]$. But then

$$\nu((0,1]) = P\left(f^{-1}((0,1])\right) = P\left(h((0,1])\right) = P\left(h([0,1])\right) > 0,$$

which contradicts Proposition 1. □

Proof of Corollary 3

Suppose that $\mathcal{I}$ does contain the Borel σ-algebra $\mathcal{B}$. Then the bijection f is Loeb measurable. The distribution ν of f is simply the restriction $\mu_{\mathcal{B}}$ of μ to $\mathcal{B}$. Proposition 1 implies that $f^{-1}(x)$ cannot be a singleton for ν-almost all $x \in [0,1]$, which contradicts the bijective hypothesis on f. □

Acknowledgments

The results were obtained while the authors were attending the Symposium on Reuniting the Antipodes – Constructive and Nonstandard Views of the Continuum at Venice, Italy on May 16-23, 1999. The authors are very grateful to Robert Anderson for helpful conversations and also to the organizers of the Venice conference, Ulrich Berger, Horst Osswald and Peter Schuster for the invitation to the conference. The research was supported in part by the National Science Foundation and the Vilas Trust Fund.

References

[AFHL] S. Albeverio, J. E. Fenstad, R. Hoegh-Krohn, and T. L. Lindstrom, Nonstandard Methods in Stochastic Analysis and Mathematical Physics, Academic Press, Orlando, Florida, 1986.

[A] R. M. Anderson, A nonstandard representation for Brownian motion and Ito integration, Israel J. Math. 25 (1976) pp. 15-46.

[FK1] S. Fajardo and H. J. Keisler, Neometric spaces, Advances in Math. 118 (1996), pp. 134-175.

[FK2] S. Fajardo and H. J. Keisler, Existence theorems in probability theory, Advances in Math. 120 (1996), pp. 191-257.

[HL] A. E. Hurd and P. A. Loeb, An Introduction to Nonstandard Real Analysis, Academic Press, Orlando, Florida, 1985.

[K1] H. J. Keisler, An infinitesimal approach to stochastic analysis, Memoirs Amer. Math. Soc. 48 (1984), No. 297.

[K2] H. J. Keisler, Rich and Saturated Adapted Spaces, Advances in Math. 128 (1997), pp. 242-288.

[K3] H. J. Keisler, Quantifier elimination for neocompact sets, J. Symbolic Logic 63 (1998), pp. 1442-1472.

[L] P. A. Loeb, Conversion from nonstandard to standard measure spaces and applications in probability theory, Trans. Amer. Math. Soc. 211 (1975), pp. 113-122.

[P] K. R. Parthasarathy, Probability Measures on Metric Spaces, Academic Press, New York, 1967.

[S1] Y. N. Sun, Hyperfinite law of large numbers, Bull. Symbolic Logic 2 (1996), pp. 189-198.

[S2] Y. N. Sun, Distributional properties of correspondences on Loeb spaces, J. Functional Analysis 139 (1996), pp. 68-93.

[S3] Y. N. Sun, Integration of correspondences on Loeb spaces, Trans. Amer. Math. Soc. 349 (1997), pp. 129-153.

[S4] Y. N. Sun, A theory of hyperfinite processes: the complete removal of individual uncertainty via exact LLN, J. Math. Econ. 29 (1998), pp. 419-503.

[S5] Y. N. Sun, The almost equivalence of pairwise and mutual independence and the duality with exchangeability, Probability Theory and Related Fields 112 (1998), pp. 425-456.

ON BROUWERIAN BAR INDUCTION

Boris A. Kushner

Department of Mathematics, University of Pittsburgh at Johnstown
Johnstown, PA 15904, USA
boris+@pitt.edu

Keywords: intuitionism, continuum, intuitionistic continuum, tree, detachable tree, founded tree, induction, bar-induction

Abstract: As is well known, Brouwer's concept of continuum is based on the notion of a choice sequence. In turn, the mathematical treatment of choice sequences is based on two fundamental ideas of Brouwer: the Principle of Continuity and the Principle of Bar Induction. The intuitive meaning of the first principle is quite clear. At the same time intuition backing the Bar Induction is much more sophisticated. This Principle is rather complicated and seems to carry on a touch of the historical development of the intuitionistic concept of a spread. We establish here a more clear version of bar induction in form of an induction over detachable founded trees.

1. As is well known, Brouwer's concept of the continuum is based on the notion of a choice sequence (a notion that can be traced back to E. Borel (see [1]). In turn, the mathematical treatment of choice sequences is based on two fundamental ideas of Brouwer: the Principle of Continuity and the Principle of Bar Induction (PBI). The intuitive meaning of the first principle is quite clear. It reflects the developing, non-completed nature of choice sequences. As only an initial segment of a choice sequence is available to an observer, only such "finite" information can be used in an intuitionistic assignment of natural numbers to choice sequences. Hence, every such assignment is necessarily continuous (cf. [1–4]). The intuition backing PBI is much more sophisticated. Methodological aspects of PBI are scarcely discussed in the literature on the subject. We will present a few considerations that show that PBI reflects in a peculiar form the "manifoldness" of choice sequences. This "manifoldness" is readily available to our intuition, but it is difficult to describe in precise terms (cf. [1–4]). Philosophical aspects of PBI and

P. Schuster, U. Berger and H. Osswald (eds.),
Reuniting the Antipodes – Constructive and Nonstandard Views of the Continuum, 119–126.

its place in Brouwer's theory of the continuum are discussed in detail in my work [5].

2. We consider here Bar Induction in its detachable version. In literature one can find also the axiom of monotone Bar Induction. In presence of some strong form of the Principle of Continuity both forms of Bar Induction are equivalent (see, e.g., [4]).

Modulo technical detail, Bar-Induction is an inductive process that enables one to "spread" a sentence A from terminal nodes of a detachable, well-founded tree M to all its nodes by means of inductive steps of the following type: if A is true on all successors of a given node t, then A is true on t. Here a tree M is said to be detachable if the following statement is intuitionistically true

$$\forall s(s \in M \vee s \notin M).$$

M is well-founded if every intuitionistic path (i.e., a choice sequence) in M arrives at a terminal node (cf. [1–8]). The problem of the admissibility of such induction closely depends on the meaning of the well-foundness condition, i.e., on the extent of the concept of a path in a given tree or, in other words, on the extent of the concept of a choice sequence. If one should replace choice sequences by classical paths (functions), one would get a classically valid inductive principle quite similar to the principle of induction over recursive well-founded trees or constructive ordinals in classical recursion theory (see [7–9]). Therefore, a study of possible restrictions on classical functions in the classical counterpart of PBI could shed new light on the rather mysterious concept of choice sequence. Fortunately, we have here remarkable results due to Kleene, namely, even the class of all hyper-arithmetic functions is not enough to express the well-foundness condition in the above principle (see, e.g., [9]). Therefore, the class of choice sequences intended in PBI is much wider that the class of general recursive (effective, computable) functions. This conclusion will hold even in the case of finitely branching trees, as follows from results of Kleene-Zaslavsky (see, e.g., [6, 9–10]).

As is well known, PBI and Principle of Continuity imply the central fact of Intuitionistic Analysis, the local uniform continuity of everywhere defined intuitionistic functions over the reals. It is worth noting that the classical counterpart of this result, namely the Heine-Borel Theorem, is based on highly non-constructive compactness properties of the classical continuum. In turn those properties are guaranteed by the width of the classical concept of a real number (or, what is the same, by a very liberal traditional concept of sequence). Well-known counterexamples of Recursive (Constructive) Analysis show that the class of general recursive (computable) reals (and so the class of general recursive sequences) is not wide enough to assure any uniform continuity properties (see [10]). Let us note that there are metamathematical arguments to support our point of view, too (see [11, 4.1.4]).

Therefore PBI is a remarkably strong inductive principle that expresses the width of the concept of a choice sequence. The consistency of this principle with standpoints of Intuitionistic Mathematics which focus on effectiveness is a significant fact from the methodological point of view.

3. A Theorem. (cf. [7–8]).

We will prove a theorem that establishes an equivalence between Bar Induction and an Induction over well-founded detachable trees. Alternative formulations of the axiom of Bar Induction in the context of formal constructive analysis can be found in the paper of Howard and Kreisel [12].

We consider finite sequences of natural numbers, including the empty sequence Λ. By a *beginning* of a finite sequence $n_1 * \ldots * n_k$ we mean either the empty sequence or any sequence $n_1 * \ldots * n_i$ where $1 \leq i \leq k$. We will write $s_1 \leq s_2$ if s_1 is a beginning of s_2. $s_1 < s_2$ means that $s_1 \leq s_2$ but $s_1 \neq s_2$.

We define the operation of concatenation of finite sequences as follows

$$K(s_1, s_2) \rightleftharpoons \begin{cases} s_1, & \text{if } s_2 \rightleftharpoons \Lambda \\ s_2, & \text{if } s_1 \rightleftharpoons \Lambda \\ s_1 * s_2, & \text{if both } s_1 \text{ and } s_2 \text{ are non-empty.} \end{cases}$$

Below we will use the simpler notation $s_1 * s_2$ instead of $K(s_1 * s_2)$ (one should adjust this notation in case one or both of s_1, s_2 are empty).

An inhabited species of finite sequences M is said to be a *tree* if

$$\forall s(s \in M \ \& \ t \leq s \supset t \in M).$$

Having in mind the evident geometrical interpretation we call finite sequences belonging to a tree M *nodes* of M.

A node $s \in M$ is said to be *terminal* if

$$\forall i(s * i \notin M).$$

Let M be a tree, A be a property of finite sequences. A is said to be *M-hereditary* if for every $s \in M$ the following holds

$$\forall i(s * i \in M \supset A(s * i)) \supset A(s). \qquad ((1))$$

In other words a property that holds on all one-member continuations in M of a given $s \in M$ is inherited by s itself.

Let α be a choice sequence. We denote by $\bar{\alpha}$ the sequence such that

$$\bar{\alpha}(0) = \Lambda,$$

$$\bar{\alpha}(n+1) = \alpha(0) * \ldots * \alpha(n).$$

We denote by U the *universal* tree that consists of all finite sequences. Let R be a unary predicate (a property) on U. s is said to be *R-secured* if

$$\forall \alpha \exists n(R(s * \bar{\alpha}(n)). \qquad ((2))$$

A predicate R is said to be *detachable* if the following statement holds (intuitionistically):

$$\forall s(R(s) \vee \neg R(s)).$$

We will now formulate PBI following Kleene-Vesley [2] (it would be more appropriate to talk about the Principle of Detachable Bar-Induction for U).

Let R and A be predicates on U, $s \in U$ such that

$$R \text{ is a detachable predicate;} \qquad ((3))$$

$$\forall t(R(t) \supset A(t)); \qquad ((4))$$

$$A \text{ is a } U\text{-hereditary property;} \qquad ((5))$$

$$s \text{ is } R\text{-secured.} \qquad ((6))$$

PBI allows one to conclude under conditions ((3))–((6)) that $A(s)$ holds.

A tree M is said to be *well-founded*, if

$$\forall \alpha \exists n(\bar{\alpha}(n) \notin M). \qquad ((7))$$

where α is a variable over choice sequences.

A tree M is said to be *detachable* if

$$\forall s(s \in M \vee s \notin M).$$

We consider the following Principle of Induction (WFI) on detachable, well-founded trees: if M is a detachable, well-founded tree and A is an M-hereditary property, then A holds on every node of M.

It is worth noting that an M-hereditary property automatically holds at every terminal node of M. Indeed, if s is a terminal node, then all the statements $s * i \in M$ are false. Hence the antecedent of ((1)) is true and thus $A(s)$ holds.

Therefore WFI is an inductive process that allows a hereditary property to be extended from terminal nodes of a detachable, well-founded tree to all its nodes.

Theorem. Principles BI and WFI are intuitionistically equivalent.

Proof.

1. From WFI to BI. Let conditions ((3))–((6)) be fulfilled for predicates R, A and a node s. We introduce a property $\bar{A}$ and a species of nodes T such that

$$\bar{A}(t) \equiv A(s * t),$$

$$t \in T \equiv \forall t_1(t_1 < t \supset \neg R(s * t_1)).$$

It is evident that T is a detachable tree. Then T is well-founded by ((6)) and ((2)). Let us prove that $\bar{A}$ is a T-hereditary property. Let $t \in T$. Since R is detachable we can consider two separate cases: $R(s*t)$ and $\neg R(s * t)$. In the first case (i.e., t is a terminal node of T) we have (by ((4))) $A(s * t)$, hence $\bar{A}(t)$. Thus, the implication ((1)) (with an appropriate adjustment of notation) holds. In the second case it is evident that all the sequences $t * i$ are in T because t is in T. Hence the condition of T-heritability ((1)) for such a t takes the following form

$$\forall i \bar{A}(t * i) \supset \bar{A}(t).$$

By definition of $\bar{A}$ this is equivalent to

$$\forall i A(s * t * i) \supset A(s * t).$$

The last statement holds by definition of U-heritability.

Thus, T-heritability of $\bar{A}$ is proved. Now applying WFI to $\bar{A}$ and T we get $\bar{A}(\Lambda)$, i.e., $A(s)$, as was required.

The above construction shows that bar-induction is an induction by a special detachable, well-founded tree T that consists of all sequences that do not intersect the "barrier" R. This tree T is peculiar because, not only is it detachable itself, but the species of its terminal nodes is detachable, too. The essence of our theorem is the elimination of this peculiarity. (Here is a simple counterexample. Consider the tree P such that the empty sequence Λ is in P, all sequences with the length greater or equal than 2 do not belong to P and the one-term sequence n is in P iff $2n+1$ is a perfect number. Then P is a detachable tree but we can not decide whether Λ is a terminal node of P or not).

2. From BI to WFI. Let M and A be, respectively, a detachable, well-founded tree and an M-hereditary property. Let $s \in M$. We extend the property A to the complement of M by declaring it to be always true there. Let us consider the property $\bar{A}$ such that

$$\bar{A}(t) \equiv (t \in M \supset A(t))$$

It is evident that $\bar{A}(t)$ is equivalent to $A(t)$ for every $t \in M$ and $\bar{A}(t)$ is true for $t \notin M$. Consider now a predicate R such that

$$R(t) \equiv t \notin M. \tag{(8)}$$

We prove that clauses ((3))–((6)) of BI are fulfilled for $\bar{A}, s$ and R. Indeed, it is clear that R is a detachable predicate. Then $R(t) \supset \bar{A}(t)$, since $\bar{A}(t)$ is true for $t \notin M$. It is easy to see that $\bar{A}$ is a U-hereditary property. Indeed, we have to prove that for every t

$$\forall i \bar{A}(t,i) \supset \bar{A}(t). \tag{(9)}$$

If $t \notin M$ then $\bar{A}(t)$ is true and so ((9)) holds. If $t \in M$ then $A(t) \equiv \bar{A}(t)$. Moreover, since A is M-hereditary we have

$$\forall i(t * i \in M \supset A(t*i)) \supset A(t).$$

Hence,

$$\forall i(t * i \in M \supset \bar{A}(t*i)) \supset \bar{A}(t).$$

Therefore, we have ((9)) since $\forall i \bar{A}(t*i)$ implies

$$\forall i(t * i \in M \supset \bar{A}(t*i)).$$

Using well-foundness of M and ((8)) it is not difficult to prove that every $s \in M$ is R-secured. We have

$$\forall \alpha \exists n(\bar{\alpha}(n) \notin M),$$

i.e.,

$$\forall \alpha \exists n R(\bar{\alpha}(n)). \qquad ((10))$$

Since $s \in M$ and ((8)) holds, $\neg R(s_1)$ holds for every beginning s_1 of s. Therefore by ((10)) we can conclude that

$$\forall \alpha \exists n R(s * \bar{\alpha}(n)),$$

i.e., s is R-secured.

Applying BI to R, s and $\bar{A}$ we get $\bar{A}(s)$. But for $s \in M$ $\bar{A}(s)$ is equivalent to $A(s)$. The theorem is proved.

The author is deeply grateful to Dr. A. Wilce for his friendly support.

References

[1] A.S. Troelstra, *On the early history of intuitionistic logic*, Mathematical Logic, P.P. Petkov, Ed., Plenum Press, New York–London, 1990, pp. 3–17.

[2] A. Heyting, *Intuitionism. An Introduction*, North-Holland, Amsterdam, 1956.

[3] A.S. Troelstra, *Choice Sequences. A Chapter of Intuitionistic Mathematics*, Clarendon Press, Oxford, 1977.

[4] A.S. Troelstra and D. van Dalen, *Constructivism in Mathematics. An Introduction. Vol. 1–2*, North-Holland, Amsterdam–New York–Oxford–Tokyo, 1988.

[5] B.A. Kushner, *Printsip bar-induktsii i teoriya kontinuuma u Brauera. Russian*, Zakonomernosti razvitiya sovremennoy matematiki, Nauka, Moscow, 1987, pp. 230–250.

[6] S.C. Kleene and R.E. Vesley, *The Foundations of Intuitionistic Mathematics. Especially in Relation to Recursive functions*, North-Holland, Amsterdam, 1965.

[7] B.A. Kushner, *Ob odnom predstavlenii Bar-Induktsii. Russian*, Voprosy Matematicheskoy Logiki i Teorii Algoritmov, Vychislitel'ny Tsentr AN SSSR, Moscow, 1988, pp. 11–18.

[8] B.A. Kushner, *A Version of Bar-Induction. Abstract*, The Journal of Symbolic Logic **57** (1992), 361.

[9] H. Rogers, Jr., *Theory of Recursive Functions and Effective Computability*, McGraw-Hill Book Company, New York, 1967.

[10] B.A. Kushner, *Lectures on Constructive Mathematical Analysis. (Translation from the Russian)*, AMS, Providence, Rhode Island, 1984.

[11] A.G. Dragalin, *Mathematical Intuitionism: Introduction to proof theory. (Translation from the Russian)*, AMS, Providence, Rhode Island, 1988.

[12] W.A. Howard and G. Kreisel, *Transfinite induction and bar induction of types zero and one, and the role of continuity in intuitionistic analysis*, The Journal of Symbolic Logic **31** (1966), 325–358 .

CURT SCHMIEDEN'S APPROACH TO INFINITESIMALS

An eye-opener to the historiography of analysis

Detlef Laugwitz †
Ahornweg 23, 64367 Mühltal, Germany

Abstract: In the early 1950s C. Schmieden developed the idea to 'adjoin' an infinitely large natural number Ω to real analysis. A formula $A(\Omega)$ was true if $A(n)$ was true for almost all finite natural n. In a paper of 1958 we used rational (and later real) sequences to establish a model in which Ω was represented by the sequence of the finite natural numbers. Clearly, this approach was less powerful with respect to new applications than Robinson's (1961) was. Yet it turned out that the Omega Calculus could serve as a means to clarify earlier uses of infinitesimals and infinitely large numbers. The sequential approach was successful in re-considering the foundational aspects of Cauchy's textbooks, and also early Fourier analysis and the use of delta functions around 1820. Adjoining an ideal element was more in the spirit of Euler (and, in some sense, of Leibniz and Bolzano). Like most of pre-Cantorian mathematics the two versions of Schmieden's approach were basically constructive.

The Cauchy continuum consists of all $r + \alpha$, r real, α infinitesimal. Any real function mapping this continuum into itself must be continuous. (This result, though attributed to Brouwer, does not depend on intuitionism. It is not valid in nonstandard continua.) Discontinuous real functions, and also distributions, can be represented by polynomials of infinite degree.

1. INTRODUCTION

In 1954 Carl Friedrich von Weizsäcker gave me a typescript of some fifty pages to report on it in his seminar. This script was a collection of strange examples, taken from introductory calculus, treated in an unusual manner, freely using infinitely large numbers and infinitesimals. I found this interesting, and Weizsäcker disclosed the name of the author. A fruitful collaboration began.

P. Schuster, U. Berger and H. Osswald (eds.),
Reuniting the Antipodes – Constructive and Nonstandard Views of the Continuum, 127–142.

The author was Curt Schmieden (1905–1992). He had contributed to hydrodynamics, aerodynamics, and to nonlinear vibrations. The examples had been collected since 1948 during his mathematics courses for students of physics and engineering. He felt that the usual Weierstrassian approach to analysis was not suitable for these. From his own research work he was used to a way of thinking in which both infinitesimals and functions assuming also infinitely large values were not yet eliminated, and he felt uneasy when he had to tell engineering students that no such things existed. (Neither of us actually used the new approach in introductory courses.)

After years of debates among the two of us, and with the referees (one of them was Paul Lorenzen), a joint paper was eventually accepted for publication in 1957 [SL58]. The approach was soon superseded by Abraham Robinson's Nonstandard Analysis [Ro61,66] which was much more powerful when it came to applications in contemporary research. Right from the beginning I had felt that our work opened new ways for understanding the historical development of analysis. Indeed what Schmieden had in mind was a concretization of Leibniz's Principle of continuity, as expressed in a letter to Varignon (2 February 1702): The rules of the finite remain valid for the infinite. Of course, one had to specify the rules.

We shall consider a few texts from the history of infinitesimals. More material can be found in my publications listed in the bibliography at the end of the paper.

2. Ω-RATIONAL NUMBERS AND FUNCTIONS: EULER

The new and fruitful idea of Schmieden was indeed a turning point in the approach to infinitesimals. He did not start from considering infinitely small numbers but adjoined to the field of rational numbers an infinitely large natural number Ω. To adjoin was meant in a more general sense than in algebra but defined by the following postulate: Whatever is true for all very large finite natural numbers will, by definition, remain true for Ω. More precisely, if $A(n)$ is a formula that is true for all sufficiently large finite natural n then $A(\Omega)$ is true. (Leibniz's Principle)

Since $n > 10^{100}$ for all large n it follows that $\Omega > 10^{100}$, etc. Thus Ω is indeed infinitely large. Since n^n and 2^{n^2} are natural numbers, and since $n^n < 2^{n^2}$ for all large n, the inequality $\Omega^\Omega < 2^{\Omega^2}$ is true, and both sides are infinitely large natural numbers. Also, $r(\Omega)$ is an Ω-rational number if $r(n)$ is rational for almost all n. The functions accessible here are polynomials of infinite degree and also quotients of such. It is easily seen that for every finite rational x

$$e_\Omega(x) := \left(1+\frac{x}{\Omega}\right)^\Omega \approx \sum_{k=0}^{\Omega}\frac{x^k}{k!}. \qquad (a \approx b \quad \text{for:} \quad a-b \quad \text{infinitesimal})$$

This is in the spirit of Euler. The elementary transcendental functions are represented by polynomials of infinite degree.

Definite integrals are obtained as infinite sums of infinitesimals, and derivatives are, again up to infinitesimal 'errors', nothing but quotients of infinitesimal differentials. Still, the advantage of what was later called the derived function was well known to Euler. He considered the indefinite integral as an antiderivative.

An early application was the logarithmic function, in particular the formula for the partial sums of the harmonic series, for $N = N(\Omega)$,

$$\begin{aligned} H(N) &= 1+\frac{1}{2}+\frac{1}{3}+\ldots+\frac{1}{N} = \log N + C_N, \\ C_N &\approx C = .577\ldots \quad \text{if} \quad N \gg 1. \end{aligned}$$

In Eu34 he finds, by manipulation of the divergent harmonic series, the values of many convergent series, as

$$\begin{aligned} 1-\frac{1}{2}+\frac{1}{3}-\frac{1}{4}\pm\ldots+\frac{1}{2N-1}-\frac{1}{2N} &= H(2N)-2\cdot\frac{1}{2}H(N) \\ &\approx \log 2N - \log N = \log 2, \end{aligned}$$

$$1+\frac{1}{3}-\frac{1}{2}+\frac{1}{5}+\frac{1}{7}-\frac{1}{4}\pm\ldots+\frac{1}{4N-3}+\frac{1}{4N-1}-\frac{1}{2N}$$

$$\begin{aligned} &= H(4N)-\frac{1}{2}H(2N)-\frac{1}{2}H(N) \approx \log 4N - \frac{1}{2}\log 2N - \frac{1}{2}\log N \\ &= \frac{3}{2}\log 2, \quad \text{and similarly} \\ \log 3 &= \log 3N - \log N \approx H(3N) - H(N) \\ &= 1+\frac{1}{2}-\frac{2}{3}+\frac{1}{4}+\frac{1}{5}-\frac{2}{6}\pm\ldots \end{aligned}$$

Such were also among Schmieden's early examples, though at the time he did not know about the paper by Euler.

With Euler this was not just an occasional trick but rather a fruitful method to which he came back every now and then [Eu48, 55]. The key-word here is *inexplicable functions,* as e. g. $H(x)$ for real x, or the gamma function.

So far everything is obtained by algebraic manipulation. But the paper [Eu34] begins, surprisingly, with a general principle on the convergence of series

$$S = p_1 + p_2 + \ldots + p_N + p_{N+1} + \ldots = S_N + R_N;$$

The series has a value (or converges) if $R_N \approx 0$ *for all* $N \gg 1$.

Euler formulates this principle for positive terms but applies it more generally. No proof is offered, and it appears that such a proof may have surpassed the level of abstraction of his time.

Obviously $S_N \approx S$, or $R_N \approx 0$ is a necessary condition. Suppose now that $R_N \approx 0$ for all $N \gg 1$. Euler was not a man of general statements or of conceptual proofs. Still we may try to find a proof by means of what he had at hand. One of these is Fermat's method of *descente infinie* which says, in modern terms, that a non-empty set of natural numbers contains a least element, or in terms of the time: if a formula (equation or inequality) $F(m)$ is true for some natural m_1, then there exists a smallest m_0 such that $F(m_0)$ is true.

Consider the formula $F(m) : k \geq m \Rightarrow |R_k| < \varepsilon$ for a fixed positive rational ε. This is, by assumption, true for all $m = N \gg 1$. So the minimal natural $m_0 = m(\varepsilon)$ must be finite. It follows that S_m for $m \geq m_0$ approximates the sum up to an error smaller than ε. (In particular, if S_{m_0} is finite then S must be finite.)

Another *Principle* frequently used by Euler is the following: If $S_m \approx \hat{S}_m$ for all finite m, and if $S_N \approx S$, $\hat{S}_N \approx \hat{S}$ for all infinite N, then $S \approx \hat{S}$. Examples of this are his proof of $\left(1 + \frac{x}{N}\right)^N \approx \sum_{k=0}^{\infty} \frac{x^k}{k!}$, and also the proof for the sine product. For the latter, Wim Luxemburg [Lu73] has completed Euler's proof by using this principle which says, in terms of *Robinson's Lemma:* If S_m is an internal sequence and if $S_m \approx 0$ for all finite m, then $S_N \approx 0$ up to some $N \gg 1$.

Again, the Lemma follows by descente infinie. Consider the formula

$$F(m) : k \geq m \Rightarrow k|a_k| \geq 1;$$

if there is no such m, then $a_m \approx 0$ for all m. Otherwise, let $N + 1$ be the least natural number. It must be infinite, and for $k \leq N$ we have

$|a_k| < \frac{1}{k} \approx 0$ if $k \gg 1$. Now the Principle follows immediately: Let $a_m = S_m - \hat{S}_m$; then $a_N = S_N - \hat{S}_N \approx 0$ for some $N \gg 1$, and since both series converge, $S \approx S_N \approx \hat{S}_N \approx \hat{S}$.

(Incidentally, a result equivalent to Robinson's Lemma was proved in a strictly constructive manner by du Bois-Reymond. See La74.)

You will have observed that some of my formulas $F(m)$ contained quantifiers in disguise, and this may appear to have been outside the conceptual realm of Euler. Still, Euler used 'it exists' and 'for all' even if he had no symbols for them.

In any case, Schmieden's approach was an eye-opener with respect to some parts of Euler's work. Up to then, the results on divergent series mentioned here and the convergence condition of Euler had been ignored. Other parts, as the proof on the exponential function, had been misunderstood as dealing with limits. Then the proof appeared based on an illicit interchanging of two limits. But Euler did not think in the conceptual framework of limits.

Euler's algebraic use of infinitesimals was criticized even by Lagrange who was primarily an algebraist. Lagrange put as a prize question of the Berlin Academy in 1785 the problem to eliminate infinitesimals (and also limits and Newtonian moments) from analysis, or else to put their use on a "firm" ground. This was achieved by Cauchy. It was an outcome of a paradigm change, from ideal elements to real variables.

3. MODELS FOR EXTENDED NUMBER SYSTEMS: CAUCHY

The referees of our paper SL58 did, like Lagrange, not accept the approach. (It was eventually published in La80, 83.) We rewrote the paper and defined the new numbers in terms of equivalence classes of rational (and later real) sequences, in other words, we supplied a model. When Abraham Robinson, in the Appendix of Ro66, drew attention to the use of infinitesimals in Cauchy's textbooks it became clear that we had simply rediscovered this approach of Cauchy on a level of abstraction that was certainly closer to Cauchy's than Robinson or the ultrapowers were. Moreover, a careful study of the research work of Fourier, Poisson, Cauchy and their contemporaries revealed that they went much farther in their practical use of infinitesimals and infinitely large numbers than Cauchy in his introductory textbooks. There was a substantial use of delta functions in Fourier analysis and partial differential equations of mathematical physics [La89].

In SL58 we discussed delta functions, such as $\delta(\xi) = \frac{1}{\pi}\frac{\omega}{\xi^2+\omega^2}$ for some infinitely small positive ω. In subsequent papers La59, La61 more general delta functions, distributions and other generalized functions were studied systematically. The conceptual framework now was the number system $\mathbb{R}^{\mathbb{N}}/F$, where F was the Fréchet filter on $\mathbb{N}$. The functions appearing here were no longer standard but closely related to standard functions. For instance, consider a normal (or internal) function $\varphi(\xi) = f(\Omega, \xi)$, $\xi = x(\Omega)$ which is differentiable and infinitely close to some standard continuous function g. Then, the k-th derivative of φ represents a distribution of order k.

Later I found that such conceptions had already been used around 1820. Cauchy had carefully proved in 1818, that the rational function δ from above had the property

$$\int_{-\infty}^{+\infty} \delta(x-k)F(k)dk \approx F(x)$$

for standard continuous bounded F. Also, he knew that

$$\frac{\omega}{\omega^2+\xi^2} = \int_0^{\infty} \cos \xi t e^{-\omega t} dt = \frac{1}{2}\int_{-\infty}^{+\infty} \cos \xi t e^{-\omega |t|} dt.$$

He obtained the following version of the Fourier Inversion Formula

$$F(x) \approx \frac{1}{2\pi}\int\int e^{-\omega |t|} \cos(x-k)t \; F(k) dt \; dk.$$

If ω is replaced by 0 the original formula of Fourier appears. To it belongs the awkward delta function

$$\delta_F(\xi) = \frac{\sin p\xi}{\pi p} = \frac{1}{2\pi}\int_{-p}^{p} e^{it\xi} dt = \frac{1}{\pi}\int_0^p \cos t\xi \; dt$$

for $p \gg 1$. Fourier hat proved the delta function property by a geometrical argument, based on the rapid oscillation of $\delta_F(\xi)$.

Smoothing (or cutoff) factors like $e^{-\omega t}$ were a natural outcome of the individual partial differential equation and of physical considerations. Cauchy also made an attempt at a general theory of such "auxiliary multipliers" $\alpha(t) \approx 1$ for finite t [Ca22].

Moreover, the delta function method was a special case of his " singular integrals".

Infinitesimals, infinitely large numbers and generalized functions were used in a matter of fact way exceeding Euler's earlier methods. On the level of introductory textbooks Cauchy supplied the rudiments of a theoretical foundation [Ca21, Ca23].

He gave his famous description of infinitely small quantities as variables converging to zero. This was open to misinterpretations. Actually his practical use of infinitesimals (and also of infinitely large numbers, when he says that a sequence s_m converges to s if $s_\mu \approx s$ for each infinitely large μ) shows that he considered these quantities as members of an extension of the real number field. From 1823 on we find occasional references to " systems of infinitesimals" having a "basis" i (corresponding to our ω); $f(i)$ was an infinitesimal if $f(x)$, considered as a real function in some positive neighborhood of $x = 0$, had limit 0 for x tending to 0. Without further comment he used $1/i$, $\log i, \ldots$

From a point of view that was generally accepted only a century after Cauchy, and which was above the level of abstraction of his contemporaries, Cauchy considered a "space" of elements that appeared represented by functions. This is similar to considering the real numbers as represented by their decimal developments.When we put Cauchy's text before this conceptual background the results in his textbooks appear clear and correct. We shall now consider only two of his famous theorems.

A *function* is, by definition, *continuous* if infinitesimal increments of the argument yield infinitesimal increments of the value. Let α, β be infinitesimals. Then,

$$F(x+\alpha,\ y+\beta) - F(x,y) =$$
$$[F(x+\alpha,\ y+\beta) - F(x,y+\beta)] + [F(x,y+\beta) - F(x,y)],$$

and it is obvious that a function of several variables is continuous in these variables together if it is continuous in each single variable. (The theorem is false if we consider β as a variable but it is correct if $y+\beta$ denotes a fixed value of the independent variable!)

A *series* is, by definition, *convergent* to s if $s_\mu \approx s$ for all partial sums with $\mu \gg 1$. Let $s_m(x)$ be the partial sums of a series of continuous functions $u_k(x)$. Suppose that the series converges everywhere (toujours) in some interval and has $s(x)$ as its sum. Let
$s(x) = s_m(x) + r_m(x)$. Suppose that $\alpha \approx 0$. Then,

$$s(x+\alpha) - s(x) = r_m(x+\alpha) + [s_m(x+\alpha) - s_m(x)] - r_m(x).$$

Here m is any finite natural number, and the middle term on the right is infinitesimal since s_m is a continuous function. For m sufficiently large the two remainder terms, though still finite, will be made arbitrarily small. Hence $|s(x + \alpha) - s(x)|$ will be smaller than any finite positive number, and must be infinitesimal. The sum function is continuous.

The critical term here is $r_m(x+\alpha)$. The argument essentially depends on $x + \alpha$ being a fixed number, so α cannot be variable. The series is assumed to converge everywhere (toujours), including numbers $x + \alpha$, x real and α infinitesimal. Such "nonstandard" numbers constitute Cauchy's continuum.

When Robinson [Ro66] had drawn attention to an interpretation of Cauchy by means of his Nonstandard Analysis this gave rise to a vivid debate, also on the question whether conceptions of the second half of the 20th century were permitted as a justification here. It seems to me that our approach is closer to the level of abstraction of the early 19th century. Also it permits an easy characterization of continuous functions which are central in Cauchy's texts. Let F be a real function which is fully defined, that is, has real values for all real x. This function is, in the obvious way, extended to all Ω-numbers. (The value of the number represented by the sequence x_n is represented by the sequence $y_n = F(x_n)$.) It is easy to see that F must be continuous if it maps the Cauchy continuum into itself. In Nonstandard Analysis this conclusion does not follow.

This theorem may remind one of a similar proposition by Brouwer. (Compare also La96, 97 and Pa96.) Still, Cauchy and his contemporaries used delta and other nonstandard (internal) functions in their research papers. These had a weaker continuity property: For each $\varepsilon > 0$ there exists some $\delta > 0 \ldots$, as usual, but now δ may be infinitesimal for finite ε. Cauchy said that the usual theorems of real analysis extended to such functions "by analogy". In our approach they can be proved rigorously. As Cauchy knew well there is some need of discontinuous functions in physics. Constructive theories admitting only functions that are continuous in the usual sense are of no interest for applications.

Like Euler, Cauchy says that $s_m \approx s_r$ for $m, r \gg 1$ is a necessary condition for convergence. Then he states verbally the famous ε-condition and concludes without proof that this condition is sufficient.

4. CONTEMPORARY BACKGROUND

Euler and Cauchy were not isolated. Their work was embedded in the manner of thinking of their contemporary scientific communities. Each such community has, in addition to published work and even as prerequisites for this work, an amount of tacit understanding, or tacit knowledge, that does not appear in print, and is difficult to trace. Also there are hidden lemmas, that are not mentioned explicitly and are taken for granted, to be made explicit and proved (that is, reduced to other assumptions) by later generations. [La87]

Leibniz, Euler and Gauss took the intermediate value theorem for continuous curves (or functions) for granted. Bolzano in 1817 and Cauchy in 1821 felt the need to prove this property from the concept of continuity as it became defined during their time. Leibniz and Euler, and to some extent the Bernoullis, used ideal elements, such as imaginary numbers, infinitesimals, and infinitely large numbers. The connection with the linear continuum (or the real line) was loose or not at all intended. Actually Euler and Lagrange banned geometry from analysis and considered the latter as algebra (of the infinite). Even in 1821 Cauchy chose "Analyse algébrique" as the subtitle for his Cours d'Analyse. Analogously, Leibniz and Euler never tried to find a geometrical model for complex numbers. Ideal numbers became a tool in the work of Kummer, and Dedekind supplied set-theoretic models for them.

The ideal infinitesimals of Euler and his predecessors were modeled by Cauchy, in the shape of variables. There was a tacit understanding about infinitesimals as variables converging to zero during the two decades preceding Cauchy's Cours of 1821. In addition to Fourier and Poisson we may mention Lacroix and Legendre. When refereeing Cauchy's prize paper on integration in 1814 they explained the discovery of a double integral whose value depended on the order of integrations by a use of infinitesimals [LL14]. In the same year 1814 Gauss wrote a note on trigonometric series in which he used infinitesimals in the same way as the French did, and used in passing the definition of continuity as it was published in Ca21. That note by Gauss was not published at the time [La89a]. But in reviews of papers by Poisson Gauss underlined that the author must take more care in considering the diverse orders of infinitesimals. And, as Stroyan has observed, the famous Surface Theory published by Gauss in 1827 rests substantially on infinitesimals.

Around 1870 variables lost their role as an undefined basic conception during the process of arithmetization of analysis (Weierstrass, Cantor, Dedekind). The linear continuum became indentified with the set of real numbers, and infinitesimals were expelled from the official world of mathematics. The tacit knowledge of the early 19th century of which Cauchy's propositions had only been like the tip of an iceberg, was no longer present, though it had left its traces which will be considered presently.

5. THE DEGENERATION OF INFINITESIMALS

The Fourier transform was soon replaced by other methods for partial differential equations such as Green's functions. Delta functions and infinitesimal methods had lost their applications in mathematical physics soon after 1830.

Also even the rudiments of an abstract theory of infinitesimals as developed in Cauchy's textbooks were on a level of abstraction that was not accessible to his contemporaries. In physics and geometry infinitesimal reasoning became a mere heuristics.

Elementary parts of proper infinitesimal analysis were preserved in a degenerate form in introductory textbooks, above all at the École polytechnique, in the shape of a rigid dogma. This was put in two Principles by Duhamel in 1841:

(i) If α, β are infinitesimals and a, b finite, then

$$\frac{a+\alpha}{b+\beta} \approx \frac{a}{b},$$

(ii) If A is a finite quantity and $\sum a_k \approx A$, and if $\alpha_k/a_k \approx 0$, then $\sum(a_k + \alpha_k) \approx A$.

Obviously these Principles are basic for differentiation and integration, respectively. As soon as the fundamental theorems of the Calculus had been established there was no need for further use of infinitesimals.

Let us leap to the year 1927 and to two outstanding mathematicians of the time, Jacques Hadamard and William Fogg Osgood, to learn about surprising facts.

Hadamard published his *Cours d'analyse professé à l'École Polytechnique.* It begins with preliminary notions, recalling highschool knowledge. The five first pages are on " infiniment petits, infiniment grands". He writes: "Les notions fondamentales du calcul infinitésimal sont celles d'infiniment petit et d'infiniment grand. Une quantité infiniment petite est une quantité variable qui tend vers zéro. Une quantité infiniment grande est une quantité variable dont la valeur absolue augmente au delà de toute limite ... "

He discusses the orders of infinitesimals and infinitely large quantities. In the subsequent section on series he says that u_n must be an infinitesimal if $\sum u_n$ converges, but convergence or divergence will depend on the order of the infinitesimal u_n if $1/n$ is taken as the principal infinitesimal (the "basis" of Cauchy's system). The series will converge if this order is greater than 1.

Thus, Hadamard avoids Weierstrassian epsilontics in his text for engineering students.

Also in 1927 there was a debate on Duhamel's Principle (ii) in the American Mathematical Monthly. At Harvard Osgood had used this Principle for decades, and had attempted to reconcile it with epsilontic methods. He emphasized the usefulness of the Principle (ii) in mathematical education. There was strong opposition, and indeed to put the Principle on a firm ground in the sense of the Weierstrassians seems next to impossible. Both of Duhamel's Principles are easy to prove in our approach. The second one appears here in a precise form, with specified lower and upper limits of the sum in terms of possibly infinitely large integers that were missing in the more informal earlier versions. Neither Hadamard nor Osgood used infinitesimals in their *research* papers.

Recently Volkov and Demidov [VD97] published two letters by the Moscow mathematician N.N. Luzin of the early 1930s, addressed to M.Y. Vygodsky, on the subject of the latter's textbook Foundations of Infinitesimal Calculus. The text used infinitesimals, and Luzin was very much in favor of this approach. The editors claim that Luzin anticipated the discovery of nonstandard analysis. An English translation, commented by myself, will appear in the American Mathematical Monthly.

6. CONCLUDING REMARKS

Duhamel's Principle (ii) is a good example for the power of our conceptions. Suppose that

$$\frac{\varepsilon_k}{\alpha_k} \approx 0 \qquad \text{for all } k, \qquad 1 \leq k = k(\Omega) \leq N = N(\Omega).$$

This means that, for each standard $p > 0$,

$$\left|\frac{\varepsilon_k}{\alpha_k}\right| \leq p$$

Suppose further, that $\sum_{k=1}^{N} \alpha_k \approx A$, and that

$$\sum_{k=1}^{N} |\alpha_k| < B \qquad \text{for some standard numbers} \quad A, B.$$

Then,

$$|\sum \varepsilon_k| \le \sum |\varepsilon_k| \le p \sum |\alpha_k| < pB.$$

Since this holds for each standard $p > 0$, we obtain $\sum \varepsilon_k \approx 0$. Finally,

$$\sum_{k=1}^{N}(\alpha_k + \varepsilon_k) = \sum_{k=1}^{N} \alpha_k + \sum_{k=1}^{N} \varepsilon_k \approx \sum_{k=1}^{N} \alpha_k \approx A.$$

This is Principle (ii).

To see what this means consider the definite integral of a continuous function on an interval $a \le x \le b$. Duhamel's Principle (ii) immediately yields that each infinitesimal subdivision of the interval gives us one and the same value for the standard part of the Riemann sum. (Existence of standard parts is needed only for numbers $\alpha = a(\Omega)$ whose representing sequence $a(n)$ is monotonic and bounded. For the existence of the Riemann integral consider upper and lower sums.)

We see here how the Principle circumvents the concept of uniform continuity. Actually the proof by Cauchy in Ca23 was quite similar.

The example shows once more that our approach suffices to clarify problems of 19th century analysis. Infinitely large integers are essential, and the new numbers cannot be identified with variables (as Osgood had attempted to do) but must be considered as members of an extension of the standard number system. Sequences α_k are to be considered as $\alpha_k = a(\Omega, k)$ where a is a function of two variables. The sequence is then also defined for infinitely large integers by $\alpha_N = a(\Omega, N(\Omega))$. (Internal sequences.) Neither Robinson's Nonstandard Analysis nor the ultrapower models are needed here.

It has turned out that Schmieden's approach is close to the historical development. This explains why this version of a non-standard mathematics is, in many parts, as constructive as mathematics up to and including Cauchy. It may be of interest that the very first contribution to history was a paper on Bolzano (La65). Not content with his proof of the intermediate value theorem of 1817 Bolzano considered from about 1830 until his death in 1848 a theory of number expressions, including infinitesimal and infinitely large numbers. This work was not published until the early 1960s. Needless to say that he failed to give a constructive proof of the theorem.

On a new level of abstraction all of the nonstandard approaches are revivals of Euler's 'expressions', but now expressions may also be built up from logical symbols in addition to symbols of standard mathematics.

Also it has turned out that many models of the linear continuum were discovered or rediscovered. Schmieden's model is, in a precise sense, the smallest possible extension to give us a sufficiently large theory of analysis. It is founded on adjoining one single infinitely large number Ω to the theory of natural numbers. If one adjoins Ω to real analysis, one obtains, among other things, what we called the Cauchy continuum, consisting of all $r + \alpha$, r real and α infinitesimal. Since oscillatory infinitesimals exist, this continuum is not strictly ordered but is infinitesimally thick. This reflects one of the intuitive properties of the continuous line. As the examples may have made clear neither the lack of a strict order nor the existence of zero divisors are obstacles for the theory and its applications. Still, some caution will be advisable.

A few examples will clarify this. A function f and its derivative are related by

$$f(x+\alpha) - f(x) = (f'(x) + o)\alpha,$$

with $o = o(x, \alpha) \approx 0$ for $\alpha \approx 0$. Similarly, denominators can be avoided. Also, $(1 + (-1)^{\Omega})(1 + (-1)^{\Omega}) = 0$, but we cannot conclude that one of the factors equals 0. Neither of them should appear as a denominator. In any ultrapower model one can conclude that either the first or the second factor vanishes. But since it is in general unknown which one equals zero one has the same situation.

In a discussion Per Martin-Löf pointed out that one may be tempted to use intuitionistic logic here. The formula

$$(-1)^{\Omega} = +1 \vee (-1)^{\Omega} = -1$$

can be proven, but we cannot infer that either $(-1)^{\Omega} = +1$ or $(-1)^{\Omega} = -1$ is valid. Similarly we have restrictions for conclusions from existential propositions. As we saw men like Euler and Cauchy were reluctant to prove existence theorems, and managed with certain 'Principles'.

When trying to prove such Principles one may introduce more general rules of inference. One such was 'descente infinie' which, in the case of statements on finite natural numbers, is equivalent to complete induction. The extension of the method to internal statements comes as close to constructivity as appears possible.

The highly non-constructive existence of the standard part can here be replaced by its weak version: Let $a(n)$ be a rational (or real) sequence which is monotonic and has a finite bound. Then some real number (or Dedekind cut) r exists such that $a(N) \approx r$ for all $N \gg 1$. Cauchy's proof of the intermediate value theorem in Ca21, Note III rests on this version.

References

[Ca21] A.-L. Cauchy: Cours d'analyse, 1821, Œuvres (2) 3, Paris 1897. Reprint of the original edition, edited and commented by U. Bottazzini, Bologna 1990.

[Ca22] A.-L. Cauchy: Mémoire sur l'intégration des équations ... Œuvres (2) 1, 275–357.

[Ca23] A.-L. Cauchy: Résumé des leçons, 1823. Œuvres (2) 4, Paris 1899.

[Du41] J.M.C. Duhamel: Cours d'analyse, Paris 1841.

[Eu34] L. Euler: De progressionibus harmonicis observationes, 1734. Opera omnia (1) 14, 73–86.

[Eu48] L. Euler: Introductio in analysin infinitorum. Lausanne 1748. Opera omnia (1) 8.

[Eu55] L. Euler: Institutiones calculi differentialis. 1755. Opera omnia (1) 10.

[La59] D. Laugwitz: Eine Einführung der δ-Funktionen. Sitzungsber. Bayer. Akad. Wiss., Math.-nat.Kl., 1959, 41–59.

[La61] D. Laugwitz: Anwendungen unendlich kleiner Zahlen I. Zur Theorie der Distributionen. J.r.u. angew. Math. 207 (1961), 53–60.

[La65] D. Laugwitz: Bemerkungen zu Bolzanos Grössenlehre. Arch. Hist. Ex. Sci. 2 (1965), 398–409.

[La74] D. Laugwitz: On Abraham Robinson's Sequential Lemma. Fachbereich Mathematik, Darmstadt, Reprint No. 167, 1974.

[La80] D. Laugwitz: Ω-Calculus as a generalization of field extension. In: Nonstandard Analysis – Recent Developments. Lecture Notes in Mathematics, 1983.

[La83] D. Laugwitz: Eulers Begründung der Analysis aus der Algebra. Mathem. Semesterber. 30 (1983), 171–193.

[La86] D. Laugwitz: Zahlen und Kontinuum. Eine Einführung in die Infinitesimalmathematik. BI. Wissenschaftsverlag, Mannheim 1986.

[La87] D. Laugwitz: Hidden lemmas in the early history of infinite series. Aequationes Math. 34 (1987), 264–276.

[La87a] D. Laugwitz: Infinitely small quantities in Cauchy's textbooks. Historia math. 14 (1987), 258–274.

[La89] D. Laugwitz: Definite values of infinite sums: Aspects of the foundations of infinitesimal analysis around 1820. Arch. Hist. Ex. Sci. 39 (1989), 195–245.

[La89a] D. Laugwitz: Grundlagen der Analysis bei C.F. Gauss: Trigonometrische Reihen. Math. Semesterber. 36 (1989), 159–174.

[La91] D. Laugwitz: Cauchy-Zahlen als Grundlage der Infinitesimalmathematik. Math. Semesterber. 38 (1991), 175–213.

[La96] D. Laugwitz: Controversies about numbers and functions. Fachb. Mathematik, Darmstadt, Preprint No 1853 (1996). – To appear in Proceedings of Two International Conferences on the History and Philosophy of Mathematics, Penn State University.

[La97] D. Laugwitz: Mathematische Modelle zum Kontinuum und zur Kontinuität. Philosophia naturalis 34 (1997), 265–313.

[LL14] G.F. Lacroix, A.M. Legendre: Report to the Institute of 7 November 1814 on Cauchy's Mémoire sur les intégrales définies. In: Cauchy, Œuvres (1) 1, 321–327.

[Lu73] W.A.J. Luxemburg: What is nonstandard analysis? Amer. Math. Monthly 80 (supplement) June–July 1973, 87–109.

[Pa96] E. Palmgren: Sheaf-theoretic nonstandard analysis: constructive aspects. Dept. of Math. Uppsala University, Report 1996: 28 (p. 26).

[Ro61] A. Robinson: Non-standard analysis. Indagationes Math. 23 (1961), 432–440.

[Ro66] A. Robinson: Nonstandard Analysis. North Holland Publ., Amsterdam 1961.

[SL58] C. Schmieden, D. Laugwitz: Eine Erweiterung der Infinitesimalrechnung. Math. Zeitschr. 69 (1958), 1–39.

[VD97] V.A. Volkov, S.S. Demidov: Two letters of N.N. Luzin to M.Ya. Vygodskii. Historical-mathematical Investigations 2 (37) (1997). 113–152. (Russian)

A SEQUENT CALCULUS FOR CONSTRUCTIVE ORDERED FIELDS

Sara Negri
Department of Philosophy
PL 24, 00014 University of Helsinki
Finland
negri@helsinki.fi

Keywords: Constructive axiomatizations, cut elimination, ordered fields, real closed fields.

Mathematical Subject Classification: 03F05, 03F65, 12J15

Abstract: The theory of constructive ordered fields, based on a relation of strict linear order, is formalized as a proof-theoretical system, a sequent calculus extended with nonlogical rules. It is proved that structural rules, the rules of cut and contraction in particular, can be eliminated from derivations. The method of extension by nonlogical rules is applied also to the theory of real closed fileds, starting from a quantifier-free axiomatization.

Introduction

The failure of cut elimination for logical systems extended with axioms for specific mathematical theories has so far limited the application of the methods of structural proof theory for the study of even elementary mathematical theories.

In previous work (cf. Negri 1999, Negri and von Plato 1998) a method was found for extending logical sequent calculi with nonlogical rules representing mathematical axioms while maintaining eliminability of the structural rules, cut especially.

Here the method is applied to an axiomatization of constructive ordered fields. The axiomatization is based on a single primitive notion of *constructive linear order* and uses constructions and free parameters instead of quantifiers. A well known problem with such an approach to field theory is that it goes beyond first-order logic. Namely, a proposition containing an inverse x^{-1} is well formed only if it is true that $x \neq 0$.

P. Schuster, U. Berger and H. Osswald (eds.),
Reuniting the Antipodes – Constructive and Nonstandard Views of the Continuum, 143–155.

However, the analysis of formal derivations can equally be performed if we treat the condition for the construction of inverses as a meta-level rule of well-formedness of sequents, similarly to other conditions of well-formedness occurring in first-order logic.

After giving the axiomatization for constructive ordered fields, we show how to convert it into a system of sequent calculus nonlogical rules. We then prove cut elimination for the system obtained by adding these rules to the sequent calculus for first-order intuitionistic logic **G3i**.

Similar methods can be applied to the study of the metamathematics of the theory of real closed fields. We eliminate quantifiers already in the axiomatization, by expressing the axioms of existence of square roots of positive elements and of zeros of polynomials of odd degree by means of constructions, rather than by $\forall\exists$-axioms.

1. AXIOMS FOR CONSTRUCTIVE ORDERED FIELDS

We assume a set R with a primitive relation $a < b$ of strict order, instead of the relation $a \leqslant b$ of weak order. This choice of basic relation is needed in order to secure computational meaning: Intuitively, a finite approximation of a and b is sufficient for the verification of $a < b$, whereas with $a \leqslant b$ the verification may lead to an infinite computation. Negation $\sim A$ is defined as $A \supset \bot$.

The axioms for constructive ordered fields can be grouped as follows:

I. Axioms for constructive linear order:

1. $\sim(a < b \,\&\, b < a)$ (*asymmetry*)
2. $a < b \supset a < c \vee c < b$ (*split*)

Given a relation $a < b$ satisfying the above axioms, the relation $a \neq b$ defined by

$$a \neq b \equiv a < b \vee b < a$$

is an apartness relation. Equality is then defined as the negation of $a \neq b$

$$a = b \equiv \sim a \neq b$$

and it is easy to see that the relation $a = b$ satisfies reflexivity, symmetry and transitivity. Clearly, the relation $a = b$ is stable, that is, it satisfies $\sim\sim a = b \supset a = b$. Substitution in the form

$$a < b \,\&\, b = c \supset a < c$$

follows from split and definition of $a = b$. We argue similarly for deriving $a < b \,\&\, a = c \supset c < b$. The usual weak order is defined by

$$a \leqslant b \equiv \sim b < a$$

Note that classically, the first axiom is equivalent to $a \leqslant b \vee b \leqslant a$. But constructively, not any two elements need be comparable. Contraposition of the second axiom expresses transitivity of weak order.

We stipulate the existence of two distinct elements 0 and 1 in R satisfying

3. $0 < 1$.

We have a binary operation $+$ (addition) and a unary operation $-$ (opposite) satisfying:

II. Axioms of additive group:

4. $a + b = b + a$ (*commutativity*)

5. $(a + b) + c = a + (b + c)$ (*associativity*)

6. $a + 0 = a$ (*zero*)

7. $a + (-a) = 0$ (*opposite*)

We assume a *principle of strong extensionality*

8. $a + b < a + c \supset b < c$

from which the principle of extensionality,

$$b = c \supset a + b = a + c$$

follows by contraposition and definition of equality. The apparently more general form of extensionality, $a = b \,\&\, c = d \supset a + c = b + d$, follows from the previous one and transitivity of equality.

Monotonicity of addition with respect to the strict linear order, i.e.,

$$a < b \supset a + c < b + c$$

is now provable: From the antecedent, by axioms 7, 8 and substitution we obtain $a+(c+(-c)) < b+(c+(-c))$, hence by associativity and substitution again we have $(a+c)+(-c) < (b+c)+(-c)$, thus by symmetry and strong extensionality $a + c < b + c$. We also have

$$0 < a + b \supset 0 < a \vee 0 < b$$

as follows: From the antecedent, by split, we have $0 < a \vee a < a + b$. By extensionality and the axiom for opposite, the second disjunct gives $0 < b$.

Multiplication of a and b is denoted by ab. For $a \neq 0$ we have an operation of inverse, denoted a^{-1}.

III. Axioms for multiplication:

9. $ab = ba$ (*commutativity*)

10. $(ab)c = a(bc)$ (*associativity*)

11. $a1 = a$ (*unit*)

12. $aa^{-1} = 1$, provided that $a \neq 0$ (*inverse*)

13. $a(b + c) = ab + ac$ (*distributivity*)

Strong extensionality for multiplication is

14. $ab < ac \supset (a < 0 \,\&\, c < b) \vee (0 < a \,\&\, b < c)$

and gives by contraposition and definition of equality the principle of extensionality in multiplication

$$b = c \supset ab = ac$$

thus by transitivity of equality also $a = b \,\&\, c = d \supset ab = cd$.

By distributivity, from $0a = (0 + 0)a$ and extensionality for addition we obtain $0a = 0$. From this we obtain

$$0 < c \supset 0 < c^{-1}$$

as follows: From $0 < 1$, by extensionality we have $0c^{-1} < cc^{-1}$, so by commutativity and strong extensionality together with the hypothesis $0 < c$, the result $0 < c^{-1}$ follows.

We are now ready to prove monotonicity of multiplication with respect to the strict linear order, i.e.,

$$a < b \,\&\, 0 < c \supset ac < bc$$

From $a < b$, by using the unit axiom and extensionality in multiplication we obtain $a(cc^{-1}) < b(cc^{-1})$ so by associativity and substitution $(ac)c^{-1} < (bc)c^{-1}$. By commutativity and strong extensionality, since $0 < c^{-1}$, the result $ac < bc$ follows.

Standard axiomatizations of constructive ordered fields are based on a primitive notion of weak order. Our axiomatization is equivalent to other constructive axiomatizations, except for the axiom of strong extensionality for multiplication. In Palmgren (1998) the theory of constructive ordered fields is presented through a predicate $P(a)$ of strict positivity that corresponds to $0 < a$. The split property, here taken as an axiom, is instead shown to follow from the axiom

$$0 < a + b \supset 0 < a \lor 0 < b$$

The latter is in turn provable in our axiomatization using split and the axiom for opposite.

In Bridges (1999) the Archimedean axiom and a constructive version of the least upper bound principle are added to constructive ordered fields for obtaining an axiomatization of real numbers. These two axioms are not expressible in our approach since the first relies on natural numbers and the second is inherently second order (see the discussion in Truss 1997).

What are known as formal reals (cf. Negri and Soravia 1999) give a model for the axioms of the theory of constructive ordered fields. More generally, constructive reals (e.g. Bishop's reals) constitute a model for the theory.

2. SEQUENT CALCULUS PROOF THEORY AND ITS EXTENSION BY NONLOGICAL RULES

We shall be using an intuitionistic multi-succedent sequent calculus in this work. Atomic formulas are denoted by $P, Q, R, \ldots$ and arbitrary formulas by $A, B, C, \ldots$. The basic expressions are *sequents* of the form $\Gamma \Rightarrow \Delta$, where Γ and Δ are finite *multisets* of formulas, i.e., sets where the multiplicity of an element is counted. The left and right parts of the sequent are the *antecedent* and the *succedent*. Derivations in sequent calculus start with *logical axioms*, i.e., zero-premiss rules of two forms:

Logical axioms:

$$P, \Gamma \Rightarrow \Delta, P \qquad \bot, \Gamma \Rightarrow \Delta$$

The first axiom corresponds to making the assumption P, the second to the rule *ex falso quodlibet.* The first axiom is restricted to atomic formulas and in the second, it is essential that the false formula $\bot$ is not considered atomic, but a zero-place logical operation.

The logical rules are divided into *right* rules that correspond to the introduction rules of natural deduction, and to *left* rules that correspond to elimination rules.

Logical rules:

$$\frac{A, B, \Gamma \Rightarrow \Delta}{A\&B, \Gamma \Rightarrow \Delta} L\& \qquad \frac{\Gamma \Rightarrow \Delta, A \quad \Gamma \Rightarrow \Delta, B}{\Gamma \Rightarrow \Delta, A\&B} R\&$$

$$\frac{A, \Gamma \Rightarrow \Delta \quad B, \Gamma \Rightarrow \Delta}{A \vee B, \Gamma \Rightarrow \Delta} L\vee \qquad \frac{\Gamma \Rightarrow \Delta, A, B}{\Gamma \Rightarrow \Delta, A \vee B} R\vee$$

$$\frac{A \supset B, \Gamma \Rightarrow A \quad B, \Gamma \Rightarrow \Delta}{A \supset B, \Gamma \Rightarrow \Delta} L\supset \qquad \frac{A, \Gamma \Rightarrow B}{\Gamma \Rightarrow \Delta, A \supset B} R\supset$$

$$\frac{A(t/x), \forall x A, \Gamma \Rightarrow \Delta}{\forall x A, \Gamma \Rightarrow \Delta} L\forall \qquad \frac{\Gamma \Rightarrow A(y/x)}{\Gamma \Rightarrow \Delta, \forall x A} R\forall$$

$$\frac{A(y/x), \Gamma \Rightarrow \Delta}{\exists x A, \Gamma \Rightarrow \Delta} L\exists \qquad \frac{\Gamma \Rightarrow \Delta, \exists x A, A(t/x)}{\Gamma \Rightarrow \Delta, \exists x A} R\exists$$

In the rules for the quantifiers, $A(t/x)$ and $A(y/x)$ denote substitution of x in A with the term t and the variable y. The usual restrictions apply: in $R\forall$ and $L\exists$, y is not free in the conclusion; in $L\forall$ and $R\exists$ the term t has to be *free for* x *in* A, that is, no variable of t is in the scope of some quantifier in A.

This calculus is designated **G3i** (after Kleene's system, of which it is a simplified version). The calculus is the same as the calculus **GHPC** of Dragalin (1988).

The *height* of a derivation is its height as a tree, that is, the length of its longest branch.

The structural rules of weakening, contraction and cut are formulated as follows:

$$\frac{\Gamma \Rightarrow \Delta}{A, \Gamma \Rightarrow \Delta} LW \qquad \frac{\Gamma \Rightarrow \Delta}{\Gamma \Rightarrow \Delta, A} RW$$

$$\frac{A, A, \Gamma \Rightarrow \Delta}{A, \Gamma \Rightarrow \Delta} LC \qquad \frac{\Gamma \Rightarrow \Delta, A, A}{\Gamma \Rightarrow \Delta, A} RC$$

$$\frac{\Gamma \Rightarrow \Delta, A \quad A, \Gamma' \Rightarrow \Delta'}{\Gamma, \Gamma' \Rightarrow \Delta, \Delta'} Cut$$

None of these structural rules need be assumed in **G3i** since they can be proved *admissible*, in the sense that if the premisses of a rule are derivable, its conclusion also is derivable. Weakening and contraction are admissible and *height preserving*, that is, if their premisses are derivable with a certain derivation height, then their conclusions are also derivable with the same derivation height (cf. Dragalin 1988). *Exchange* rules, permitting the permutation of order in a list of assumptions as in Gentzen's original calculus, are absent due to the use of multisets. We refer to Negri and von Plato (2001), where the above calculus is called **G3im**, for a detailed proof of admissibility of the structural rules.

A rule is said to be *invertible* if whenever the conclusion is derivable, then the premisses are derivable.

All logical propositional rules of **G3i** except $L\supset$ and $R\supset$ are invertible. The rules for implications are partially invertible in the following sense: $L\supset$ is only invertible with respect to the second premiss, i.e., from $A \supset B, \Gamma \Rightarrow \Delta$, the sequent $B, \Gamma \Rightarrow \Delta$ follows. The rule $R\supset$ is invertible when the context Δ is empty. All the quantifier rules are invertible except $R\forall$. As for $R\supset$, the rule $R\forall$ is invertible when the context Δ is empty.

Starting with the calculus **G3i**, the structural rules remain admissible also in extensions of **G3i** by suitably formulated nonlogical rules. Given any formula A, we can check through a formula decomposition via the invertible rules of **G3i** whether the formula is equivalent in the system to a conjunction of formulas of the form $P_1\&\ldots\&P_m \supset Q_1 \vee \cdots \vee Q_n$, where the consequent is $\perp$ if $n = 0$. (This is a slightly modified version of conjunctive normal form.) It is shown in sections 2 and 3 of Negri and von Plato (1998) that such formulas may be replaced by sequent calculus rules without losing the admissibility of the structural rules of **G3i**. Formulas are now converted into rules through the following general

Rule-scheme:

$$\frac{Q_1, \Gamma \Rightarrow \Delta \quad \ldots \quad Q_n, \Gamma \Rightarrow \Delta}{P_1, \ldots, P_m, \Gamma \Rightarrow \Delta}\, \scriptstyle{Reg}$$

where Γ and Δ are arbitrary multisets. Addition of a rule following the scheme will make sequents $\Rightarrow P_1\&\ldots\&P_m \supset Q_1 \vee \cdots \vee Q_n$ derivable by the logical rules.

Special cases of the rule-scheme are obtained when n is 0 and the zero-premiss rule *Reg* becomes a *nonlogical axiom*, with inference line omitted,

$$P_1, \ldots, P_m, \Gamma \Rightarrow \Delta$$

which translates a Hilbert-style axiom of the form $\sim(P_1\&\ldots\&P_m)$.

As explained in Negri and von Plato (1998), the *principal formulas* $P_1, \ldots, P_m$ of the scheme must be repeated in the antecedent of each premiss, but we leave this out for better readability. Such repetition is needed for proving the rule of contraction admissible. We also note the following subtlety:

Closure condition: *Given a system with nonlogical rules, if it has a rule with an instance of form*

$$\frac{Q_1, \Gamma \Rightarrow \Delta \quad \dots \quad Q_n, \Gamma \Rightarrow \Delta}{P_1, \dots, P_{m-2}, P, P, \Gamma \Rightarrow \Delta} \; Reg$$

then also the rule

$$\frac{Q_1, \Gamma \Rightarrow \Delta \quad \dots \quad Q_n, \Gamma \Rightarrow \Delta}{P_1, \dots, P_{m-2}, P, \Gamma \Rightarrow \Delta} \; Reg$$

has to be included in the system.

It can happen that a substitution in the atoms of a rule produces duplications of formulas the contraction of which requires the condition. But it is in principle unproblematic, since the number of rules to be added to a given system of nonlogical rules is bounded. Often the closure condition is superfluous, for example, as noted in Negri (1999), the rule expressing irreflexivity in the theory of strict linear order is derivable from the other rules. However, the rule of irreflexivity can be added to the system if height-preserving admissibility (and not just admissibility) of contraction is needed.

Theorem 3.1: *The structural rules are admissible in extensions of* **G3i** *following the rule-scheme and satisfying the closure condition. Weakening and contraction are admissible and height preserving.*

This result is proved in Negri and von Plato (1998, section 3).

The immediate subformula property fails for nonlogical rules, but a *weak subformula property*, stating that in a derivation only subformulas of the endsequent or atomic formulas occur, is enough for proof-analysis. In particular, we have the

Corollary 3.2: *If a derivable sequent* $\Gamma \Rightarrow \Delta$ *has no logical operator, then its derivation uses only logical axioms and nonlogical rules, but no logical or structural rules.*

3. SEQUENT CALCULUS FOR CONSTRUCTIVE ORDERED FIELDS

In usual sequent calculi, the rules for well-formed sequents are meta-level rules, not explicit rules in derivations. For instance, the variable conditions in quantifier rules, the conditions on terms for substitution, and even the syntactic rules for well-formed formulas are such rules. We shall handle the conditions arising in the theory of constructive ordered fields in the same way. If a sequent $\Gamma \Rightarrow \Delta$ contains a term x^{-1} we assume that the condition $x \neq 0$ has been established.

Another example of conditions is found in elementary geometry. The construction of a line through two points a and b can only be made if a and b are distinct points. In treating the conditions of wellformedness of sequents as meta-level rules we have followed von Plato (1998).

The rules for constructive ordered fields are grouped as in the axiomatic presentation. For simplicity of notation we have omitted the repetition of the principal formula (or formulas) of each rule in its premisses. The rule of irreflexivity has to be added in order to satisfy the closure condition.

I Rules for constructive linear order:

$$a < a, \Gamma \Rightarrow \Delta \ {}_{\mathit{irref}} \qquad a < b, b < a, \Gamma \Rightarrow \Delta \ {}_{\mathit{asym}}$$

$$\frac{a < c, \Gamma \Rightarrow \Delta \quad c < b, \Gamma \Rightarrow \Delta}{a < b, \Gamma \Rightarrow \Delta}\ {}_{\mathit{split}}$$

The axiom $0 < 1$ becomes the rule

$$\frac{0 < 1, \Gamma \Rightarrow \Delta}{\Gamma \Rightarrow \Delta}\ {}_{\mathit{nondeg}}$$

multiplication in the form of inference rules we first have to formulate the axioms for defined equality in terms of the primitive notion of strict linear order. In each axiom involving equality, the atom $a = b$ is replaced by $\sim (a < b \vee b < a)$ which is in turn replaced by its logical equivalent $\sim a < b \,\&\, \sim b < a$. The latter form of axiom is rendered into two inference rules with zero premisses, namely $a < b, \Gamma \Rightarrow \Delta$ and $b < a, \Gamma \Rightarrow \Delta$. Following the same general method we obtain:

II Rules for additive group:

$$a + b < b + a, \Gamma \Rightarrow \Delta \ {}_{+\text{-}\mathit{comm1}} \qquad b + a < a + b, \Gamma \Rightarrow \Delta \ {}_{+\text{-}\mathit{comm2}}$$

$$(a+b)+c < a+(b+c), \Gamma \Rightarrow \Delta \ {}_{+\text{-}\mathit{ass1}} \qquad a+(b+c) < (a+b)+c, \Gamma \Rightarrow \Delta \ {}_{+\text{-}\mathit{ass2}}$$

$$a + 0 < a, \Gamma \Rightarrow \Delta \ {}_{\mathit{zero1}} \qquad a < a + 0, \Gamma \Rightarrow \Delta \ {}_{\mathit{zero2}}$$

$$a + (-a) < 0, \Gamma \Rightarrow \Delta \ {}_{\mathit{opp1}} \qquad 0 < a + (-a), \Gamma \Rightarrow \Delta \ {}_{\mathit{opp2}}$$

$$\frac{b < c, \Gamma \Rightarrow \Delta}{a + b < a + c, \Gamma \Rightarrow \Delta} \ {\scriptstyle +\text{-}ext}$$

III Rules for multiplication:

$$ab < ba, \Gamma \Rightarrow \Delta \ {\scriptstyle \times\text{-}comm1} \qquad ba < ab, \Gamma \Rightarrow \Delta \ {\scriptstyle \times\text{-}comm2}$$

$$(ab)c < a(bc), \Gamma \Rightarrow \Delta \ {\scriptstyle \times\text{-}ass1} \qquad a(bc) < (ab)c, \Gamma \Rightarrow \Delta \ {\scriptstyle \times\text{-}ass2}$$

$$a1 < a, \Gamma \Rightarrow \Delta \ {\scriptstyle unit1} \qquad a < a1, \Gamma \Rightarrow \Delta \ {\scriptstyle unit2}$$

$$aa^{-1} < 1, \Gamma \Rightarrow \Delta \ {\scriptstyle inv1} \qquad 1 < aa^{-1}, \Gamma \Rightarrow \Delta \ {\scriptstyle inv2}$$

$$a(b+c) < ab + ac, \Gamma \Rightarrow \Delta \ {\scriptstyle distr1} \qquad ab + ac < a(b+c), \Gamma \Rightarrow \Delta \ {\scriptstyle distr1}$$

Observe that by symmetry, the rules $+$-*comm2* and $\times$-*comm2* can be dropped.

The axiom of strong extensionality for multiplication gives rise, by using the regular decomposition described in the previous section, to the following four rules:

$$\frac{a < 0, \Gamma \Rightarrow \Delta \quad 0 < a, \Gamma \Rightarrow \Delta}{ab < ac, \Gamma \Rightarrow \Delta} \ {\scriptstyle \times\text{-}ext1} \qquad \frac{c < b, \Gamma \Rightarrow \Delta \quad b < c, \Gamma \Rightarrow \Delta}{ab < ac, \Gamma \Rightarrow \Delta} \ {\scriptstyle \times\text{-}ext2}$$

$$\frac{a < 0, \Gamma \Rightarrow \Delta \quad b < c, \Gamma \Rightarrow \Delta}{ab < ac, \Gamma \Rightarrow \Delta} \ {\scriptstyle \times\text{-}ext3} \qquad \frac{c < b, \Gamma \Rightarrow \Delta \quad 0 < a, \Gamma \Rightarrow \Delta}{ab < ac, \Gamma \Rightarrow \Delta} \ {\scriptstyle \times\text{-}ext4}$$

It follows as in Negri and von Plato (1998, pp. 421–422) that these four rules above can be reduced to the single rule

$$\frac{a < 0, c < b, \Gamma \Rightarrow \Delta \quad 0 < a, b < c, \Gamma \Rightarrow \Delta}{ab < ac, \Gamma \Rightarrow \Delta} \ {\scriptstyle \times\text{-}ext}$$

In the rules for the inverse the condition $a \neq 0$ is required. We notice that substitutions in the atomic formulas may produce duplications only in the rule containing two principal formulas in the conclusion, namely asymmetry. In this case the closure condition requires the rule of irreflexivity. Since all of our rules follow the rule-scheme, we conclude from theorem 3.1 the

Corollary 4.1: *The structural rules of weakening, contraction and cut are admissible in the sequent calculus for constructive ordered fields. Weakening and contraction are admissible and height preserving.*

It follows that our system is *complete*, in the sense that if A is derivable in the Hilbert-style axiomatization given in section 2, then $\Rightarrow A$ is derivable in the sequent calculus system.

With logical systems, proofs of cut elimination have as a trivial consequence a proof of consistency for the system. With the extension by nonlogical rules, the situation is different. Here a purely proof-theoretic consistency proof is complicated by the rule *nondeg*: In a root-first analysis of derivations, the rule *nondeg* introduces atoms in the antecedents of sequents. All the other nonlogical rules have atoms in the antecedent of their conclusion and therefore, in absence of *nondeg*, cannot produce a derivation of $\Rightarrow\perp$.

4. CONCLUDING REMARKS

We have found a cut-free sequent calculus for the theory of constructive ordered fields. Therefore the methods of proof analysis which are typical of structural proof theory become applicable. For pure logical systems standard applications of cut elimination include consistency proofs as direct consequences of the subformula property. In systems with nonlogical axioms as those studied here the subformula property does not hold, but only a weaker form of the subformula property. Nevertheless the control of the structure of derivations in a cut-free system permits to answer non-trivial questions of conservativity. For instance, the methods developed in Negri (1999) for showing conservativity of apartness over equality can be applied also in the context of constructive ordered fields. By these methods it is possible to show that the theory of constructive ordered fields is conservative over the theory of ordered fields based on a relation of weak linear order, defined as the negation of strict linear order.

Our method of treating constructive ordered fields can be applied also to *real closed fields* (see Delzell 1996 for references on the literature on real closed fields). The axioms asserting the existence of square roots of positive elements and the existence of zeros of polynomials of odd degree will be first formulated in terms of two constructions, $sqr(a)$ and $z(a_0, \ldots, a_{2n+1})$ satisfying the quantifier-free axioms

$$\begin{aligned} sqr(a)sqr(a) &= a, \quad \text{for } 0 \leqslant a \\ \textstyle\sum_{i=0}^{2n+1} a_i z(a_0, \ldots, a_{2n+1})^i &= 0, \quad \text{for } a_{2n+1} \neq 0 \end{aligned}$$

By adding the two constructions and the respective axioms to the theory of constructive ordered fields, we obtain a quantifier-free axiomatization for real closed fields, with all axioms conforming to the rule-scheme. Consequently, we obtain a sequent calculus for real closed fields with all structural rules admissible.

In the standard theory of real closed fields, the law of excluded middle is assumed for atoms. By a result of Tarski (see Delzell 1996), the law holds for arbitrary formulas. The assumption means that atoms are decidable, but this has constructive sense only if a field is discrete. In our proof-theoretical approach, decidability for atoms can be included in the cut-free system by adding the rule of *Gentzen excluded middle for atoms*

$$\frac{a < b, \Gamma \Rightarrow \Delta \qquad \sim a < b, \Gamma \Rightarrow \Delta}{\Gamma \Rightarrow \Delta}\ {\scriptstyle Gem\text{-}at}$$

By arguing as in von Plato (1998a) it can be shown that the corresponding excluded middle rule for arbitrary quantifier-free formulas is admissible in the theory of real closed fields with decidable atoms, thus showing that, at least for the quantifier-free fragment, the theory behaves like a classical theory. Palmgren (1998) proves the decidability for all formulas using Tarski quantifier elimination and the Dragalin-Friedman translation.

References

Bridges, D.S. (1999) Constructive mathematics: a foundation for computable analysis, Theoretical Computer Science, vol. 219, pp. 95–109.

Delzell, C. N. (1996) Kreisel's unwinding of Artin's proof, in P. Odifreddi (ed.), *Kreiseliana: about and around Georg Kreisel.* Wellesley, A.K. Peters, pp. 113–246.

Dragalin, A. (1988) *Mathematical Intuitionism: Introduction to Proof Theory*, American Mathematical Society, Providence, Rhode Island.

Negri, S. (1999) Sequent calculus proof theory of intuitionistic apartness and order relations, *Archive for Mathematical Logic*, vol. 38, pp. 521–547.

Negri, S. and J. von Plato (1998) Cut elimination in the presence of axioms, *The Bulletin of Symbolic Logic*, vol. 4, pp. 418–435.

Negri, S. and J. von Plato (2001) *Structural Proof Theory*, Cambridge University Press.

Negri, S. and D. Soravia (1999) The continuum as a formal space, *Archive for Mathematical Logic*, vol. 38, pp. 423–447.

Palmgren, E. (1998) A note on constructive real closed fields, 3 pages. Available at `http://www.math.uu.se/ palmgren/publicat.html`.

von Plato, J. (1998) A structural proof theory of geometry, manuscript.

von Plato, J. (1998a) Proof theory of full classical propositional logic, manuscript.

Troelstra, A. and H. Schwichtenberg (1996) *Basic Proof Theory*, Cambridge University Press.
Truss, J. (1997) *Foundations of Mathematical Analysis*, Oxford University Press.

THE PURITZ ORDER AND ITS RELATIONSHIP TO THE RUDIN-KEISLER ORDER

Siu-Ah Ng*
School of Mathematics, University of Natal
Pietermaritzburg, 3209 South Africa
ngs@nu.ac.za

Hermann Render
Fachbereich Mathematik, Gerhard-Mercator-Universität
47048 Duisburg, Germany
render@math.uni-duisburg.de

Keywords: Constellation, Sky, Rudin-Keisler order, Countable compactness

Abstract: We show that there is an intimate connection between the Rudin-Keisler order on the Stone-Čech compactification βX of the discrete space X and the so-called Puritz order on a nonstandard model ${}^*X$. Furthermore it is shown that the S-topology of an ultrapower model with respect to a free ultrafilter over the natural numbers is pseudocompact but not countably compact.

1. INTRODUCTION

Let ${}^*X$ be a nonstandard model of the set X. It is a well known fact that ${}^*X$ resembles the *Stone-Čech compactification* βX of the discrete space X in many ways. In our context βX is defined as the space of all ultrafilters on X. Indeed, it is a matter of fact that βX can be constructed by nonstandard methods in various ways, provided that ${}^*X$ is an *enlargement* in the sense that standard subsets of ${}^*X$ have the finite intersection property, see [17], [12], [6], [3], [4] and [16]. For a

*Supported by South Africa NRF 2039556.

P. Schuster, U. Berger and H. Osswald (eds.),
Reuniting the Antipodes – Constructive and Nonstandard Views of the Continuum, 157–166.

discussion of common features of βX and $^*X$ in descriptive set theory we refer to [15] and [7] and the references given there.

One aim of this paper is to discuss the relationship between the Rudin-Keisler order on βX and an order $\leq_{\mathrm{P}}$ on a given nonstandard model $^*X$ which we will call according to the work in [13] the *Puritz order*: define $x \leq_{\mathrm{P}} y$ for $x, y \in {}^*X$ to mean that there exists a function $f : X \to X$ with $x = {}^*f(y)$. Formally analogous, the *Rudin-Keisler order* $\leq_{\mathrm{RK}}$ on βX is defined in the following way: for $p, q \in \beta X$ write $p \leq_{\mathrm{RK}} q$ if there exists a function $f : X \to X$ such that $p = \overline{f}(q)$, where $\overline{f}(q)$ denotes the ultrafilter $\{A \subset X \mid f^{-1}[A] \in q\}$ or equivalently, the filter generated by $\{f(A) | A \in q\}$. (We define $\overline{f}$ similarly for the more general case $f : I \to X$.) From the vast literature about the Rudin-Keisler order we mention only [1], [2], [5], [8], [10], [11]. For any order $\leq$ (i.e. a reflexive and transitive relation, also called pre-order) one can introduce an equivalence relation $\approx_{\leq}$ by defining $p \approx_{\leq} q$ iff $p \leq q$ and $q \leq p$. The $\approx_{\leq}$-equivalence class of p is called the *type of* p and will be denoted by $T_{\leq}(p)$. A point p is called $\leq$-*minimal* if $q \leq p$ implies $q \approx_{\leq} p$. For the type $T_{\leq_{RK}}(p)$ of $p \in \beta X$ with respect to the Rudin-Keisler order we write simply $T_{RK}(p)$. We also write $\approx_{\mathrm{RK}}$ for $\approx_{\leq_{\mathrm{RK}}}$. A type of $N \in {}^*X$ with respect to the Puritz order is called a *constellation* according to [14]. We shall use in this paper the more suggestive notion of the *Puritz type* of $N \in {}^*X$. Further we write $x \approx_{\mathrm{P}} y$ if $x \leq_{\mathrm{P}} y$ and $y \leq_{\mathrm{P}} x$. The set ${}^*X_\infty := {}^*X \setminus \{{}^*x \mid x \in X\}$ is called the set of all infinitely large elements. We call $^*X$ *proper* when ${}^*X_\infty \neq \emptyset$.

The formal similarity between βX and $^*X$ can be converted into a precise statement by considering the following map

$$\mathbf{u} : {}^*X \to \beta X, \quad x \mapsto \mathbf{u}_x := \{A \subset X \mid x \in {}^*A\} \tag{14.1}$$

where $\mathbf{u}_x$ is clearly an ultrafilter over X. In Section 2 we show that this map is continuous and open (with respect to the relative topology on the image) and that this map is order preserving, i.e.

$$x \leq_{\mathrm{P}} y \text{ implies } \mathbf{u}_x \leq_{\mathrm{RK}} \mathbf{u}_y. \tag{14.2}$$

Already A. Robinson has observed in [17] that for an enlargement $^*X$ the map $\mathbf{u}$ is always surjective. We show that the map $\mathbf{u}$ is surjective if and only if the S-topology on $^*X$ is compact. The situation is different if we consider nonstandard models with weaker saturation assumptions. Then the map $\mathbf{u}$ is not surjective and the description of the image depends strongly on the nonstandard models. In this context we discuss *ultrapowers*: assume that q be a free ultrafilter over an (index) set I and define the *ultrapower of* X with respect to q by

$$^*X_q := X^I/q \tag{14.3}$$

where X^I is the set of all functions $f: I \to X$ and functions $f, g: I \to X$ are identified by the following equivalence relation: define $f \sim_q g$ if $\{i \in I \mid f(i) = g(i)\} \in q$. By

$$[f] = \{g \in X^I \mid f \sim_q g\} \tag{14.4}$$

we define the equivalence class of the element $f \in X^I$. C. Puritz has shown that for $X = I = \mathbb{N}$ there exists exactly one type in $^*X_\infty$ if and only if the ultrafilter q is selective. We generalize this result for the case that $X = I$ in the context of uniform ultrafilters. Further in Section 3 (Theorem 8) we give a characterization of the injectivity of $\mathbf{u}$ for the case when $^*X$ is an ultrapower.

In Section 4 we discuss compactness properties of the S-topology of nonstandard models with weak saturation properties. For ultrapowers over a countable index set I simple cardinality considerations show that the map $\mathbf{u}$ is not surjective, hence $^*X_q$ is not compact with respect to the S-topology. Therefore the question arises whether weaker compactness properties are satisfied. We show that a proper nonstandard model $^*X$ is pseudocompact. More surprisingly is the fact that ultrapowers over a countable index set are not countably compact with respect to the S-topology.

Finally we recall some definitions. The space βX is equipped with the usual topology generated by the sets $\widehat{A} := \{p \in \beta X \mid A \in p\}$ with $A \subset X$ arbitrary. Let now $^*X$ be a nonstandard model of X. Then the topology generated by $\{^*A | A \subset X\}$ is called the *S-topology*. Further the "copy" of X in $^*X$ is denoted by $^\sigma X := \{^*x \mid x \in X\}$.

Acknowledgments

The authors wish to thank W. Rinkewitz for very helpful comments. In particular, he showed us the proof of Theorem 2 (b).

2. PROPERTIES OF THE MAP u.

In this section we collect some results which might be part of mathematical folklore. In order to hold the paper self-contained we include the proofs. In this section we only assume that $^*X$ is a nonstandard model of a given set X. In the next section we shall discuss more specialized models.

Lemma 1 (a) *Let $f : X \to X$ be a function and $y \in {}^*X$. Then*

$$\overline{f}(\mathbf{u}_y) = \mathbf{u}_{{}^*f(y)}.$$

(b) *For all $x, y \in {}^*X$: $\mathbf{u}_x \leq_{\mathrm{RK}} \mathbf{u}_y$ iff there exists a function $f : X \to X$ such that $\mathbf{u}_x = \mathbf{u}_{{}^*f(y)}$.*

(c) *$x \leq_{\mathrm{P}} y$ implies $\mathbf{u}_x \leq_{\mathrm{RK}} \mathbf{u}_y$.*

(d) *$x \approx_{\mathrm{P}} y$ implies $\mathbf{u}_x \approx_{\mathrm{RK}} \mathbf{u}_y$.*

Proof: (a) We have

$$\begin{aligned} \overline{f}(\mathbf{u}_y) \quad &= \{A \subset X \mid f^{-1}[A] \in \mathbf{u}_y\} \\ &= \{A \subset X \mid y \in {}^*(f^{-1}[A])\} \\ &= \{A \subset X \mid y \in {}^*f^{-1}[{}^*A]\} \\ &= \{A \subset X \mid {}^*f(y) \in {}^*A\} \\ &= \mathbf{u}_{{}^*f(y)}. \end{aligned}$$

(b) Immediate consequence of (a) and the definition of $\leq_{\mathrm{RK}}$.
(c) Obvious, by (b) and the definition of $\leq_{\mathrm{P}}$.
(d) Immediate from (c). □

Recall that a space is *completely regular* if for a closed subset F and $x \notin F$, there is a continuous function $f : X \to \mathbb{R}$ such that $f \equiv 1$ on F and $f(x) = 0$. It is well-known that the S-topology on $^*X$ is completely regular.

Theorem 2 *The following statements are equivalent:*

(a) *$\mathbf{u}$ is injective.*

(b) *For all $x, y \in {}^*X$: $x \leq_{\mathrm{P}} y$ iff $\mathbf{u}_x \leq_{\mathrm{RK}} \mathbf{u}_y$.*

(c) *The S-topology is Hausdorff.*

Proof: (a) $\Rightarrow$ (b). Let $x, y \in {}^*X$. Since $\mathbf{u}$ is injective, the implication $\mathbf{u}_x \leq_{\mathrm{RK}} \mathbf{u}_y \Rightarrow x \leq_{\mathrm{P}} y$ is an obvious consequence of Lemma 1 (b). According to Lemma 1 (c), the reverse implication holds always.

(b) $\Rightarrow$ (a). Fix $x, y \in {}^*X$ with $\mathbf{u}_x = \mathbf{u}_y$. In particular, $\mathbf{u}_x \leq_{\mathrm{RK}} \mathbf{u}_y$. So by (b), $x \leq_{\mathrm{P}} y$ holds. Hence there exists a function $f : X \to X$ with $x = {}^*f(y)$. Then $\mathbf{u}_{{}^*f(y)} = \mathbf{u}_x = \mathbf{u}_y$. By Lemma 1 (a), we have therefore $\mathbf{u}_y = \overline{f}(\mathbf{u}_y)$. Now, Theorem 9.2 (a) in [5] gives us $\{f = id\} \in \mathbf{u}_y$. Hence $y \in {}^*\{f = id\}$. This shows $y = {}^*f(y) = x$.

(a) $\Leftrightarrow$ (c). Let $x \neq y$. If the S-topology is Hausdorff then there exist disjoint sets A, B in X such that $x \in {}^*A$ and $y \in {}^*B$. Since ${}^*A \cap {}^*B = \emptyset$ we have $\mathbf{u}_x \neq \mathbf{u}_y$. Conversely, if $\mathbf{u}_x \neq \mathbf{u}_y$ then there exists $A \in \mathbf{u}_x \setminus \mathbf{u}_y$. Then $x \in {}^*A$ and $y \in {}^*(X \setminus A)$. □

Lemma 3 *Let $x, y \in {}^*X$. If $x \approx_{\mathrm{P}} y$ then there exists a bijection $G : X \to X$ with $y = {}^*G(x)$.*

Proof: Assume that $x \approx_{\mathrm{P}} y$. In the first case assume that $x \in {}^\sigma X$. Then there exists a function $f : X \to X$ with $y = {}^*f({}^*x)$. Hence $y \in {}^\sigma X$. But it is clear that there exists a bijective function $F : X \to X$ with $F(x) = y$. In the second case we assume that $x \in {}^*X_\infty$. Let $f : X \to X$ and $g : X \to X$ be functions such that $x = {}^*f(y)$ and $y = {}^*g(x)$. Then $x = {}^*f(y) = {}^*f({}^*g(x)) = {}^*h(x)$ with $h := f \circ g$. Set $A := \{s \in X \mid h(s) = s\}$. Then $x \in {}^*A$ and clearly g is injective on A. Since $x \in {}^*X_\infty$ the set A must be infinite. Decompose A into two disjoint sets A_0 and A_1 which have the same cardinality as A. Without loss of generality we can assume that $x \in {}^*A_0$. It is easy to see that there exists a bijective extension $G : X \to X$ of $g|A_0$. Since $x \in {}^*A_0$ it follows that $y = {}^*G(x)$. □

Lemma 4 *$\mathbf{u} : {}^*X \to \beta X$ is a continuous open map and the image $\mathbf{u}({}^*X)$ is dense in βX.*

Proof: Note that $\mathbf{u}_x \in \widehat{A}$ iff $A \in \mathbf{u}_x$ iff $x \in {}^*A$. It follows that $\mathbf{u}^{-1}(\widehat{A}) = \{x \in {}^*X \mid \mathbf{u}_x \in \widehat{A}\} = {}^*A$ and $\mathbf{u}({}^*A) = \{\mathbf{u}_x \mid x \in {}^*A\} = \widehat{A} \cap \mathbf{u}({}^*X)$ showing that $\mathbf{u}$ is continuous and open. Let us prove that the image is dense: Let $q \in \beta X$ be arbitrary and $\widehat{A}$ with $A \subset X$ be a neighborhood of q. Let $x \in {}^*A$. Then $\mathbf{u}_x \in \widehat{A}$ and the proof is complete. □

By purely elementary topological arguments we can show that

Proposition 5 *The map $\mathbf{u}$ is surjective if and only if ${}^*X$ is compact in the S-topology.*

Proof: Assume that $\mathbf{u}$ is surjective. Let $(O_j)_{j \in J}$ be an open covering of ${}^*X$ in the S-topology. One can assume that O_j are standard sets, i.e. there exist subsets A_j of X such that $O_j = {}^*A_j$. Then $\mathbf{u}({}^*A_j) = \widehat{A_j}$ is open. By compactness of βX there exists a finite subcovering, say $\widehat{A_{j1}}, ..., \widehat{A_{jn}}$. Furthermore we have $\mathbf{u}^{-1}(\mathbf{u}({}^*A_j)) = \mathbf{u}^{-1}(\widehat{A_j}) = {}^*A_j$. Hence ${}^*A_{j1}, ..., {}^*A_{jn}$ is a finite subcovering of ${}^*X$. For the converse assume that ${}^*X$ is compact. Then the image $\mathbf{u}({}^*X)$ is a dense, compact and therefore closed subset of the Hausdorff space βX. Hence $\mathbf{u}({}^*X)$ is equal to βX. □

3. THE PURITZ ORDER ON ULTRAPOWERS

Let I be a set. A filter q over I is said to be *uniform* if $|A| = |I|$ for all $A \in q$. A uniform ultrafilter q is *selective* if for each function $f : I \to I$ there exists $U \in q$ such that $f|U$ is either injective or constant.

Proposition 6 *Let q be a free ultrafilter over the set I and let X be a set. Then $\mathbf{u}({}^*X_q) = \{\overline{f}(q) \mid f : I \to X$ a function $\}$. For $I = X$ we have $\mathbf{u}({}^*X_q) = \{p \in \beta X \mid p \leq_{\text{RK}} q\}$.*

Proof: Let $x \in {}^*X_q$ and let $f : I \to X$ be a representation of the equivalence class $x \in {}^*X_q$. Then

$$\begin{aligned} \mathbf{u}_x &= \{A \subset X \mid x \in {}^*A\} \\ &= \Big\{A \subset X \mid \{i \in I \mid f(i) \in A\} \in q\Big\} \\ &= \{A \subset X \mid f^{-1}[A] \in q\} \\ &= \overline{f}(q). \end{aligned}$$

□

The following result generalizes Theorem 4.2 (c) in [13]. See [18] for the related notion of $P-$points.

Theorem 7 *Let q be a uniform ultrafilter over X. Then for ${}^*X := {}^*X_q = X^X/q$ the following statements are equivalent:*

(a) *${}^*X_\infty$ consists of exactly one Puritz type (=constellation);*

(b) *for all $x, y \in {}^*X_\infty$ there exists a function $f : X \to X$ with $x = {}^*f(y)$;*

(c) *for all $x, y \in {}^*X_\infty$ there exists a bijection $f : X \to X$ with $x = {}^*f(y)$;*

(d) *q is RK-minimal in $\beta X \setminus X$;*

(e) *q is selective.*

Proof: Clearly (a) and (b) are equivalent and by Lemma 3, (b) and (c) are equivalent. For (b) $\Rightarrow$ (d), let $p \in \beta X \setminus X$ with $p \leq_{\mathrm{RK}} q$. Then there exists $N \in {}^*X_\infty$ with $p = \mathbf{u}_N$ and $M \in {}^*X_\infty$ with $q = \mathbf{u}_M$. Then $N \approx_{\mathrm{P}} M$ by assumption (b) and we have $\mathbf{u}_N \approx_{\mathrm{RK}} \mathbf{u}_M$ by Lemma 1. Hence $p \approx_{\mathrm{RK}} q$ and q is RK-minimal.

The implication (d) $\Rightarrow$ (e) follows from Theorem 9.6. in [5].

For (e) $\Rightarrow$ (a), it suffices to show that $N \approx_{\mathrm{P}} [\mathrm{id}]$ for each $N \in {}^*X_\infty$. Let $h : X \to X$ be a function such that $N = [h]$. Since q is selective there exists $U \in q$ such that $h|U$ is either injective or constant. If $h|U$ is constant then $[h] \in X$, a contradiction to $N \in {}^*X_\infty$. If h is injective on $U \in q$, it is easy to see that there exists $A_0 \in q$ such that $h|A_0$ possesses a bijective extension $H : X \to X$ (cf. the proof of Lemma 3). Let $f := H^{-1}$. Then ${}^*f[h] = [\mathrm{id}]$, since $\{i \in I \mid f(h(i)) = i\} \supset A_0 \in q$. □

Theorem 8 *Let q be an ultrafilter over X. Then for ${}^*X := {}^*X_q = X^X/q$ the following statements are equivalent for the mapping $\mathbf{u} : {}^*X \to \beta X$:*

(a) *$\mathbf{u}$ is injective.*

(b) *$\mathbf{u}_x = \mathbf{u}_y$ implies $x \approx_{\mathrm{P}} y$ for all $x, y \in {}^*X_\infty$.*

(c) *$x \leq_{\mathrm{P}} y \Leftrightarrow \mathbf{u}_x \leq_{\mathrm{RK}} \mathbf{u}_y$ for all $x, y \in {}^*X_\infty$.*

(d) *$\overline{g_1}(q) = \overline{g_2}(q)$ implies $[g_1] = [g_2]$ for all functions $g_1, g_2 : X \to X$.*

Proof: The equivalence of (a) and (c) follows from Theorem 2. (c) $\Rightarrow$ (b) is clear. The proof of (b) $\Rightarrow$ (a) is the same as the proof of (b) $\Rightarrow$ (a) in Theorem 2.

For (b)$\Rightarrow$(d) let $g_1, g_2 : X \to X$ be functions as in (d) and $x := [g_1]$ and $y := [g_2]$ be the corresponding equivalence class in the nonstandard model. By Proposition 6 we have $\mathbf{u}_x = \overline{g_1}(q)$ and $\mathbf{u}_y = \overline{g_2}(q)$. By (b) there exists a function $F : X \to X$ with $x = {}^*F(y)$. Hence $A_0 := \{s \in X : g_1(s) = F(g_2(s))\} \in q$. It is now easily checked that $\overline{F}(\overline{g_1}(q)) = \overline{g_1}(q)$ using our assumption $\overline{g_1}(q) = \overline{g_2}(q)$. Hence $\overline{F}(\overline{g_2}(q)) = \overline{g_2}(q)$. Then Theorem 9.2 in [5] implies that $\{s \in X : F(s) = s\} \in \overline{g_2}(q)$. So $A_1 := g_2^{-1}(\{s \in X \mid F(s) = s\}) \in q$. It follows that $A_0 \cap A_1 \subset \{s \in X \mid g_1(s) = g_2(s)\}$ is contained in q proving condition (d).

The implication (d) $\Rightarrow$ (a) is easy: $\mathbf{u}_x = \mathbf{u}_y$ implies $\overline{g_1}(q) = \overline{g_2}(q)$ (notation as above) and (d) means that $x = [g_1] = [g_2] = y$. □

4. COMPACTNESS PROPERTIES OF THE S-TOPOLOGY

A topological space X is *countably compact* if each sequence $(x_n)_{n\in\mathbb{N}}$ of X has a cluster point. Alternatively, a space X is countably compact if and only if the system $\mathcal{F}(X)$ of all closed subsets has the *countable compactness property*: if $(A_n)_{n\in\mathbb{N}}$ is a sequence of descending non-empty sets in $\mathcal{F}(X)$ then the intersection $\bigcap_{n=1}^{\infty} A_n$ is non-empty. Note that under ω_1-saturation the set $\mathcal{I}({}^*X)$ of all internal subsets of ${}^*X$ has the countable compactness property. On the other hand $\mathcal{I}({}^*X)$ contains only a base of the S-topology. We will show in this section that the S-topology of models of the form $\mathbb{N}^{\mathbb{N}}/q$ is not countably compact! Our proof uses arguments given by Frolik for showing that Rudin-Keisler types are not countably compact subspaces of βX.

Nonetheless, a weaker form of compactness can be proved as the first result shows. Recall that a topological space X is *pseudocompact* if each continuous function $f : X \to \mathbb{R}$ is bounded.

Proposition 9 *Let ${}^*X$ be a proper nonstandard model. Then ${}^*X$ is pseudocompact with respect to the S-topology.*

Proof: Suppose that there exists a continuous unbounded function $\phi : {}^*X \to \mathbb{R}$. Then $U_n := \{x \in {}^*X \mid |\phi(x)| > n\}$ is a non-empty open set. Since X is dense in ${}^*X$ with respect to the S-topology there exists $x_n \in X$ with ${}^*x_n \in U_n$, i.e., that $|\phi({}^*x_n)| > n$ for all $n \in \mathbb{N}$. Let $N \in {}^*\mathbb{N}_\infty$ be fixed. By continuity of ϕ, there exists $O \subset X$ with $x_N \in {}^*O$ such that $\phi({}^*O)$ is bounded, say by C. Let $F_m := \{x_n \mid n \geq m\}$. Then $x_N \in {}^*F_m \cap {}^*O$ for all $m \in \mathbb{N}$. By transfer we have $F_m \cap O \neq \emptyset$ for all $m \in \mathbb{N}$. Choose $x_m \in F_m \cap O$ for some $m > C$. Then ${}^*x_m \in {}^*O$ by transfer and therefore $|\phi({}^*x_m)| \leq C$. On the other hand we know that ${}^*x_m \in U_m$, hence $|\phi({}^*x_m)| > m > C$, a contradiction. □

Theorem 10 *Let ${}^*\mathbb{N} := \mathbb{N}^{\mathbb{N}}/q$ be an ultrapower with respect to a free ultrafilter over $\mathbb{N}$. Then the S-topology is not countably compact.*

Proof: Suppose that ${}^*\mathbb{N}$ be countably compact. Then the continuous image is countably compact as well. It follows that

$$T := \{p \in \beta\mathbb{N} \mid p \leq_{\mathrm{RK}} q\} \supset T_{RK}(q) \tag{14.5}$$

is countably compact. It is well known that $T_{RK}(q)$ is an infinite set. Hence there exists a sequence $(x_n)_{n\in\mathbb{N}}$ in $T_{RK}(q)$ which is discrete in the Hausdorff space $\beta\mathbb{N}$ (cf. [9], p. 5), i.e. that there exist pairwise disjoint open subsets U_n of $\beta\mathbb{N}$ with $x_n \in U_n$. By the definition of the topology in $\beta\mathbb{N}$ we can assume that U_n is of the form

$$U_n = \{p \in \beta\mathbb{N} \mid A_n \in p\} \tag{14.6}$$

where A_n are pairwise disjoint subsets of $\mathbb{N}$ (since U_n are pairwise disjoint). Without loss of generality one can assume that $\cup_{n=1}^{\infty} A_n = \mathbb{N}$. By countably compactness of T there exists an adherent point $x \in T$ of the sequence $(x_n)_{n\in\mathbb{N}}$. Since $x \in T$ we have $x \leq_{RK} q$, and since $x_n \in T_{RK}(q)$ we infer $x \leq_{RK} x_n$. Hence there exist functions $h_n\colon \mathbb{N} \to \mathbb{N}$ with $x = \overline{h_n}(x_n)$. Define a function $H\colon \mathbb{N} \to \mathbb{N}$ by $H(t) = h_n(t)$ for $t \in A_n$. We claim that $\overline{H}(x) = x$: clearly $x_n \in U_n$ implies $A_n \in x_n$. Now let $A \in x_n$. Then $A \cap A_n \in x_n$. Hence

$$H(A) \supset H(A \cap A_n) = h_n(A \cap A_n) \in x \tag{14.7}$$

since $\overline{h_n}(x_n) = x$. It follows that $\overline{H}(x_n) \subset x$. Since $\overline{H}(x_n)$ is an ultrafilter we infer $\overline{H}(x_n) = x$. The continuity of $\overline{H}$ now implies that $\overline{H}(x) = x$. By [[5], Theorem 9.2 (a)], it follows that

$$V := \{p \in \beta\mathbb{N} \mid \overline{H}(p) = p\} = \Big\{p \in \beta\mathbb{N} \mid \{H = id\} \in p\Big\} = \{\widehat{H = id}\}. \tag{14.8}$$

Since V is open, $x \in V$ and x is an adherent point of $(x_n)_n$ it follows that $x_n \in V$ for infinitely many indices n. Hence $x_n = \overline{H}(x_n) = x$ for infinitely many indices n which is a contradiction. □

References

[1] A. Blass, *The Rudin-Keisler ordering of P-points*, Trans. Amer. Math. Soc. 179 (1973), 145–166.

[2] A. Blass, *Selective ultrafilters and homogeneity*, Annals of Pure and Applied Logic 38 (1988), 215–255.

[3] M. Canjar, *Countable ultraproducts without CH*, Annals of Pure and Applied Logic 37 (1988), 1-79.

[4] G. Cherlin and J. Hirshfeld, *Ultrafilters and Ultraproducts in Non-Standard Analysis*, in "Contributions to Non-Standard Analysis",

ed. W.A.J. Luxemburg and A. Robinson, North Holland 1972, 261-279.

[5] W.W. Comfort and S. Negrepontis, *The Theory of Ultrafilters*, Die Grundlehren der mathematischen Wissenschaften 211, Springer, Berlin 1974.

[6] J.C. Dyre, *Nonstandard characterizations of ideals in C(X)*, Math. Scand. 50 (1982), 44–54.

[7] W. Eck, *Über Ultrafilter und Nichtstandardmodelle mit vorgeschriebener Verteilung der Konstellationen*, Dissertation 1976, Freie Universität Berlin.

[8] S. Garcia-Ferreira, *Three orderings on* $\beta(\omega)\setminus\omega^*$, Topology and its Applications 50 (1993), 199–216.

[9] L. Gillman and M. Jerison, *Rings of Continuous Functions,* Van Nostrand, Princeton 1960.

[10] J. Ginsburg and V. Saks, *Some applications of ultrafilters in topology*, Pacific Journal Math. 57 (1975), 403–418.

[11] N. Hindman, D. Strauss, *Algebra in the Stone-Čech compactification*, Walter de Gruyter, Berlin 1998.

[12] W.A.J. Luxemburg, *A general theory of monads,* in "Applications of Model theory to Algebra, Analysis and Probability". Holt, Rinehart and Winston, New York 1969, 18–86.

[13] Ch. Puritz, *Ultrafilters and Standard Functions in Non-Standard Arithmetic*, Proc. London Math. Soc. 22 (1971), 705–733.

[14] Ch. Puritz, *Skies, Constellations and Monads*, in "Contributions to Non-Standard Analysis", ed. W.A.J. Luxemburg and A. Robinson, North Holland 1972, 215–243.

[15] H. Render, *Countably determined sets and a conjecture of C.W. Henson,* Math. Ann. 303 (1995), 307–324.

[16] H. Render, *Topologies on the nonstandard model,* Bolyai Society Mathematical Studies **4**, Topology with Applications, Szekszárd (Hungary) (1993), 413–418.

[17] A. Robinson, *Compactification of groups and rings and nonstandard analysis,* J. Symb. Logic 34 (1969), 576–588.

[18] N.I. Rosen, *P-Points with countably many constellations.* Trans. Amer. Math. Soc. 290 (1985), 585–596.

UNIFYING CONSTRUCTIVE AND NONSTANDARD ANALYSIS

Erik Palmgren *

Department of Mathematics, Uppsala University
P.O. Box 480, SE-751 06 Uppsala, Sweden
palmgren@math.uu.se

Keywords: Constructive mathematics, nonstandard analysis, sheaves.

Abstract: This paper reviews Moerdijk's constructive method for obtaining a nonstandard model of analysis, and exemplifies the use of the model. Two new results about the model, countable saturation and a characterization of internal functions, are established. It is shown that the model supports the construction of Loeb measure, and a weak principle of omniscience (LLPO). For so called internal statements there is a transfer principle to the constructive metatheory.

Introduction

The familiar models for nonstandard analysis are obtained using nonconstructive tools, e.g. the compactness theorem or ultrapowers. However, there are other methods for building models which comply to strict constructivism. One may use a filter of cofinite sets instead of an ultrafilter in the ultrapower construction, with ensuing restrictions on the transfer principle. This approach was, in effect, taken already in the Schmieden–Laugwitz model for infinitesimal calculus [20] from 1956. It has more recently been put forward, in extended form, as a suitable foundation for constructive nonstandard analysis by Martin-Löf [10]. Some subsequent development can be found in [13, 14]. A rich class of constructive models for infinitesimals emerges from topos theory and the synthetic differential geometry of Lawvere; see Kock [6] and Moerdijk and Reyes [12]. These models have in general no strong transfer principle. Moerdijk [11] succeeded in giving a constructive nonstandard model

*Research supported by a grant from the Swedish Natural Science Research Council (NFR).

P. Schuster, U. Berger and H. Osswald (eds.),
Reuniting the Antipodes – Constructive and Nonstandard Views of the Continuum, 167–183.

of arithmetic with a full transfer principle. This opened up the possibility to develop a constructive version of nonstandard analysis where some of the most important tools are available. This was realized to a large extent in [15, 16, 17], by using a slightly generalized version of Moerdijk's original sheaf model. In the present paper we show that an important principle that was absent in these articles — the countable saturation principle — is indeed valid in the sheaf model.

The paper is organized as follows. In Section 1 we review the construction of the nonstandard universe $\mathcal{N}$. Section 2 discusses the internal sets and functions of $\mathcal{N}$. Here we present two new results for $\mathcal{N}$: the $\aleph_1$-saturation principle and a characterization of internal functions between nonstandard versions of standard sets. As a consequence, $\mathcal{N}$ will be a constructive model of LLPO (but not of LPO). We also briefly indicate how to make the Loeb measure construction over hyperfinite sets. Section 3 discusses the relation between nonstandard real numbers and the canonical real numbers of $\mathcal{N}$. In the final section we exemplify the use of the model to prove results in the calculus of several variables, e.g. the Implicit Function Theorem.

1. THE SHEAF MODEL AS A NONSTANDARD UNIVERSE

As is well-known from topos theory [7] the category of sheaves over a (generalized) topological space is a universe of "variable sets" — a Grothendieck topos — which obeys the laws of intuitionistic logic rather than classical logic. We briefly review Moerdijk's sheaf model construction from [15, 16] (see also the seminal paper by Moerdijk [11]). To motivate the construction recall Łos's fundamental theorem for ultrapowers, which states that for any ultrafilter $\mathcal{U}$ on a set I, any L-structure M, and any L-formula $\Phi(x_1, \ldots, x_n)$,

$$M^I/\mathcal{U} \models \Phi(\alpha_1, \ldots, \alpha_n) \Leftrightarrow \{i \in I : M \models \Phi(\alpha_1(i), \ldots, \alpha_n(i))\} \in \mathcal{U}. \tag{15.1}$$

That the filter is an ultrafilter, i.e. maximal among proper filters on I, is crucial for proving the equivalence (15.1) when Φ involves the logical constants $\vee$ and $\neg$. Moerdijk [11] gave a constructive analogue of fundamental theorem (Theorem 1.2 below) by changing the notion of model to a sheaf model over a category of filters. The idea of shifting to a nonstandard semantics occurs also in Martin-Löf [10].

A *filter base* $\mathcal{F} = (F, \{F_i\}_{i \in I})$ consists of a non-void (i.e. inhabited) index set I and a family $\{F_i\}_{i \in I}$ of subsets of an underlying set F, called *base sets*, satisfying the filtering condition: for all $i, j \in I$, there exists $k \in I$ such that $F_k \subseteq F_i \cap F_j$. The filter generated is then $\{S \subseteq F : (\exists i \in I)\, F_i \subseteq S\}$. For constructive reasons, it will be better to work only with the bases of filters. So in the sequel we shall abuse the language and simply call them *filters*. Let $\mathcal{F} = (F, \{F_i\}_{i \in I})$ and $\mathcal{G} = (G, \{G_j\}_{j \in J})$ be filters. A *continuous map* α from $\mathcal{F}$ to $\mathcal{G}$, in symbols $\alpha : \mathcal{F} \to \mathcal{G}$, is a partial function $\alpha : F \to G$ which is totally defined on some base set of $\mathcal{F}$, and satisfies the *continuity* condition, $(\forall j \in J)\, (\exists i \in I)\, \alpha[F_i] \subseteq G_j$. Two such morphisms are equivalent if they agree on some base set of $\mathcal{F}$. The filters together with the continuous maps then form a category $\mathbb{B}$ with terminal object and all pullbacks, see [11] and [15]. For each set A there is a trivial filter $\overline{A} = (A, \{A\})$. In this way the category of sets can be considered as a (full) subcategory of $\mathbb{B}$. Note that the set of morphisms from $\mathcal{F}$ to $\overline{A}$, $\mathrm{Hom}_{\mathbb{B}}(\mathcal{F}, \overline{A})$, can be identified with the reduced power $A^F/\mathcal{F}$. As for any category, its hom-sets give a contravariant functor

$$^*A =_{\mathrm{def}} \mathrm{Hom}_{\mathbb{B}}(-, \overline{A}) : \mathbb{B}^{\mathrm{op}} \to \mathbf{Sets}.$$

We now define a Grothendieck topology K on $\mathbb{B}$ so that $^*A$ becomes a sheaf. Let $\mathcal{G}^{(k)} = (G^{(k)}, \{G_j^{(k)}\}_{j \in J^{(k)}})$, $k = 1, \ldots, n$, and $\mathcal{F} = (F, \{F_i\}_{i \in I})$ be filters. A finite set of continuous maps $\{\beta_k : \mathcal{G}^{(k)} \to \mathcal{F}\}_{k=1}^n$ is called a *K-cover* of $\mathcal{F}$ if for all $j_1 \in J^{(1)}, \ldots, j_n \in J^{(n)}$ there exists $i \in I$ for which

$$\beta_1[G_{j_1}^{(1)}] \cup \cdots \cup \beta_n[G_{j_n}^{(n)}] \supseteq F_i. \tag{15.2}$$

In case $\{\beta : \mathcal{G} \to \mathcal{F}\}$ is a cover, we say that β is a *covering map* (which is the same as an epimorphism in $\mathbb{B}$). The following is a typical example of a cover and is used when verifying the disjunctive case for the constructive fundamental theorem.

Example 1.1 Let $\mathcal{F} = (F, \{F_i\}_{i \in I})$ be a filter. Suppose $S_1 \cup \cdots \cup S_n = F_i$ for some $i \in I$. Then the inclusions

$$\{(\mathcal{F}|S_k) \hookrightarrow \mathcal{F}\}_{k=1}^n$$

form a K-cover of $\mathcal{F}$, where the base sets of $\mathcal{F}|S_k$ are $\{F_i \cap S_k\}_{i \in I}$.

The category of sheaves over the topology $(\mathbb{B}, K)$ — denoted $\mathcal{N}$, for *nonstandard universe* — is a universe containing both standard and nonstandard objects. With this topology each presheaf of the form $^*A$ becomes a sheaf. This sheaf is the *nonstandard version* of A. The sheaf of locally constant functions $\Delta(A)$ is a subsheaf of $^*A$, denoted by $^\sigma A$, which constitute the *standard elements* of $^*A$. In particular, each constant element $^*a = x \mapsto a \in {}^*A(\mathcal{F})$ is standard. For each relation $R \subseteq A_1 \times \cdots \times A_n$ define a subsheaf $^*R$ of $^*A_1 \times \cdots \times {}^*A_n$ by

$$(\alpha_1, \ldots, \alpha_n) \in {}^*R(\mathcal{F}) \Leftrightarrow_{\mathrm{def}} (\exists i \in I)\,(\forall u \in F_i)\,(\alpha_1(u), \ldots, \alpha_n(u)) \in R,$$

where $\mathcal{F} = (F, \{F_i\}_{i \in I})$. We assume that L is a first order language including symbols for all sets, relations, functions and constants of interest to us here (this can be made precise using universes of sets, see [16]). $^*L$ denotes the language where all symbols have been decorated with $*$. For any L-formula Φ, we define its $*$-transform, $^*\Phi$, to be the $^*L$-formula where all symbols have been replaced by their starred counterparts. Thus for instance

$$(\exists y \in {}^*M)(\forall x \in {}^*S)\,{}^*f(x) \,{}^*\!< y$$

is the $*$-transform of $(\exists y \in M)(\forall x \in S) f(x) < y$. A formula which is a $*$-transform of an L-formula is called *internal.* The language $^*L$ can be regarded as a sublanguage of the language $L(\mathcal{N})$ of the topos $\mathcal{N}$. We now use ordinary sheaf semantics to interpret $L(\mathcal{N})$. Corresponding to the fundamental theorem for ultrapowers we have the following.

Theorem 1.2 (Moerdijk) *Let $\Phi(x_1, \ldots, x_n)$ be an L-formula, where $x_1, \ldots, x_n$ vary over $S_1, \ldots, S_n$ respectively. Then for $\alpha_1 \in {}^*S_1(\mathcal{F}), \ldots, \alpha_n \in {}^*S_n(\mathcal{F})$:*

$$\mathcal{F} \Vdash {}^*\Phi(\alpha_1, \ldots, \alpha_n) \quad \textit{iff} \quad (\exists i \in I)\,(\forall u \in F_i)\,\Phi(\alpha_1(u), \ldots, \alpha_n(u)).$$

We list the main principles valid for the model $\mathcal{N}$ but refer to [15, 16] for more background:

(Transfer) For any L-formula Φ: Φ is true iff $^*\Phi$ holds in $\mathcal{N}$. (This is a direct consequence of Theorem 1.2.)

(Idealization) The following is true in $\mathcal{N}$ for any L-formula Φ: If for any standard n and any sequence $a_0, \ldots, a_n \in {}^\sigma S$ there exists $z \in {}^*T$ such that $^*\Phi(x, a_k, z)$ for $k = 0, \ldots, n$, then there is some $z \in {}^*T$ such that for all $y \in {}^\sigma S$: $^*\Phi(x, y, z)$.

(Overspill) The following holds in $\mathcal{N}$ for any L-formula Φ: If $^*\Phi(x, n)$ for all standard $n \in {}^*\mathbb{N}$, then there is some infinite n with $^*\Phi(x, n)$.

(Underspill) The following holds in $\mathcal{N}$ for any L-formula Φ: If ${}^*\Phi(x, n)$ for all infinite $n \in {}^*\mathbb{N}$, then there is some standard n with ${}^*\Phi(x, n)$.

In this paper we establish the $\aleph_1$-*saturation principle*, see Theorem 2.1 below. Thereby all the main principles of nonstandard analysis are available to us. Note, however, that the transfer principle is weaker than the usual since the interpretation of the logical constants is nonstandard in $\mathcal{N}$. As a consequence the standard part map does not take its customary form (see Section 3). Induction and dependent choice is valid in $\mathcal{N}$ for the set of standard natural numbers ${}^\sigma\mathbb{N}$; see [15, 18]. This means that the results of constructive analysis [2, 3] can be reused within the model. To prove results in constructive analysis one need to use the transfer principle. For some examples of this, see Section 4.

2. INTERNAL SETS AND FUNCTIONS

The common notion of a constructive set [2, 3] is that of a type, or preset, equipped with a chosen equivalence relation. Power sets can be handled within type theories by introducing a set Ω of truth values with a distinguished element **true**. The power set of A is then the set of functions $\wp(A) = (A \to \Omega)$, and the membership predicate is defined by $x \epsilon S \Leftrightarrow_{\text{def}} (\text{ev}(S, x) = \textsf{true})$ for $x \in A$ and $S \in \wp(A)$. We require $(\Omega, \textsf{true})$ to satisfy, for any formula $\Phi(x_1, \ldots, x_n, y)$ of the language L,

$$(\forall x_1 \in A_1) \cdots (\forall x_n \in A_n) \qquad (15.3)$$
$$(\exists S \in \wp(B))(\forall y \in B)[y \epsilon S \Leftrightarrow \Phi(x_1, \ldots, x_n, y)].$$

In a topos such a pair $(\Omega, \textsf{true})$ is called a subobject classifier, and L can be taken to be the language of the topos. Within predicative theories, such as Martin-Löf type theory Ω can be taken to be a universe U of types [9], where any two types which are logically equivalent as propositions are identified, and Φ must then be restricted to sets and relations that belong to U. (It is however possible to introduce tiers of truth values as in Russell's type theory.)

The nonstandard version ${}^*A$ of A is the basic example of an internal set. Its elements are called *internal objects.* Thus the elements of ${}^*(A \to B)$ are *internal functions* and we have an evaluation functional ${}^*\text{ev} : {}^*(A \to B) \times {}^*A \to {}^*B$. A subset X of ${}^*A$ is *internal* if there is some

$S \in {}^*(\wp(A))$ such that for all $x \in {}^*A$: ${}^*(x \mathrel{\epsilon} S)$ iff $x \in X$. Note that ${}^*(x \mathrel{\epsilon} S)$ is ${}^*\mathrm{ev}(S, x) = {}^*\mathbf{true}$. By the transfer principle the *-transform of (15.3) holds in $\mathcal{N}$ and thus yields Keisler's *internal definition principle* [4].

Theorem 2.1 ($\aleph_1$-saturation) *Let A be a non-void set. Then in $\mathcal{N}$: For every $f : {}^\sigma\mathbb{N} \to {}^*A$, then there exists $g : {}^*(\mathbb{N} \to A)$ such that $f(n) = g(n)$ for all $n \in {}^\sigma\mathbb{N}$. This g is called an* internal extension *of f.*

Proof. Let $H = [{}^\sigma\mathbb{N} \to {}^*A]$. Suppose $\varphi \in H(\mathcal{F})$. By the construction of function spaces in sheaves φ is a natural transformation ${}^\sigma\mathbb{N}\times\mathbf{y}(\mathcal{F}) \dot{\to} {}^*A$, where $\mathbf{y}$ is the Yoneda embedding given by $\mathbf{y}(\mathcal{F}) = \mathrm{Hom}_{\mathbb{B}}(-, \mathcal{F})$. For each $n \in \mathbb{N}$, we let

$$\alpha_n =_{\mathrm{def}} \varphi({}^*n, 1_{\mathcal{F}}) \in {}^*A(\mathcal{F}),$$

where ${}^*n \in {}^\sigma\mathbb{N}(\mathcal{F})$. Write $\mathcal{F} = (F, \{F_p\}_{p \in P})$. Using the principle of dependent choice, choose in turn $p_0, p_1, p_2, \ldots \in P$ such that α_n is defined on F_{p_n} and $F_{p_{n+1}} \subseteq F_{p_n}$, for all $n \in \mathbb{N}$. We now form a filter $\mathcal{G} = (G, \{G_r\}_{r \in R})$ which is essentially a product of $\mathcal{F}$ and the Fréchet filter. More precisely, let $G = \{(n, x) : n \in \mathbb{N}, x \in F_{p_n}\}$ and $G_{(m,p)} = \{(n, x) : n \geq m \text{ and } x \in F_p\}$ for $(m, p) \in R$, where $R = \{(k, p) \in \mathbb{N} \times P : F_q \subseteq F_{p_k}\}$. It is readily checked that $\mathcal{G}$ is a filter and that the second projection $\pi_2 : \mathcal{G} \to \mathcal{F}$ is a continuous covering map. Now we construct the extension at $\mathcal{G}$. Define $\gamma \in {}^*(\mathbb{N} \to A)(\mathcal{G})$ by

$$\gamma(n, x) = k \mapsto \begin{cases} \alpha_k(x) & \text{if } n \geq k, \\ a_0 & \text{if } n < k, \end{cases}$$

where a_0 is a fixed element of A. Then γ is defined on $G_{(m,p_m)}$ for each m. We need to check that for each $n \in \mathbb{N}$,

$$\mathcal{G} \Vdash \mathrm{ev}(H(\pi_2)(\varphi), {}^*n) = {}^*\mathrm{ev}(\gamma, {}^*n).$$

By definition of evaluation and the naturality of φ we have

$$\begin{aligned} \mathrm{ev}(H(\pi_2)(\varphi), {}^*n) &= H(\pi_2)(\varphi)({}^*n, 1_{\mathcal{G}}) \\ &= \varphi({}^*n, \pi_2) \\ &= \varphi({}^\sigma\mathbb{N}({}^*n), \mathbf{y}(\pi_2)(1_{\mathcal{F}})) \\ &= {}^*A(\pi_2)(\varphi({}^*n, 1_{\mathcal{F}})) \\ &= \alpha_n \pi_2 \end{aligned}$$

On the other hand $\alpha_n \pi_2 = {}^*\mathrm{ev}(\gamma, {}^*n)$ since for $(m,x) \in G_{(n,p_n)}$ we have $\alpha_n \pi_2(m,x) = \alpha_n(x)$ and ${}^*\mathrm{ev}(\gamma, {}^*n)(m,x) = \gamma(m,x)({}^*n(m,x)) = \gamma(m,x)(n) = \alpha_n(x)$. □

Corollary 2.2 *Let* $X : {}^\sigma\mathbb{N} \to {}^*\wp(A)$ *be a decreasing sequence of non-void internal sets. Then* $\bigcap_{n \in {}^\sigma\mathbb{N}} X_n$ *is non-void.*

Proof. Let $Y : {}^*(\mathbb{N} \to \wp(A))$ be an internal extension of X. By the assumption we have for each $n \in {}^\sigma\mathbb{N}$:

$$(\exists y \in {}^*A)\,(\forall k < n)\, y \,\epsilon\, Y_k.$$

By overspill (a special case of idealization) there is some infinite n for which this holds. Hence there exists $y \in {}^*A$ such that $y \,\epsilon\, Y_n = X_n$ for every $n \in {}^\sigma\mathbb{N}$. □

The above saturation principle may be used to show a weak principle of omniscience. Recall from [3] that the *lesser limited principle omniscience* (LLPO) states that if the infinite 0-1-sequence $(a_n)_n$ contains at most one 1, it can be decided whether $a_{2n} = 0$ for all n, or whether $a_{2n+1} = 0$ for all n. Within constructive mathematics this principle is equivalent to the principle that for all real numbers x and y: $x \leq y$ or $y \leq x$. For systematic account of the consequences of LLPO, see Ishihara [5].

Corollary 2.3 *LLPO is valid in the nonstandard universe* $\mathcal{N}$.

Proof. Suppose that $a : {}^\sigma\mathbb{N} \to {}^\sigma\{0,1\}$ is a sequence containing at most one 1. Let $\hat{a}$ be its internal extension $\hat{a} : {}^*(\mathbb{N} \to \{0,1\})$. Then for all $k \in {}^\sigma\mathbb{N}$:

$$(\forall m \leq k)\hat{a}_{2m} = 0 \vee (\forall m \leq k)\hat{a}_{2m+1} = 0.$$

Since the displayed statement is internal, we obtain by overspill an infinite k for which it holds. Thus $(\forall m \in {}^\sigma\mathbb{N})\, \hat{a}_{2m} = 0$ or $(\forall m \in {}^\sigma\mathbb{N})\, \hat{a}_{2m+1} = 0$. □

In [16] we showed, using only overspill, that LLPO holds for the more restricted set of functions, ${}^\sigma(\mathbb{N} \to \mathbb{N})$, a function space which does not supply enough choice functions. There is a stronger omniscience principle, LPO, which states that any 0-1 sequence is constant zero or has some term of value 1. From [16] follows that this principle is not valid $\mathcal{N}$ unless it is valid in the metatheory:

Proposition 2.4 *LPO holds, if LPO is valid in* $\mathcal{N}$.

We briefly indicate how to construct Loeb measures on hyperfinite sets. Consider the set $\mathbb{HF}$ of decidable, hereditarily finite sets. This set as well as the membership predicate and all finite boolean operations can be coded using the natural numbers and primitive recursive operations. We now consider measure spaces $(\Omega, \mathcal{A}, \mu)$, where the sample space Ω and the measure $\mu : {}^*(\mathbb{HF} \to [0, \infty))$, defined on the algebra $\mathcal{A} \in {}^*\mathbb{HF}$, satisfy the $*$-version of the conditions for a finitely additive measure space. Moreover we assume that $\mu(\Omega)$ is finite. Then we say that a complemented subset [1] $A = (A^1, A^0)$ of Ω is *Loeb measurable* if for every standard rational $\varepsilon > 0$ there are $S_1, S_0 \in \mathcal{A}$ such that $S_1 \subseteq A^1$, $S_0 \subseteq A^0$ and $\mu(S_0^c \backslash S_1) < \varepsilon$. The *Loeb measure* of this set is then the Dedekind cut given by the lower and upper standard rational approximations to $\mu(S_1)$ and $\mu(S_0^c)$, respectively, for $S_1, S_0 \in \mathcal{A}$, $S_1 \subseteq A^1$ and $S_0 \subseteq A^0$. Using these definitions it is possible to deduce the Bishop–Cheng conditions [2, Def. 6.10.13] for a constructive measure space. These results will be presented in a forthcoming paper.

The model $\mathcal{N}$ has the remarkable property that every function between two internal sets of the form ${}^*A$ and ${}^*B$ is internal, i.e. is in ${}^*(A \to B)$. This stems from the following general result in topos theory. Recall [7] that a site $(\mathbb{D}, J)$ is called *subcanonical* if each representable presheaf

$$\mathbf{y}(C) = \mathrm{Hom}_{\mathbb{D}}(-, C)$$

is a sheaf over the site.

Theorem 2.5 *[7, Exercise III.11]. Let $(\mathbb{D}, J)$ be a subcanonical site. Then the Yoneda embedding*

$$\mathbf{y} : \mathbb{D} \to \mathrm{Sh}(\mathbb{D}, J)$$

preserves the category theoretic exponentials (function spaces) and limits that exist in $\mathbb{D}$.

The site $(\mathbb{B}, K)$ is indeed subcanonical and $\mathbf{y}(\overline{A}) = {}^*A$ so we need only to establish the following easy lemma.

Lemma 2.6 *In the category of filters $\mathbb{B}$, the exponent of two trivial filters $\overline{A}$ and $\overline{B}$ is the trivial filter $\overline{A^B}$.*

Proof. Let $\varphi : \mathcal{C} \times \overline{B} \to \overline{A}$ be a morphism in $\mathbb{B}$, where $\mathcal{C} = (C, \{C_i\}_{i \in I})$ is an arbitrary filter. Then $\mathcal{C} \times \overline{B}$ is $(C \times B, \{C_i \times B\}_{i \in I})$, so φ is defined on some $C_i \times B$. Let $\hat{\varphi} : C_i \to B^A$ be the exponential transpose given by $\hat{\varphi}(u)(v) = \varphi(u, v)$. Thus $\hat{\varphi} : \mathcal{C} \to \overline{B^A}$ is the unique morphism such that $\varphi = \mathrm{ev} \circ (\hat{\varphi} \times 1)$, where $\mathrm{ev}(f, x)$ is $f(x)$. □

Corollary 2.7 *The exponential of* $^*A$ *and* $^*B$ *is* $^*(A^B) = {}^*(B \to A)$, *and thus the principle of unique choice holds in* $\mathcal{N}$ *in the following form: If for every* $x \in {}^*B$ *there exists a unique* $y \in {}^*A$ *such that* $\Phi(x, y)$, *then there exists an* $f \in {}^*(A^B)$ *such that for all* $x \in {}^*B$, $\Phi(x, f(x))$.

This result may seem contradictory since one might expect to be able to define a non-internal function $\chi : {}^*\mathbb{N} \to {}^*\mathbb{N}$ by

$$\chi(n) = \begin{cases} 1 & \text{if } n \text{ is standard,} \\ 0 & \text{if } n \text{ is nonstandard.} \end{cases}$$

However, this function is not definable, since it is not decidable whether a natural number is standard or nonstandard. We have in fact by [11]:

$$\neg(\forall x \in {}^*\mathbb{N})\,(x \in {}^\sigma\mathbb{N} \vee x \notin {}^\sigma\mathbb{N}). \tag{15.4}$$

One might compare Corollary 2.7 with the result that in some sheaf models all functions on $\mathbb{R}$ are continuous. This does, of course, not preclude that discontinuous or external functions may be dealt with as partial functions.

3. THE STANDARD PART MAP AND REAL NUMBERS

In this section we examine the notions of real number, standard part map and nonstandard hull within the context of the model $\mathcal{N}$. In an arbitrary sheaf model the structure of real numbers represented as Cauchy sequences will usually differ from that defined using Dedekind cuts. Fortunately, these two notions coincide in our model $\mathcal{N}$, since the principle of dependent choice is valid here [18]. A Dedekind cut is a pair (L, R) of non-void subsets of ${}^\sigma\mathbb{Q}$ satisfying the well-known conditions of consistency, openness and locatedness; see [7] or [18] for a formal definition. We denote the set of such pairs by $\mathbb{R}_d$. The nonstandard version of $\mathbb{R}$ is ${}^*\mathbb{R}$, the set of *hyperreals*. A hyperreal ξ is called *finite* if $-n < \xi < n$ for some $n \in {}^\sigma\mathbb{N}$; it is *infinitesimal* if $-2^{-n} < \xi < 2^{-n}$ for each $n \in {}^\sigma\mathbb{N}$. Denote the set of finite elements in ${}^*\mathbb{R}$ by $\mathrm{Fin}({}^*\mathbb{R})$ and the set of infinitesimals by $\mathrm{Inf}({}^*\mathbb{R})$. Define for $\alpha, \beta \in {}^*\mathbb{R}$ the relation that α is *appreciably* *smaller than* β (in symbols $\alpha \prec \beta$) to hold iff there exists $n \in {}^{\sigma}\mathbb{N}$ with $\alpha + 2^{-n} < \beta$. For $\xi \in {}^{*}\mathbb{R}$ define the sets

$$L_{\xi} = \{q \in {}^{\sigma}\mathbb{Q} : q \prec \xi\}, \quad R_{\xi} = \{r \in {}^{\sigma}\mathbb{Q} : \xi \prec r\}. \tag{15.5}$$

The pair (L_{ξ}, R_{ξ}) is a Dedekind cut for any finite hyperreal ξ, a result that follows easily using the transfer principle; see [18]. Thus

$$\xi \mapsto (L_{\xi}, R_{\xi})$$

is indeed a *standard part map* from $\mathrm{Fin}({}^{*}\mathbb{R})$ to $\mathbb{R}_d$. This map also induces a well-defined function

$$\phi : \mathrm{Fin}({}^{*}\mathbb{Q})/\mathrm{Inf}({}^{*}\mathbb{Q}) \to \mathbb{R}_d \tag{15.6}$$

given by $\phi([\xi]) = (L_{\xi}, R_{\xi})$. By constructing a certain covering map in the category $\mathbb{B}$ one can show [18] that ϕ is indeed an isomorphism of ordered rings. Thus $\mathbb{R}_d$ may be viewed as the nonstandard hull of the rational numbers, just as in the classical setting. In [16] an infinitesimal calculus was developed using this nonstandard hull and it displayed several classical features. For instance, the Cauchy–Peano theorem and the intermediate value theorem are true in $\mathcal{N}$. Using the isomorphism (15.6), and further results of [16], we can prove the following.

Proposition 3.1 ([18]) *The set of reals* $\mathbb{R}_d$ *has the following properties in the model* $\mathcal{N}$.

(i) $\mathbb{R}_d$ *is an integral domain.*

(ii) For all $x, y \in \mathbb{R}_d$: $x \leq y$ *or* $y \leq x$.

(iii) The intermediate value theorem holds for $\mathbb{R}_d$.

Remark 3.2 The set ${}^{\sigma}\mathbb{R}$ of standard reals in ${}^{*}\mathbb{R}$ does not correspond to the Dedekind reals since one can show [15] that $(\forall x \in {}^{*}\mathbb{R})\,[x \text{ finite} \to (\exists u \in {}^{\sigma}\mathbb{R})\, x \simeq u]$ is not valid in $\mathcal{N}$.

We now consider nonstandard hulls of standard metric spaces. First note that the inclusion ${}^{*}\mathbb{Q} \hookrightarrow {}^{*}\mathbb{R}$ yields a monomorphism

$$\psi : \mathrm{Fin}({}^{*}\mathbb{Q})/\mathrm{Inf}({}^{*}\mathbb{Q}) \to \mathrm{Fin}({}^{*}\mathbb{R})/\mathrm{Inf}({}^{*}\mathbb{R}). \tag{15.7}$$

By transfer we have $(\forall x \in {}^{*}\mathbb{R})\,(\forall n \in {}^{*}\mathbb{N})\,(\exists q \in {}^{*}\mathbb{Q})\,|x - q| < 2^{-n}$, which shows that ψ is onto, and hence an isomorphism. Let $d : X \times X \to \mathbb{R}$ be a metric. Define $\mathrm{Fin}({}^*X)$ to be the set of $x \in {}^*X$ such that ${}^*d(x,a)$ is finite, for some $a \in {}^\sigma X$. Introduce an equivalence relation $\sim$ on this set by letting $x \sim y \Leftrightarrow {}^*d(x,y)$ is infinitesimal. Thus ${}^*d : {}^*(X \times X \to \mathbb{R})$ defines a metric ρ on $\mathrm{Fin}({}^*X)/\sim$ with values in $\mathrm{Fin}({}^*\mathbb{R})/\mathrm{Inf}({}^*\mathbb{R})$ by $\rho([x],[y]) = [{}^*d(x,y)]$.

Proposition 3.3 $(\mathrm{Fin}({}^*X)/\sim, \rho)$ *is a complete metric space.*

Proof. Consider a Cauchy sequence $a : {}^\sigma\mathbb{N} \to \mathrm{Fin}({}^*X)/\sim$. Hence there is some $c \in {}^\sigma X$ and some $K \in {}^\sigma\mathbb{N}$ such that ${}^*d(a_m, c) < K$ for all $m \in {}^\sigma\mathbb{N}$. Let $b : {}^*(\mathbb{N} \to X)$ be an internal extension of the sequence a, according to the $\aleph_1$-saturation principle. Thus there is some infinite $\mu \in {}^*\mathbb{N}$ such that ${}^*d(b_m, c) < K$ for each $m \leq \mu$. That a is a Cauchy sequence implies that for some $f : {}^\sigma\mathbb{N} \to {}^\sigma\mathbb{N}$ we have

$$(\forall k, m, n \in {}^\sigma\mathbb{N})\,[m \geq f(k), n \geq f(k) \Rightarrow {}^*d(b_m, b_n) < 2^{-k}]. \qquad (15.8)$$

By $\aleph_1$-saturation, we can replace f in (15.8) by an internal extension $g : {}^*(\mathbb{N} \to \mathbb{N})$. Hence overspill is applicable to (15.8), so there is some infinite $\nu \in {}^*\mathbb{N}$ such that ${}^*d(b_m, b_n) < 2^{-k}$ for all $k, m, n \leq \nu$ with $m, n \geq f(k)$. Let $\eta = \min(\mu, \nu)$. Then clearly $b_\eta \in \mathrm{Fin}({}^*X)$ is the limit of the sequence a. □

Given a Dedekind cut (L, R) in $\mathcal{N}$ it is possible to extract some information about concrete rational approximations. For each filter $\mathcal{F}$ define

$$L_\mathcal{F} = \{q \in \mathbb{Q} : {}^*q \in L(\mathcal{F})\}, \quad R_\mathcal{F} = \{q \in \mathbb{Q} : {}^*q \in R(\mathcal{F})\}. \qquad (15.9)$$

Then one can show [18] that $(L_\mathcal{F}, R_\mathcal{F})$ is a Dedekind cut for any ultrafilter $\mathcal{F}$. In this case the combination of (15.5) and (15.9) gives the usual standard part map associated with an ultrapower model over $\mathcal{F}$. For a general filter $\mathcal{F}$, $(L_\mathcal{F}, R_\mathcal{F})$ may fail to be located unless there is further information about the convergence of (L, R).

Remark 3.4 For results on a standard part map for the Schmieden–Laugwitz model, see Schuster [21].

4. NONSTANDARD PROOF OF THE IMPLICIT FUNCTION THEOREM

While results like Proposition 3.1 can clearly not be transferred from $\mathbb{R}_d$ in $\mathcal{N}$ to $\mathbb{R}$ in the real world, some of its consequences may, if expressible by internal formulas. In this section we shall make use of the model, though not the results of Section 3, to prove two standard theorems from the calculus several variables. We claim that their nonstandard proofs are more intuitive than their standard proofs. For further examples of this kind, see [15, 16].

Let $I_1, \ldots, I_m \subset \mathbb{R}$ be compact intervals, and put $I = I_1 \times \cdots \times I_m$. Let f and g be a uniformly continuous functions $I \to \mathbb{R}$. Then g is the *ith partial derivative* of f in case for every k there is some n such that for all p if $2^{-p} < |x_i - u| \leq 2^{-n}$, then

$$\left|\frac{f(\mathbf{x}) - f(\mathbf{y})}{x_i - u} - g(\mathbf{x})\right| \leq 2^{-k},$$

whenever $\mathbf{x} = (x_1, \ldots, x_m)$ and $\mathbf{y} = (x_1, \ldots, x_{i-1}, u, x_{i+1}, \ldots, x_m)$.

Lemma 4.1 *g is the ith partial derivative of f iff the following implication is valid in the model:*

$$2^{-p} < |x_i - u| \simeq 0 \Rightarrow \frac{{}^*f(\mathbf{x}) - {}^*f(\mathbf{y})}{x_i - u} \simeq {}^*g(\mathbf{x}),$$

whenever $\mathbf{x} = (x_1, \ldots, x_m)$ and $\mathbf{y} = (x_1, \ldots, x_{i-1}, u, x_{i+1}, \ldots, x_m)$.

Proof. See [16]. □

In constructive analysis we have only approximate versions of the mean value theorem. Here are two variants that will be used below.

Lemma 4.2 *Let $f : I \to \mathbb{R}$ be a differentiable function on the compact interval I. Let $a \neq b$ in I. Then for each $\varepsilon > 0$ there exists $c \in (a, b)$ for which*

$$\left|f'(c) - \frac{f(a) - f(b)}{a - b}\right| < \varepsilon.$$

Lemma 4.3 *Let $f : [a, b] \to \mathbb{R}$ be a uniformly continuous function. Then for each $\varepsilon > 0$ there exists $c \in (a, b)$*

$$\left|\int_a^b f(x)dx - f(c)(b - a)\right| < \varepsilon.$$

Proposition 4.4 *Let $f : I = I_1 \times \cdots \times I_m \to \mathbb{R}$ be a uniformly continuous function, which has uniformly continuous partial derivatives of first and second order. Then*

$$\frac{\partial^2 f}{\partial x_i \partial x_j} = \frac{\partial^2 f}{\partial x_j \partial x_i}.$$

Proof. We work inside the model. Let $(a_1, \ldots, a_m) \in I$. Write

$$g(x, y) = {}^*f(a_1, \ldots, a_{i-1}, x, a_{i+1}, \ldots, a_{j-1}, y, a_{j+1}, \ldots, a_m),$$

and $\frac{\partial g}{\partial x}$ for $\frac{\partial f}{\partial x_i}$ and so on. Let $a = a_i$ and $b = a_j$. Pick infinitesimals h and k such that $|h|, |k| > 2^{-n}$, for some infinite n, and so that $a + h \in I_i$ and $b + k \in I_j$. Consider the expression

$$\frac{1}{hk}[g(a+h, b+k) - g(a+h, b) - g(a, b+k) + g(a, b)]. \tag{15.10}$$

The mean-value theorem applied to $\varphi(t) = g(t, b+k) - g(t, b)$ with $\varepsilon = |hk|$ gives some $c \in (a, a+h)$ and some δ with $|\delta| < \varepsilon$, such that (15.10) equals

$$\begin{aligned} \frac{1}{k}\Big[\frac{\partial g}{\partial x}(c, b+k) - \frac{\partial g}{\partial x}(c, b) + \delta\Big] &\simeq \frac{1}{k}\Big[\frac{\partial g}{\partial x}(c, b+k) - \frac{\partial g}{\partial x}(c, b)\Big] \\ &\simeq \frac{\partial^2 g}{\partial y \partial x}(c, d) \simeq \frac{\partial^2 g}{\partial y \partial x}(a, b), \end{aligned}$$

where $d \in (b, b+k)$. The second member follows by applying the mean-value theorem with ε infinitesimal. The last member follows by uniform continuity. Analogously, if we apply the mean-value theorem to $\psi(t) = g(a+h, t) - g(a, t)$ we find that (15.10) equals

$$\frac{1}{h}\Big[\frac{\partial g}{\partial y}(a+h, d') - \frac{\partial g}{\partial y}(a, d')\Big] \simeq \frac{\partial^2 g}{\partial x \partial y}(c', d') \simeq \frac{\partial^2 g}{\partial x \partial y}(a, b). \square$$

As is well-known the exact version of the intermediate-value theorem does not hold in constructive analysis unless some further restrictions are put on the continuous function.

Lemma 4.5 *Let $f : [a, b] \to \mathbb{R}$ be differentiable such that for some subinterval $[s, t]$ of $[a, b]$ we have $f'(u) \neq 0$ for all $u \in [s, t]$. If $f(s) < 0 < f(t)$, then f has a zero in the interval $[s, t]$.*

In $\mathbb{R}^n$, let $\overline{B}(\mathbf{x}, \delta)$ denote the closed ball of radius δ centered at $\mathbf{x}$.

Theorem 4.6 (Implicit Function Theorem) *Let $F : K \to \mathbb{R}$ be a differentiable function on the compact subspace $K \subseteq \mathbb{R}^{n+1}$, whose partial derivatives are uniformly continuous. Let $(\mathbf{x}_0, z_0) \in K$ be an inner point of K such that*

$$F(\mathbf{x}_0, z_0) = 0 \text{ and } \frac{\partial F}{\partial z}(\mathbf{x}_0, z_0) \neq 0.$$

Then there are $\delta, \eta > 0$ such that $\overline{B}(\mathbf{x}_0, \delta) \times \overline{B}(z_0, \eta) \subseteq K$ and a unique uniformly continuous function $g : \overline{B}(\mathbf{x}_0, \delta) \to \overline{B}(z_0, \eta)$ such that

$$F(\mathbf{x}, g(\mathbf{x})) = 0 \qquad (\mathbf{x} \in \overline{B}(\mathbf{x}_0, \delta))$$

with continuous partial derivatives given by

$$\frac{\partial g}{\partial x_i}(\mathbf{x}) = -\frac{\partial F}{\partial x_i}(\mathbf{x}, g(\mathbf{x}))\Big/\frac{\partial F}{\partial z}(\mathbf{x}, g(\mathbf{x})) \quad (i = 1, \ldots, n;\ \mathbf{x} \in \overline{B}(\mathbf{x}_0, \delta)).$$

Proof. Suppose $\frac{\partial F}{\partial z}(\mathbf{x}_0, z_0) > 0$ (the negative case is analogous). Let M be a bound of the partial derivatives of F on K. We may find positive numbers a and b such that the set $\overline{B}(\mathbf{x}_0, a) \times \overline{B}(z_0, a)$ is included in K, and for $(\mathbf{x}, z)$ in this set $\frac{\partial F}{\partial z}(\mathbf{x}, z) > b$. First choose $0 < \eta < a$ and then $\delta < \eta$ with $\delta < b\eta(2M\sqrt{n})^{-1}$.

We now reason inside the model. Using the mean-value theorem we obtain the following. For any $(\mathbf{x}, z), (\mathbf{u}, w) \in K$ and each $\varepsilon > 0$ there are some θ and ϕ in $[0, 1]$ such that

$$|{}^*F(\mathbf{x}, z) - {}^*F(\mathbf{u}, z) - D_{\mathbf{x}}F(\theta\mathbf{x} + (1-\theta)\mathbf{u}, z) \cdot (\mathbf{x} - \mathbf{u})| < \varepsilon, \quad (15.11)$$

$$|{}^*F(\mathbf{u}, z) - {}^*F(\mathbf{u}, w) - {}^*\frac{\partial F}{\partial z}(\mathbf{u}, \phi z + (1-\phi)w)(z - w)| < \varepsilon \quad (15.12)$$

where $D_{\mathbf{x}}F(\mathbf{y}, v) = ({}^*\frac{\partial F}{\partial x_1}(\mathbf{y}, v), \ldots, {}^*\frac{\partial F}{\partial x_n}(\mathbf{y}, v))$. Writing ${}^*F(\mathbf{x}, z) = ({}^*F(\mathbf{x}, z) - {}^*F({}^*\mathbf{x}_0, z)) + ({}^*F({}^*\mathbf{x}_0, z) - {}^*F({}^*\mathbf{x}_0, {}^*z_0))$, we find using (15.11) and (15.12) that for some θ and ϕ

$$\Big|{}^*F(\mathbf{x}, z) - D_{\mathbf{x}}F(\theta\mathbf{x} + (1-\theta){}^*\mathbf{x}_0, z) \cdot (\mathbf{x} - {}^*\mathbf{x}_0) \quad (15.13)$$
$$-{}^*\frac{\partial F}{\partial z}({}^*\mathbf{x}_0, \phi z + (1-\phi){}^*z_0)(z - {}^*z_0)\Big| < 2\varepsilon.$$

By the Cauchy-Schwarz inequality it follows that for $||\mathbf{x} - {}^*\mathbf{x}_0|| \leq \delta$,

$$\begin{aligned} &|D_{\mathbf{x}}F(\theta\mathbf{x} - (1-\theta)^*\mathbf{x}_0, z) \cdot (\mathbf{x} - {}^*\mathbf{x}_0)| \\ &\leq \ ||D_{\mathbf{x}}F(\theta\mathbf{x} - (1-\theta)^*\mathbf{x}_0, z)|| \cdot ||\mathbf{x} - {}^*\mathbf{x}_0|| \\ &\leq \ ({}^*M\sqrt{{}^*n})\delta < {}^*b^*\eta/2. \end{aligned}$$

Using this inequality and putting $z = {}^*z_0 + {}^*\eta$ with $\varepsilon = {}^*b^*\eta/4$ in (15.13) yields

$$|{}^*F(\mathbf{x}, {}^*z_0 + {}^*\eta) - {}^*\frac{\partial F}{\partial z}({}^*\mathbf{x}_0, \phi z + (1-\phi)^*z_0)^*\eta| < \frac{2^*b^*\eta}{4} + \frac{{}^*b^*\eta}{2} = {}^*b^*\eta.$$

Since ${}^*\frac{\partial F}{\partial z}$ exceeds ${}^*b$ on K, we have ${}^*F(\mathbf{x}, {}^*z_0 + {}^*\eta) > 0$. Similarly, we obtain ${}^*F(\mathbf{x}, {}^*z_0 - {}^*\eta) < 0$. According to Lemma 4.5 there is a unique $z \in [{}^*z_0 - {}^*\eta, {}^*z_0 + {}^*\eta]$ such that $F(\mathbf{x}, z) = 0$. We show that z depends uniformly continuously on $\mathbf{x}$. Let $\mathbf{x}, \mathbf{u} \in \overline{B}(\mathbf{x}_0, \delta)$ and $z, w \in \overline{B}(z_0, \eta)$ be such that

$$\mathbf{x} \simeq \mathbf{u} \quad \text{and} \quad F(\mathbf{x}, z) = F(\mathbf{u}, w) = 0. \tag{15.14}$$

We need only to show $z \simeq w$. From (15.11) it follows by choosing $\varepsilon > 0$ infinitesimal that for some $\mathbf{y} \simeq \mathbf{x}$

$${}^*F(\mathbf{x}, z) - {}^*F(\mathbf{u}, z) \simeq D_{\mathbf{x}}(\mathbf{y}, z) \cdot (\mathbf{x} - \mathbf{u}) \tag{15.15}$$

and from (15.12) we obtain for some $v = \phi z + (1-\phi)w$,

$${}^*F(\mathbf{u}, z) - {}^*F(\mathbf{u}, w) \simeq {}^*\frac{\partial F}{\partial z}(\mathbf{u}, v)(z - w). \tag{15.16}$$

Thus (15.14) gives $D_{\mathbf{x}}F(\mathbf{y}, z)(\mathbf{x} - \mathbf{u}) \simeq -{}^*\frac{\partial F}{\partial z}(\mathbf{u}, v)(z - w)$. Since $||D_{\mathbf{x}}||$ is finite, $\mathbf{x} \simeq \mathbf{u}$ and ${}^*\frac{\partial F}{\partial z} > {}^*b$ it follows from this equality that $z \simeq w$. We now temporarily return to the standard world, and retrieve a function ${}^*g$ such that

$${}^*F(\mathbf{x}, {}^*g(\mathbf{x})) = 0 \qquad (\mathbf{x} \in {}^*B({}^*\mathbf{x}_0, \delta))$$

(this g may be constructed as a uniform limit using Weierstrass's approximation theorem).

Let ${}^*e_i = (0, \ldots, 0, 1, 0, \ldots, 0)$ (1 in the ith coordinate). Let $|h| > 0$ be infinitesimal such that $\mathbf{x}$ and $\mathbf{u} = \mathbf{x} + h\mathbf{e}_i$ lie in $\overline{B}(\mathbf{x}_0, \delta)$. Then for $\varepsilon \in$

$(0, h^2)$ we obtain, using (15.11) and (15.12), some $(\mathbf{y}, v) \in \overline{B}({}^*\mathbf{x}_0, \delta) \times \overline{B}(z_0, \eta)$ such that $\mathbf{y} \simeq \mathbf{x}$, $v \simeq {}^*g(\mathbf{x})$ and

$$|{}^*\frac{\partial F}{\partial x_i}(\mathbf{y}, {}^*g(\mathbf{x}))h - {}^*\frac{\partial F}{\partial z}(\mathbf{u}, v)({}^*g(\mathbf{x}) - {}^*g(\mathbf{u}))| < 2\varepsilon.$$

Dividing by $\frac{\partial F}{\partial z}(\mathbf{u}, v)h$ and using ${}^*\frac{\partial F}{\partial z} > {}^*b$ gives

$$\left| \frac{{}^*\frac{\partial F}{\partial x_i}(\mathbf{y}, g(\mathbf{u}))}{{}^*\frac{\partial F}{\partial z}(\mathbf{u}, v)} - \frac{{}^*g(\mathbf{x}) - {}^*g(\mathbf{u})}{h} \right| < \frac{2\varepsilon}{h^*b} < \frac{2h}{{}^*b}.$$

Thus by uniform continuity of the partial derivatives and g

$$\frac{{}^*g(\mathbf{u}) - {}^*g(\mathbf{x})}{h} \simeq -\frac{{}^*\frac{\partial F}{\partial x_i}(\mathbf{y}, g(\mathbf{u}))}{{}^*\frac{\partial F}{\partial z}(\mathbf{u}, v)} \simeq -\frac{{}^*\frac{\partial F}{\partial x_i}(\mathbf{x}, g(\mathbf{x}))}{{}^*\frac{\partial F}{\partial z}(\mathbf{x}, g(\mathbf{x}))}. \square$$

Remark 4.7 The above proof is modeled on the standard proof in [8].

References

[1] E. Bishop. *Foundations of Constructive Analysis.* McGraw-Hill 1967.

[2] E. Bishop and D.S. Bridges. *Constructive Analysis.* Springer, Berlin 1985.

[3] D.S. Bridges and F. Richman. *Varieties of Constructivism.* Cambridge University Press 1987.

[4] C.C. Chang and H.J. Keisler. *Model Theory,* third ed., North-Holland 1990.

[5] H. Ishihara. An omniscience principle, the König lemma and the Hahn-Banach Theorem. *Zeitschr. f. math. und Grundlagen d. Math.* 36(1990), 237 – 240.

[6] A. Kock. *Synthetic Differential Geometry.* Cambridge University Press 1981.

[7] S. Mac Lane and I. Moerdijk. *Sheaves in Geometry and Logic.* Springer 1992.

[8] J.E. Marsden and A.S. Tromba. *Vector Calculus.* Second ed. Freeman 1981.

[9] P. Martin-Löf. *Intuitionistic Type Theory.* Bibliopolis 1984.

[10] P. Martin-Löf. Mathematics of infinity. In: P. Martin-Löf and G.E. Mints (eds.) *COLOG-88 Computer Logic,* Lecture Notes in Computer Science, vol. 417, Springer, Berlin 1990, 146 – 197.

[11] I. Moerdijk. A model for intuitionistic nonstandard arithmetic. *Ann. Pure Appl. Logic* 73(1995), 37 – 51.

[12] I. Moerdijk and G.E. Reyes. *Models of Smooth Infinitesimal Analysis.* Springer 1991.

[13] E. Palmgren. A constructive approach to nonstandard analysis. *Ann. Pure Appl. Logic* 73(1995), 297 – 325.

[14] E. Palmgren. Constructive nonstandard analysis. In: A. Pétry (ed.) *Méthodes et analyse non standard,* Cahiers du Centre de Logique, vol. 9, Academia, Louvain-la-Neuve 1996, 369 – 397.

[15] E. Palmgren. Sheaf-theoretic foundation for nonstandard analysis. *Ann. Pure Appl. Logic* 85(1997), 69 – 86.

[16] E. Palmgren. Developments in constructive nonstandard analysis. *Bulletin of Symbolic Logic* 4(1998), 233 – 272.

[17] E. Palmgren. Constructive nonstandard representations of generalized functions. *Indagationes Mathematicae* N.S. 11(2000), 129 – 138.

[18] E. Palmgren. Real numbers in the topos of sheaves over the category of filters. *Journal of Pure and Applied Algebra,* to appear.

[19] J. Ruokolainen. *Some Constructive Hyperfinite Nonstandard Analysis.* Licentiate Thesis. University of Helsinki 1999.

[20] C. Schmieden and D. Laugwitz. Eine Erweiterung der Infinitesimalrechnung. *Math. Zeitschr.* 69(1958), 1 – 39.

[21] P. Schuster. A constructive look at generalized Cauchy reals. *Math. Logic Quarterly* 46(2000), 125 – 134.

POSITIVE LATTICES

Jan von Plato
Department of Philosophy
PL 24, 00014 University of Helsinki
Finland
vonplato@helsinki.fi

Abstract: From a constructive point of view, a weak linear order is a negative concept, to be replaced by the positive notion of strict order. It turns out that for a similar positive theory of partial order and lattices, one basic relation is sufficient. The treatment is elementary throughout and uses only free parameters and constructions. In lattice theory, the partial order relation can be characterized through algebraic equalities for lattices. A constructivization of this characterization is given, through use of an apartness relation instead of an equality.

Introduction

The idea of an apartness relation is that if a and b are apart, $a \neq b$, the disjunction $a \neq c \vee b \neq c$ is constructively justified. An example of such a 'positive' (we could also say 'observable') relation is the one between distinct points of the geometrical plane. We shall carry over the idea of a positive relation to the theory of partial order. An axiomatization of *positive partial order* is given through a refinement of an apartness relation, written for two elements a and b as $a \nleq b$, and read as, *a exceeds b*. The weak partial order will be the negation of the excess relation. Apartness is a symmetrization of the excess relation. As a motivating example, we shall first briefly review the theory of positive linear order.

Lattice operations are introduced as constructions, similarly to the minimum and maximum constructions of a linear order. There are axioms that state their ideal properties, as well as constructive uniqueness axioms. The treatment is thoroughly elementary, using free parameters and constructions instead of quantifier structures. The usual characterization of partial order through a set of algebraic laws for lattice operations is formulated in positive terms. The excess relation turns out

P. Schuster, U. Berger and H. Osswald (eds.),
Reuniting the Antipodes – Constructive and Nonstandard Views of the Continuum, 185–197.

to be indispensable, for, with apartness only, the constructive equivalent of an algebraic characterization cannot be carried through.

It is shown in Negri (1999) that the positive structures are conservative over the usual structures expressed in terms of weak partial order. This conservativity holds for formulas of the positive structures that have all their atomic formulas negated.

A remark on intuitionistic notation: The traditional notation for an apartness relation is $a \# b$. Bishop (1967) and Scott (1968) depart from this notation and write instead $a \neq b$ that in the intuitionistic literature stands for $\sim a = b$, so $\sim\sim a \# b$. We follow Bishop and Scott, and adopt the notation also for order relations. Our reasons are as follows: There is no need for a specific 'intuitionistic notation' in mathematics. In the cases that the intuitionistic basic relation is decidable, a totally unnecessary doubling of notation is created. The notation with the single slash is standard, and the thing to remember is whether negation is put in front of the unslashed (classical) relation, as in the definition $a \neq b \equiv \sim a = b$ or the slashed (intuitionistic) relation, as in $a = b \equiv \sim a \neq b$.

1. MOTIVATION THROUGH LINEAR ORDER

The constructive theory of linear order is obtained in a natural way through a *refinement* of an apartness relation, as follows. We postulate a relation $a < b$ such that

PLO1. $\vdash \sim(a < b \;\&\; b < a)$,
PLO2. $\vdash a < b \supset a < c \vee c < b$.

A set L with such a relation is called a *positive linear order*. Next define apartness, equality and weak order by

$$a \neq b \equiv a < b \vee b < a \tag{2.1}$$
$$a = b \equiv \sim a \neq b. \tag{2.2}$$
$$a \leqslant b \equiv \sim b < a, \tag{2.3}$$

One now derives easily all the usual constructive properties of these relations. The contraposition of PLO2 gives transitivity of the weak order relation. Using classical logic, PLO1 gives $a \leqslant b \vee b \leqslant a$, thus PLO1 is a constructive form of the law of dichotomy.

Apartness and strict order are positive or 'observable' properties, whereas equality and weak order are negative, or as we can also say, 'ideal' properties.

Minimum and maximum:

We assume a set L with a positive linear order. There are two constructions, the *minimum* and the *maximum* of two elements of L.

Constructions:

$min(a,b)$, *the minimum of a and b,*
$max(a,b)$, *the maximum of a and b.*

Axioms:

MNI $\vdash min(a,b) \leqslant a, \quad \vdash min(a,b) \leqslant b,$
MXI $\vdash a \leqslant max(a,b), \quad \vdash b \leqslant max(a,b),$

MNU $\vdash min(a,b) < c \supset a < c \vee b < c,$
MXU $\vdash c < max(a,b) \supset c < a \vee c < b.$

MNI and MXI give the ideal properties of the two constructions, MNU and MXU are constructive forms of stating their uniqueness. The axioms of constructive linear order with minimum and maximum are given in von Plato (1999).

In the axiomatization of tree and lattice structures, we add to partial order construction postulates for meet and join, their ideal properties, and uniqueness principles for the constructions. If the partial order collapses to a linear order, meet and join become the minimum and maximum constructions of linear order. Axiomatization through constructions makes provable algebraic laws the postulation of which seems somewhat arbitrary without such constructions.

Every reasonable axiom system for the constructive continuum has to include positive linear order with minimum and maximum. In Negri and Soravia (1999) it is shown that formal reals, i.e., reals in the sense of constructive pointfree topology, satisfy the axioms of a constructive linear order with minimum and maximum. The structure is further used in Cederquist (1998) for organizing the computer implementation of proofs in formal topology. In the constructive axiomatization of reals in Bridges (1999), the properties of minimum and maximum are somewhat hidden, in a more general constructive least upper bound principle.

2. POSITIVE PARTIAL ORDER

We have a set with elements $a, b, c, \ldots$, and a two-place relation $a \nleqslant b$, satisfying the axioms of *irreflexivity* and *splitting*:

PPO1. $\vdash \sim a \nleqslant a,$

PPO2. $\vdash a \nleq b \supset a \nleq c \vee c \nleq b$.

The second axiom is sometimes called 'cotransitivity.' We define six other relations, given here together with their readings:

$$\begin{array}{llll}
a \neq b & \equiv & a \nleq b \vee b \nleq a, & \textit{a and b are apart or distinct,} \quad (3.1)\\
a = b & \equiv & \sim a \neq b, & \textit{a and b are equal,} \quad (3.2)\\
a \leqslant b & \equiv & \sim a \nleq b, & \textit{a is not greater than b,} \quad (3.3)\\
a < b & \equiv & b \nleq a \;\&\; \sim a \nleq b, & \textit{a is less than b,} \quad (3.4)\\
a \nparallel b & \equiv & a \nleq b \;\&\; b \nleq a, & \textit{a and b are incomparable,} \quad (3.5)\\
a \parallel b & \equiv & \sim a \nparallel b, & \textit{a and b are comparable,} \quad (3.6)
\end{array}$$

The following theorem, easily proved, collects some of the essential properties of these relations:

Theorem 3.1:
(i) $\sim a \neq a, \quad a \neq b \supset a \neq c \vee b \neq c,$
(ii) $a \leqslant a, \quad a \leqslant b \;\&\; b \leqslant c \supset a \leqslant c,$
(iii) $\sim a < a, \quad a < b \;\&\; b < c \supset a < c,$
(iv) $\sim a \nparallel a, \quad a \nparallel b \supset b \nparallel a, \quad a \nparallel b \supset b \neq c \vee a \nparallel c.$

Thus, the symmetrization (3.1) gives an apartness relation. It follows that equality as defined in (3.2) is an equivalence relation. The negation of excess in (3.3) is a weak partial order. It is also antisymmetric: $a \leqslant b \;\&\; b \leqslant a \supset a = b$. Note that comparability only obeys the following, constructively weak substitution principle:

$$a \parallel b \;\&\; b = c \supset a \parallel c. \quad (3.7)$$

The reason why one does not get a strong, disjunctive conclusion as in theorem 3.1(iv) for incomparability, is that comparability is a negative notion.

We can give definition (3.4) also as $a < b \equiv a \leqslant b \;\&\; a \neq b$. However, if we only had weak partial order and apartness as basic relations, transitivity of strict order would not follow. (A proof of this fact using Kripke models is given in Greenleaf 1979.)

In the classical axiomatization of partial order, equality is taken as a second primitive, with the added axioms

$$a = b \supset a \leqslant b, \quad (3.8)$$
$$a \leqslant b \;\&\; b \leqslant a \supset a = b. \quad (3.9)$$

The reason why we prefer to define equality is that we only consider equality with respect to the order relation, not equality in the sense of identity. The second of the above two principles, (3.9), is not valid for the strict notion of identity.

It should be kept in mind that a positive partial order is not an order relation, for it need not be transitive. If transitivity is added, it becomes a positive linear order. Equivalently, it can be assumed that no two elements of a positive partial order are incomparable. This is expressed by $\sim (a \nleq b \ \& \ b \nleq a)$, and we see that the excess relation obeys in this case precisely the axioms PLO1 and PLO2 of a constructive linear order.

3. AN EXAMPLE: POINTS ON PARALLEL LINES

Let R be a set with a positive linear order with a strict order $a < b$ and apartness $a \neq b$ defined as in Section 2. Define excess for pairs in $R \times R$ by

$$(a, b) \nleq (c, d) \equiv \ a \neq c \vee d < b. \tag{4.1}$$

In the following, we shall use the same relation symbols in R and $R \times R$. We now show that definition (4.1) gives an excess relation in $R \times R$: If $(a, b) \nleq (a, b)$, we have by definition $a \neq a \vee b < b$ which is impossible, therefore the relation defined is irreflexive. Next assume $(a, b) \nleq (c, d)$, so $a \neq c \vee d < b$. Case 1. If $a \neq c$, then $a \neq e \vee e \neq c$. If $a \neq e$, then $(a, b) \nleq (e, f)$, if $e \neq c$, then $(e, f) \nleq (c, d)$. Case 2. If $d < b$, then $d < f \vee f < b$. If $d < f$, then $(e, f) \nleq (c, d)$, if $f < b$, then $(a, b) \nleq (e, f)$.

The general definitions (3.1)–(3.6) in Section 3 lead to the following equivalences:

$$(a, b) \neq (c, d) \supset\subset \ a \neq c \ \vee b \neq d, \tag{4.2}$$
$$(a, b) = (c, d) \supset\subset \ a = c \ \& \ b = d, \tag{4.3}$$
$$(a, b) \leqslant (c, d) \supset\subset \ a = c \ \& \ b \leqslant d, \tag{4.4}$$
$$(a, b) < (c, d) \supset\subset \ a = c \ \& \ b < d. \tag{4.5}$$
$$(a, b) \nparallel (c, d) \supset\subset \ a \neq c, \tag{4.6}$$
$$(a, b) \parallel (c, d) \supset\subset \ a = c, \tag{4.7}$$

In the example, a definition of partial order through classical concepts corresponds to the third equivalence (4.4). Reflexivity and transitivity of the weak partial order follow. If we define apartness componentwise as in (4.2), strict order becomes definable:

$$(a, b) < (c, d) \equiv \ (a, b) \leqslant (c, d) \ \& \ (a, b) \neq (c, d), \tag{4.8}$$

and the equivalence (4.5) follows. Transitivity of strict partial order becomes provable in this case, but as mentioned we cannot in general derive transitivity of strict order from a weak partial order that admits an apartness relation.

For the set R, any variety of constructive reals can be taken. Incomparability of (a, b) and (c, d) means that the points are on different vertical lines. We only used the following property of constructive reals in the example: The reals have a positive linear order relation. Since this order is undecidable, we have an example where excess, and therefore all the defined relations such as comparability, are undecidable.

4. POSITIVE LATTICES

We introduce meet and join of two elements in a positive partial order by two constructions, $a{\wedge}b$ and $a{\vee}b$. We add to the previous two axioms PPO1 and PPO2 lattice axioms that state the ideal properties of meet and join, as well as constructive uniqueness axioms.

The meet and join constructions:

$a{\wedge}b$ *the meet of a and b,*
$a{\vee}b$ *the join of a and b,*

The axioms are

MTI $\vdash a{\wedge}b \leqslant a, \quad \vdash a{\wedge}b \leqslant b.$
JNI $\vdash a \leqslant a{\vee}b, \quad \vdash b \leqslant a{\vee}b.$

MTU $\vdash c \nleqslant a{\wedge}b \supset c \nleqslant a \vee c \nleqslant b,$
JNU $\vdash a{\vee}b \nleqslant c \supset a \nleqslant c \vee b \nleqslant c.$

Contraposition of the uniqueness axioms gives the familiar classical uniqueness principles for lattice operations.

Lemma 5.1: (i) $a{\wedge}b \nleqslant c{\wedge}d \supset a \nleqslant c \vee b \nleqslant d$, (ii) $a{\vee}b \nleqslant c{\vee}d \supset a \nleqslant c \vee b \nleqslant d$.

Proof: (i) Let $a{\wedge}b \nleqslant c{\wedge}d$. By MTU, $a{\wedge}b \nleqslant c \vee a{\wedge}b \nleqslant d$. If $a{\wedge}b \nleqslant c$, axiom PPO2 gives $a{\wedge}b \nleqslant a \vee a \nleqslant c$, and axiom MTI leaves us only $a \nleqslant c$. Similarly if $a{\wedge}b \nleqslant d$, $b \nleqslant d$. (ii) is proved similarly. QED.

Since excess is a refinement of apartness, strong extensionality for meet and join follow at once: We have

Theorem 5.2: (i) $a{\wedge}b \neq c{\wedge}d \supset a \neq c \vee b \neq d$,
(ii) $a{\vee}b \neq c{\vee}d \supset a \neq c \vee b \neq d$.

5. AN EXAMPLE: THE LATTICE OF OPEN INTERVALS

Let again $R \times R$ be as in example 4. This time we define a relation for intervals (a, b) with quite a concrete sense in which one interval exceeds another:

$$(a, b) \nleqslant (c, d) \equiv a < b \;\&\; (a < c \vee d < b). \tag{6.1}$$

It is straightforward to verify that (6.1) gives a positive partial order. All the defined relations for intervals will turn out right (for details, see our 1999).

Next we define the meet and join of two intervals:

$$(a, b) \wedge (c, d) \equiv \; (max(a, c), min(b, d)), \tag{6.2}$$
$$(a, b) \vee (c, d) \equiv \; (min(a, c), max(b, d)). \tag{6.3}$$

The positive lattice axioms are satisfied: For the axioms

$$(a, b) \wedge (c, d) \leqslant (a, b) \quad \text{and} \quad (a, b) \wedge (c, d) \leqslant (c, d)\,,$$

assume $(a, b) \wedge (c, d) \nleqslant (a, b)$. Then

$$max(a, c) < min(b, d) \quad \text{and} \quad max(a, c) < a \vee b < min(b, d)\,,$$

but the latter is impossible since

$$a \leqslant max(a, c) \quad \text{and} \quad min(b, d) \leqslant b\,.$$

The second meet law and the corresponding laws for join are verified similarly. For constructive uniqueness, the assumption states that

$$a < b \;\&\; (a < max(c, e) \vee min(d, f) < b).$$

We have to prove that

$$(a < b \;\&\; (a < c \vee d < b)) \vee (a < b \;\&\; (a < e \vee f < b)).$$

If $a < max(c, e)$, axiom MXU gives $a < c \vee a < e$, and conclusion follows in both cases. If $min(d, f) < b$, axiom MNU gives $d < b \vee f < b$ and conclusion follows similarly. The uniqueness principle for join is verified analogously.

Since meet and join for intervals satisfy the positive lattice axioms, strong extensionality of these lattice operations follows by theorem 5.2:

$$(a_1, b_1) \wedge (c_1, d_1) \neq (a_2, b_2) \wedge (c_2, d_2) \supset (a_1, b_1) \neq (a_2, b_2) \vee (c_1, d_1) \neq (c_2, d_2),$$
$$(a_1, b_1) \vee (c_1, d_1) \neq (a_2, b_2) \vee (c_2, d_2) \supset (a_1, b_1) \neq (a_2, b_2) \vee (c_1, d_1) \neq (c_2, d_2).$$

No new principles have been assumed, in addition to those for constructive linear order with minimum and maximum. Therefore open intervals of the formal reals of Negri and Soravia (1999) give a concrete example of a positive lattice.

6. ALGEBRAIC CHARACTERIZATION OF POSITIVE LATTICES

In usual lattice theory, one shows that the meet and join operations obey certain algebraic laws, and that conversely, the partial order can be characterized algebraically. We shall briefly review these well known results, and then give a constructive algebraic characterization of positive lattices.

From lattice axioms to algebraic laws:

Assume a reflexive and transitive partial order, with equality defined as

$$a = b \equiv a \leqslant b \ \&\ b \leqslant a, \tag{7.1}$$

and meet and join operations satisfying

$$a \wedge b \leqslant a, \quad a \wedge b \leqslant b, \qquad a \leqslant a \vee b, \quad b \leqslant a \vee b, \tag{7.2}$$

$$c \leqslant a \ \&\ c \leqslant b \supset c \leqslant a \wedge b, \qquad a \leqslant c \ \&\ b \leqslant c \supset a \vee b \leqslant c, \tag{7.3}$$

$$b = c \supset a \wedge b = a \wedge c, \qquad b = c \supset a \vee b = a \vee c. \tag{7.4}$$

It is essential that substitution of equals (7.4) in the meet and join operations is postulated.

From the above assumptions, it follows that equality as defined in (7.1) is an equivalence relation:

Ref $a = a$,
Trans $a = b \ \&\ a = c \supset b = c$.

Meet and join obey the algebraic laws of idempotence, commutativity, associativity and cancellation:

Idem $a \wedge a = a, \ a \vee a = a$,
Comm $a \wedge b = b \wedge a, \ a \vee b = b \vee a$,
Assoc $a \wedge (b \wedge c) = (a \wedge b) \wedge c, \ a \vee (b \vee c) = (a \vee b) \vee c$,
Canc $a \wedge (a \vee b) = a, \ a \vee (a \wedge b) = a$.

From algebraic laws to lattice axioms:

Assume the laws Ref and Trans of equality, and meet and join operations obeying the algebraic laws Idem, Comm, Assoc, Canc, as well as the substitution principles

Substmeet $b = c \supset a \wedge b = a \wedge c$.
Substjoin $b = c \supset a \vee b = a \vee c$.

Now define

$$a \leqslant b \;\equiv\; a \wedge b = a. \tag{7.5}$$

It follows that the definition gives a reflexive and transitive partial order, and that meet and join obey the lattice axioms (7.2)–(7.4).

Definition (7.5) can be equivalently given in terms of join: If $a \wedge b = a$, then $a \vee b = (a \wedge b) \vee b = b$ and the other way round.

From positive lattice axioms to algebraic laws:

In Section 3 we showed that weak partial order, apartness and equality are definable and their properties provable in a positive partial order. The first lattice axioms (7.2) are already given by MTI, and the classical uniqueness principles (7.3) are the contrapositions of MTU. Substitution of equals is provable in the form of strong extensionality. Therefore the algebraic laws Idem, Comm, Assoc and Canc follow.

From algebraic laws to positive lattice axioms:

Assume an apartness relation $a \neq b$ with

Irref $\sim a \neq a$,
Split $a \neq b \supset a \neq c \vee b \neq c$.

Equality $a = b$ is defined as $\sim a \neq b$, and substitution of equals in an apartness follows. Next assume meet and join operations satisfying the algebraic laws Idem, Comm, Assoc and Canc, as well as the principle of strong extensionality:

Extmeet $a \wedge b \neq c \wedge d \supset a \neq c \vee b \neq d$.
Extjoin $a \vee b \neq c \vee d \supset a \neq c \vee b \neq d$.

Substmeet and Substjoin follow from these by contraposition.

Define excess by

$$a \nleqslant b \;\equiv\; a \wedge b \neq a. \tag{7.6}$$

We now come to our main result for positive lattices:

Theorem 7.1: Definition (7.6) gives an excess relation.

Proof: Irreflexivity: Assume $a \nleqslant a$, so $a \wedge a \neq a$, but this is impossible by Idem and Irref. Splitting: Let $a \nleqslant b$, so by definition $a \wedge b \neq a$. Then by Split,

$$a \wedge b \neq a \wedge c \vee a \neq a \wedge c.$$

In the latter case, $a \nleqslant c$. The former gives, again by Split,

$$a \wedge b \neq (a \wedge c) \wedge (b \wedge c) \vee a \wedge c \neq (a \wedge c) \wedge (b \wedge c).$$

In the first case, we get by the algebraic laws easily that

$$(a \wedge c) \wedge (b \wedge c) = (a \wedge c) \wedge b\,,$$

so that

$$a \wedge b \neq (a \wedge c) \wedge b.$$

Extmeet and Irref now give $a \neq a \wedge c$, so $a \nleqslant c$. In the second case, Extmeet gives,

$$a \neq a \wedge c \vee c \neq b \wedge c,$$

so by definition, $a \nleqslant c \vee c \nleqslant b$. QED.

As with definition (7.5), also (7.6) can be given in terms of join: If $a \wedge b \neq a$, Split gives us $a \wedge b \neq a \wedge (a \vee b) \vee a \neq a \wedge (a \vee b)$. Latter disjunct is impossible according to Canc and Irref, the former gives $b \neq a \vee b$ by Extmeet.

Finally, we check that the positive lattice axioms MTI, JNI, MTU and JNU are satisfied:

Theorem 7.2: Definition (7.6) gives the lattice laws MTI, JNI, MTU and JNU:

(i) $a \wedge b \leqslant a, \quad a \wedge b \leqslant b$, (ii) $a \leqslant a \vee b, \quad b \leqslant a \vee b$,

(iii) $c \nleqslant a \wedge b \supset c \nleqslant a \vee c \nleqslant b$, (iv) $a \vee b \nleqslant c \supset a \nleqslant c \vee b \nleqslant c$.

Proof: (i) Assume $a \wedge b \nleqslant a$, so $(a \wedge b) \wedge a \neq a \wedge b$ which leads by Assoc and Idem to $a \wedge b \neq a \wedge b$. Therefore $a \wedge b \leqslant a$. Similarly $a \wedge b \leqslant b$. (ii) Proof dual to previous. (iii) Let $c \nleqslant a \wedge b$, so $c \wedge (a \wedge b) \neq c$. Since $c \wedge (a \wedge b) = (c \wedge a) \wedge (c \wedge b)$ and $c = c \wedge c$, our assumption gives $(c \wedge a) \wedge (c \wedge b) \neq c \wedge c$, and Extmeet leads to the desired conclusion $c \wedge a \neq c \vee c \wedge b \neq c$. (iv) is proved similarly. QED.

The usual definition (7.5) of weak partial order is obtained by negating both sides in definition (7.6).

7. CONCLUDING REMARKS

If one should only assume a usual weak partial order that admits an apartness relation, but not its refinement with excess, the axioms would be as follows: a reflexive and transitive partial order, and an apartness relation. Strict partial order is defined by

$$a < b \equiv a \leqslant b \,\&\, a \neq b,$$

but as mentioned in Section 4, transitivity of strict order must be assumed, as we cannot get $a \neq c$ from $a \neq b$ and $b \neq c$. This approach is taken in Greenleaf (1979). In the example of Section 4, with points on parallel lines, we noted that the componentwise definition of apartness (4.2) gives transitivity of strict order. But (4.2) is just the principle of strong extensionality for the pair forming operation. A further concept that would need to be added to usual partial order is incomparability, together with the principles (iv) of theorem 3.1. One would also need the principle

$$a \parallel b \,\&\, a \neq b \supset\subset a < b \vee b < a.$$

Still a basic fault will remain, as strong extensionality of the lattice operations does not follow. If, on the other hand, one starts with a strict partial order and an apartness and an incomparability, the resulting natural axiomatization will be equivalent to the one with weak order.

The notion of excess we introduced is the right constructive counterpart to weak partial order, as is shown by the constructivization of the algebraic characterization of partial order through lattice axioms. This new notion also makes it clear where a principle of strong extensionality comes from: It is a consequence of a uniqueness axiom formulated positively. This seems to be a general phenomenon. For more examples and discussion, see our (1996).

Classically, a function $f : A \to B$ has the property

$$x = y \supset f(x) = f(y).$$

The positive form of this principle is called 'strong extensionality' in Troelstra and van Dalen (1988, p. 386), and 'pull-back' in Greenleaf (1979, p. 12):

$$f(x) \neq f(y) \supset x \neq y.$$

A principle to this effect can be found in Weyl (1926, p. 22), where it is referred to Helmholtz. In the present terms, we can express it as follows: If two different things are observed, here $f(x)$ and $f(y)$, then, we can at least infer that the arguments x and y were different (even if, we could complete Weyl's thought, x and y themselves should be unobservable).

A more detailed study of the lattice of open intervals of the intuitionistic continuum is given in our (1999). This paper also contains a discussion of earlier work on constructive order relations. In particular, we found that our two axioms for linear order PLO1 and PLO2 are already presented in Scott (1968), and later rediscovered by Bridges (1989).

In the equational reasoning that leads to the algebraic characterization of usual lattices, we typically have to establish some equality $a = b$. To derive it, we try to find a 'middle term' c, or sequence of such terms, such that $a = c \;\&c = b$, whence the desired equality follows by transitivity. In the corresponding constructive inference with apartnesses, we start from some apartness $a \neq b$, infer $a \neq c \vee b \neq c$, and then eliminate the disjunction. The difficulty consists in finding the right 'auxiliary term' c, one that permits us to conclude the desired apartness by strong extensionality. In equational inference, the middle terms come stepwise, whereas with apartnesses, the auxiliary terms seem to be more global, making the latter kind of inferences more difficult to perform.

References

Bishop, E. (1967) *Foundations of Constructive Analysis*, McGraw-Hill, New York.

Bridges, D. (1989) The constructive theory of preference relations on a locally compact space, *Proceedings of the Koninklijke Nederlandse Akademie van Wetenschappen*, Ser. A92, pp. 141–165.

Bridges, D. (1999) Constructive mathematics: a foundation for computable analysis, *Theoretical Computer Science*, vol. 219, pp. 95–109.

Cederquist, J. (1998) An implementation of the Heine–Borel covering theorem in type theory, in E. Gimenez and C. Paulin, eds, *Types for Proofs and Programs*, pp. 46–65 (LNCS, vol. 1512).

Greenleaf, N. (1979) Linear order in lattices: a constructive study, *Advances in Mathematics Supplementary Studies*, vol. 1, pp. 11–30.

Negri, S. (1999) Sequent calculus proof theory of intuitionistic apartness and order relations, *Archive for Mathematical Logic*, vol. 38, pp. 521–547.

Negri, S. and D. Soravia (1999) The continuum as a formal space, *Archive for Mathematical Logic*, vol. 38, pp. 423–447.

von Plato, J. (1996) Organization and development of a constructive axiomatization, in S. Berardi and M. Coppo, eds, *Types for Proofs and Programs*, pp. 288–296 (LNCS, vol. 1158), Springer, Berlin.

von Plato, J. (1999) Order in open intervals of computable reals, *Mathematical Structures in Computer Science*, vol. 9, pp. 103–108.

Scott, D. (1968) Extending the topological interpretation to intuitionistic analysis, *Compositio Mathematica*, vol. 20, pp. 194–210.

Troelstra, A. and D. van Dalen (1988) *Constructivism in Mathematics*, vol. 2, North-Holland, Amsterdam.

Weyl, H. (1926) *Philosophie der Mathematik und Naturwissenschaft*, Oldenbourg, Munich and Berlin. English version *Philosophy of Mathematics and Natural Science*, Princeton 1949.

CONSTRUCTIVE MATHEMATICS WITHOUT CHOICE

Fred Richman
Florida Atlantic University
Boca Raton, FL 33431
richman@fau.edu

Keywords: axiom of choice, constructive mathematics.

Abstract: What becomes of constructive mathematics without the axiom of (countable) choice? Using illustrations from a variety of areas, it is argued that it becomes better.

Despite the apparent unanimity among schools of constructive mathematics with respect to the acceptance of the countable axiom of choice, I believe it to be one of the central problems, or mysteries, of constructive mathematics. The axiom may be written as

$$\forall m \, \exists n \, P(m, n) \Longrightarrow \exists \alpha \, \forall m \, P(m, \alpha_m) \tag{CC}$$

where P is a binary predicate, m and n are in $\mathbf{N}$, and $\alpha \in \mathbf{N}^{\mathbf{N}}$.

Brouwer's choice sequences constitute one approach to this problem. Such sequences may be thought of as finite sequences that have the potential of going on indefinitely. That is, they embody the hypothesis of CC (or even the axiom of dependent choices): the ability to go on. We could say that Brouwer solved the problem of CC by defining the conclusion to be the hypothesis.

Other justifications of CC, including some by intuitionists, consider how we could know that the hypothesis was true, and conclude that we would have to possess an infinite sequence α as in the conclusion. We could say that this solves the problem of CC by defining the hypothesis to be the conclusion.

I would like to suggest that the way to approach the problem is in accordance with Bishop's fourth principle of constructivism, that meaningful distinctions deserve to be maintained [2, page 5]. That is, we should reject CC.

P. Schuster, U. Berger and H. Osswald (eds.),
Reuniting the Antipodes – Constructive and Nonstandard Views of the Continuum, 199–205.

It's not a question of whether you *believe* in CC or not—it is a question of whether you prefer doing mathematics with CC or without CC. I will argue that the mathematics becomes better if you reject this axiom, in much the same way that the mathematics becomes better if you reject the law of excluded middle. In particular, one consequence (or cause) of the acceptance of CC is the predominance of sequences in constructive practice. When we reject CC we find ourselves forced to formulate our theorems in nonsequential terms, which in many cases are more natural.

Three case studies are considered: the Hilbert syzygy theorem, the theory of trace-class operators on a Hilbert space, and the fundamental theorem of algebra.

1. THE HILBERT SYZYGY THEOREM

This theorem was a watershed in my thinking about countable choice. The first time I had even thought to question countable choice was when Wim Ruitenburg asked if we would use it in our book, *A course in constructive algebra*. We agreed that there were already enough problems to face, so we made no systematic effort to eliminate choice from our arguments. I didn't worry too much about it until I was explaining a constructive Hilbert syzygy theorem to an audience of algebraists.

The (abstract) Hilbert syzygy theorem relates the projective dimension of modules over a ring R to modules over the ring of polynomials $R[X]$. A module P is **projective** if whenever you have a diagram

$$\begin{array}{ccc} & & P \\ & & \downarrow \\ A & \longrightarrow & B \end{array}$$

where $A \to B$ is onto, you can lift to a map $P \to A$. The simplest nontrivial example of a projective module is the ring R viewed as a module over itself. This is a free module on one generator—to define a map from R to another R-module, you just have to say where 1 goes. It is easy to see that finite-rank free modules (direct sums of finitely many copies of R) are also projective. An example of an infinite-rank free R-module is $R[X]$, which is free on the generators $1, X, X^2, \ldots$.

Free modules are projective in the presence of the full axiom of choice. Note that the axiom of choice says that every set is projective in the category of sets (just interpret the diagram above as a diagram of sets and functions). To show that $R[X]$ is projective requires some form of countable choice.

The *projective dimension* of a module A is 0, written $\operatorname{pd} A = 0$, if A is projective. Inductively, we say that $\operatorname{pd} B \leq n+1$ if $B \cong P/A$, where P is projective and $\operatorname{pd} A \leq n$.

The abstract Hilbert syzygy theorem says that

> $\operatorname{pd} A \leq n$ *for all* R*-modules* A, *if and only if* $\operatorname{pd} A \leq n+1$ *for all* $R[X]$*-modules* A

The concrete theorem has to do with the projective dimension of finitely presented modules over $k[X_1, \ldots, X_n]$ where k is a (discrete) field. A module is finitely presented if it can be written as the quotient a finite-rank free module by a finitely generated submodule. Passing from the abstract to the concrete is already a problem for $n = 0$ because not every module over a field is free, nor is every free module projective. I pointed this out to my audience, noting that the full axiom of choice implies the law of excluded middle. So I had to restrict the class of R-modules that occur in the abstract theorem. The natural choice is finitely presented modules, the modules of interest in the concrete case.

Because the proof used the fact that $R[X]$ is projective, I had to explain to my audience, and myself, that although the full axiom of choice was unacceptable, the countable axiom of choice was okay. This is a tough sell. Accordingly, I was motivated to look for a choice-free development.

In fact, avoiding countable choice leads to a better treatment. One is led to the conclusion that the real theorem is about *flat* dimension, a much more tractable concept than projective dimension from a constructive point of view. The difference is illustrated by considering modules over the ring $\mathbf{Z}$ of integers, that is, abelian groups. In the presence of the full axiom of choice, a $\mathbf{Z}$-module is projective if and only if it is free. Flat $\mathbf{Z}$-modules are exactly the torsion-free abelian groups—groups in which the only element of finite order is the identity. The additive group of rational numbers is an example of a flat $\mathbf{Z}$-module that is not free.

It is easy to prove that all free modules, and all projective modules, are flat. In addition, every module over a field is flat. Replacing "projective" by "flat", the abstract Hilbert syzygy theorem becomes

> $\operatorname{fd} A \leq n$ *for all* R*-modules* A, *if and only if* $\operatorname{fd} A \leq n+1$ *for all* $R[X]$*-modules* A

(see [6]). One need not restrict to finitely presented modules A, as seems to be the case for projective dimension. Moreover, if the ring R is coherent, then the projective dimension of a finitely presented module is the same as its flat dimension, so the concrete Hilbert syzygy theorem follows immediately from this version of the abstract theorem.

2. TRACE-CLASS OPERATORS

Bishop defined a Hilbert space to be separable, and remarked in the preface to [1] that he was avoiding "pseudogenerality." With this convention, a closed subspace of a Hilbert space need not be a Hilbert space. Separable Hilbert spaces have orthonormal bases, but you use countable choice to prove this. Indeed, without countable choice, the separability hypothesis is generally rather useless. This is a side benefit of rejecting countable choice: you eliminate separability hypotheses from your theorems.

A basis is essentially an aligned family of *frames*—finite sets of orthogonal unit vectors. By concentrating on frames in general, and not on some particular aligned family, we get basis-free approaches. As an example, we consider trace-class operators.

In most treatments of trace-class operators, one assumes that the operator T has an adjoint, or at least an absolute value (a positive operator $|T|$ such that $\langle |T|\, x, |T|\, x\rangle = \langle Tx, Tx\rangle$ for all x). Then one takes a basis $e_1, e_2, \ldots$ of the Hilbert space and considers the sum

$$\sum_{i=1}^{\infty} |\langle |T|\, e_i, e_i\rangle|$$

If this sum converges, then T is of **trace class**. Of course one must show that the definition is independent of the particular basis.

Without countable choice we must formulate a definition that does not involve a basis, unless we want to restrict ourselves to Hilbert spaces that have bases. One might have wanted a basis-free definition of trace class in any event, but rejecting countable choice makes it imperative. In an analogous situation, anyone might prefer direct (or constructive) proofs, but rejection of the law of excluded middle requires them.

One suitable definition for an operator T to be of trace class is for the supremum

$$\|T\|_1 = \sup \sum_{i=1}^{n} |\langle Te_i, f_i\rangle|$$

to exist, where $e_1, \ldots, e_n$ and $f_1, \ldots, f_n$ range over orthonormal sets of size n, and n ranges over the positive integers [7]. So T is of trace class if a certain set of real numbers has a supremum. This elegant, basis-free definition comes about as a direct consequence of rejecting countable choice.

We can also pick out the singular values of the operator T from the expression for $\|T\|_1$. If we let $\|T\|_1^{(n)}$ denote the supremum for fixed n, then the n-th singular value is given by

$$\sigma_n = \|T\|_1^{(n)} - \|T\|_1^{(n-1)}$$

provided that the two suprema exist. (It is true, but not obvious, that if $\|T\|_1$ exists, then so does each $\|T\|_1^{(n)}$.)

3. COMPLETENESS

Hilbert spaces are complete. What does that mean? One uses countable choice to show that the Cauchy reals are sequentially complete [9, page 267 N.B.]. This indicates that Cauchy reals are the wrong notion in the absence of countable choice. In general, we rely on CC to show that the sequential completion of a metric space is sequentially complete. So sequential completeness is a concept of dubious value in the absence of CC.

We can define real numbers to be located Dedekind cuts in the rational numbers, but for an arbitrary metric space we need a more general notion. The idea is that an element of the completion $\hat{X}$ of a metric space X is something you can approximate with elements of X. Technically, to specify an element θ of $\hat{X}$, you construct, for each positive rational number r, a nonempty subset S_r which is to consist of r-approximations to θ. The defining property is if that $x \in S_r$ and $y \in S_t$, then $d(x, y) \leq r + t$. There is a natural isometry of X into $\hat{X}$, and X is complete if this isometry is onto.

We will use this notion of a completion to describe the fundamental theorem of algebra in a choiceless environment.

4. THE FUNDAMENTAL THEOREM OF ALGEBRA

It came as a bit of a surprise to find out that the fundamental theorem of algebra requires countable choice in the absence of the law of excluded middle. At least that is the case for complex numbers based on Dedekind reals. For complex numbers based on Cauchy reals, Ruitenburg [8] showed that the fundamental theorem of algebra holds.

For complex numbers based on Dedekind reals there is already a problem with square roots, that is, with the polynomial $X^2 - a$. The problem is with small values of a, and is a reflection of the fact that the function

X^2 does not admit a continuous cross-section in a neighborhood of zero. If we ever find out that a is nonzero, then we have to choose between the two square roots of a. With the law of excluded middle, we could simply consider the two cases $a = 0$ and $a \neq 0$ separately. In fact, the following weak countable choice principle [3], which is implied both by CC and by LEM, suffices:

> *Given a sequence A_n of nonempty sets, at most one of which is not a singleton, then there is a choice sequence $a_n \in A_n$.*

What is the story without choice? Let $M_n(\mathbf{C})$ denote the set of n-multisets of complex numbers, with a metric given by

$$\inf_{\sigma \in S_n} \sup_i |r_i - s_{\sigma i}| .$$

We identify two elements r and s of $M_n(\mathbf{C})$ when the distance between them is zero. One can show that the correspondence between the element $r_1, \ldots, r_n$ of $M_n(\mathbf{C})$ and the coefficients of the polynomial $(X - r_1) \cdots (X - r_n)$ is uniformly continuous on bounded subsets (in both directions).

Given a monic polynomial $p(X)$ of degree n, we can construct Gaussian numbers $r_1, \ldots, r_n$ that approximate the n roots of $p(X)$ in the sense that $(X - r_1) \cdots (X - r_n)$ is close to $p(X)$. What we are constructing is an element of the completion $\widehat{M_n}(\mathbf{Q}[i])$, where $\mathbf{Q}[i]$ denotes the Gaussian numbers. It is natural to call this element the **spectrum** of $p(X)$.

Note that the spectrum need not necessarily be in $M_n(\mathbf{C})$, which is a subset of $\widehat{M_n}(\mathbf{Q}[i])$ in a natural way. An element of $M_n(\mathbf{C})$ constructed from $p(X)$ in this way is a complete set of roots for $p(X)$, the multiset of a complete factorization. The spectrum of $p(X)$ only allows us to find polynomials near $p(X)$ that admit complete factorizations. In practice, this is what algorithms that implement the fundamental theorem of algebra do—for a given ε, they construct an n-multiset of Gaussian numbers that approximate the roots of $p(X)$ within ε.

References

[1] Bishop, Errett (1967). *Foundations of constructive analysis.* McGraw-Hill.

[2] ―――――― (1985). *"Schizophrenia in contemporary mathematics,"* in Errett Bishop: Reflections on him and his research, Contemp. Math, AMS.

[3] Bridges, Douglas, Fred Richman and Peter Schuster (2000). A weak countable choice principle, *Proc. Amer. Math. Soc.*, **128**, 2749–2752.

[4] Mines, Ray, Fred Richman and Wim Ruitenburg (1988). *A course in constructive algebra*, Springer-Verlag.

[5] Richman, Fred (2000). The fundamental theorem of algebra: a constructive development without choice, *Pac. J. Math.*, **196**, 213–230

[6] ———— (1997). Flat dimension and the Hilbert syzygy theorem, *New Zealand J. Math.*, **26**, 263–273.

[7] Richman, Fred, Douglas Bridges and Peter Schuster (preprint). Trace-class operators, www.math.fau.edu/Richman/html/docs.htm.

[8] Ruitenburg, Wim B. G. (1990). Constructing roots of polynomials over the complex numbers. *Computational aspects of Lie group representations and related topics* Amsterdam, 107–128, CWI Tract, **84**, Math. Centrum, Centrum Wisk. Inform., Amsterdam, 1991.

[9] Troelstra, Anne S., and Dirk van Dalen (1988), *Constructivism in mathematics: an introduction*, North-Holland

POINTWISE DIFFERENTIABILITY

Fred Richman
Florida Atlantic University
Boca Raton, FL 33431
richman@fau.edu

Keywords: pointwise differentiability, constructive mathematics.

Abstract: We study what can be done with pointwise properties as opposed to uniform properties on compact intervals.

Bishop defines continuity and differentiability of functions in terms of uniform behavior on compact intervals. So "continuity" is really uniform continuity on compact intervals, and "differentiability" entails uniform convergence of the difference quotient on compact intervals. Bishop says that "the concept of a pointwise continuous function is not relevant." This is a sentiment with which I have a lot of sympathy. Nevertheless, the purpose of this note is to see what can be done with the pointwise definitions.

One feature of the pointwise approach is the prominent role played by countable choice. If, like me, you prefer to reject countable choice, then you seem to be forced to use Bishop's approach. If, indeed, Bishop's approach is the right one, then this is an argument for rejecting countable choice, for by so doing you are forced to do the right thing.

1. THE MEAN VALUE THEOREM

A function f, defined on a nontrivial interval $[a, b]$, has a **derivative** m at a point x in $[a, b]$ if for each $\varepsilon > 0$ there exists $\delta > 0$ so that for all y in $[a, b]$, if $|x - y| \leq \delta$, then $|f(x) - f(y) - m(x - y)| \leq \varepsilon |x - y|$. This is the pointwise version of Bishop's definition. It suffices to require this condition only for $y \neq x$, which is the the usual definition.

If $x \neq y$ are real numbers, define the difference quotient by

$$\Delta_f(x, y) = \frac{f(x) - f(y)}{x - y}.$$

P. Schuster, U. Berger and H. Osswald (eds.),
Reuniting the Antipodes – Constructive and Nonstandard Views of the Continuum, 207–210.

If I is the interval $[x, y]$, we also write $\Delta_f(I)$ for $\Delta_f(x, y)$. Note that $\Delta_f(x, y)$ is symmetric in x and y.

An immediate consequence of these definitions is

Lemma 1 *Let f be defined on $[a, b]$ and differentiable at x. If I_n is a sequence of subintervals of $[a, b]$ containing x, whose lengths converge to 0, then $\Delta_f(I_n) \to f'(x)$.*

Proof. Suppose $a \leq c \leq x \leq d \leq b$. It suffices to show that if $d - c > 0$ is small, then so is $\Delta_f(c, d) - f'(x)$. If $d - c < \delta$, then

$$\left|f(x) - f(c) - f'(x)(x - c)\right| \leq \varepsilon(x - c)$$

and $|f(d) - f(x) - f'(x)(d - x)| \leq \varepsilon(d - x)$, so

$$\left|f(d) - f(c) - f'(x)(d - c)\right| \leq \varepsilon(d - c),$$

so $|\Delta_f(c, d) - f'(x)| \leq \varepsilon$. ■

With this lemma in hand, we can prove a mean value theorem (using countable choice) that suffices for the usual applications. In the theorem, the indicated supremum need not exist, but it is clear what it means for it to exceed a real number.

Theorem 2 *Let f be differentiable on the nontrivial interval $[a, b]$. Then*

$$\sup_{x \in [a,b]} f'(x) \geq \Delta_f(a, b).$$

Proof. The conclusion means that, for each $\varepsilon > 0$, there exists x in $[a, b]$ such that $f'(x) \geq \Delta_f(a, b) - \varepsilon$. We will construct a sequence of compact intervals $[a, b] = I_0, I_1, I_2, \ldots$ such that I_n is one of the two pieces obtained by bisecting I_{n-1}, and $\Delta_f(I_n) > \Delta_f(I_{n-1}) - \varepsilon/2^n$. It follows from the lemma that $f'(x) \geq \Delta_f(a, b) - \varepsilon$ for $x \in \bigcap I_n$.

Set $I_0 = [a, b]$. If I_{n-1} has been constructed, let L and R be its left and right halves. Then $\Delta_f(L) + \Delta_f(R) = 2\Delta_f(I_{n-1})$, so either $\Delta_f(L)$ or $\Delta_f(R)$ exceeds $\Delta_f(I_{n-1}) - \varepsilon/2^n$. Therefore we can choose I_n as desired. ■

Corollary 3 *Let f be differentiable on the nontrivial interval $[a, b]$. If $f'(x) \geq 0$ for each x in $[a, b]$, then $f(b) \geq f(a)$. If, in addition, $f'(x) > 0$ for some $x \in [a, b]$, then $f(b) > f(a)$.*

Proof. For the first claim we have

$$\Delta_f(a,b) = -\Delta_{-f}(a,b) \geq -\sup_{x\in[a,b]} -f'(x) = \inf_{x\in[a,b]} f'(x) \geq 0.$$

For the second claim, there exist $c < d$ in (a,b) such that $f(c) < f(d)$. So, using the first claim, we have $f(a) \leq f(c) < f(d) \leq f(b)$. ∎

Kushner [3, Theorem 1, page 202] shows that if $f'(x) > 0$ for each $x \in (a,b)$, then $f(b) > f(a)$. (Keep in mind that, for Kushner, all functions are pointwise continuous by Tseĭtin's theorem. Actually, he appeals to the nondiscontinuity theorem, rather than to Tseĭtin's theorem.) The key argument is his proof of [3, Lemma 1, page 200], that if f is increasing in a neighborhood of each point of (a,b), then f is increasing on $[a,b]$. This argument is essentially the same as that given in the proof of Theorem 2 above.

Strictly increasing pointwise continuous functions satisfy the intermediate value theorem (no choice necessary).

Theorem 4 *Let f be a strictly increasing, pointwise continuous function on the nontrivial interval $[a,b]$. If $f(a) \leq c \leq f(b)$, then there exists x in $[a,b]$ such that $f(x) = c$.*

Proof. Let $a = x_0, x_1, \ldots, x_n = b$ be equally spaced points in $[a,b]$. As f is strictly increasing, there exists i such that $f(x_i) \leq c \leq f(x_{i+2})$, and there may be up to three such i, but not four. The set of such intervals $[x_i, x_{i+2}]$, as n varies, has the finite intersection property, and contains arbitrarily small intervals. Thus, by the completeness of $[a,b]$, the intersection of this set contains a unique number x. The conclusion follows from the continuity of f at x. ∎

2. EXAMPLE

Let $0 = q_0 < q_1 < q_2 < \cdots$ be a Specker sequence, that is, a sequence of rational numbers in $[0,1]$ that is eventually bounded away from any real number (see [2], for example). Let a_n be a sequence of rational numbers such that $q_n < a_n < q_{n+1}$ and $n(a_n - q_n) \leq 2^{-n}$. Define $f(x)$ on $B = \bigcup[0,q_n]$ to be the piecewise linear function determined by $f(q_n) = f(a_n) = 0$ and $f((q_n + a_n)/2) = n$. Then

$$\int_{q_n}^{q_{n+1}} f(x)dx \leq \frac{1}{2^{n+1}}.$$

Note that $\bigcup[q_n, a_n]$ is located. Let C be its metric complement in $[0,1]$. As $f(x) = 0$ for x in $B \cap C$, we can extend f to $[0,1]$ by defining f to be 0 on C.

The function $F(x) = \int_0^x f(t)dt$ exists for each $x \in [0,1]$, and $F'(x) = f(x)$. The function f is pointwise continuous but not uniformly continuous on $[0,1]$ (it is unbounded), so the function F has a continuous pointwise derivative on $[0,1]$, but does not have a derivative in the sense of Bishop.

In what sense does the $\int_0^x f(t)dt$ exist? The supremum of $\int_0^x g(t)dt$ exists, where g ranges over uniformly continuous nonnegative functions that are dominated by f. It exists in the sense of Bishop-Cheng: There is a sequence f_n of uniformly continuous functions with compact support such that $\sum_n \int_0^x |f_n(x)|\,dx$ converges, and $f(x) = \sum_n f_n(x)$ whenever $\sum_n |f_n(x)|$ converges (which it always does). (See [1, page 222].) Indeed, f_n is the restriction of f to $[0, q_n]$.

References

[1] Bishop, Errett and Douglas Bridges (1985). *Constructive analysis.* Springer-Verlag.

[2] Bridges, Douglas and Fred Richman (1985). *Varieties of constructive mathematics.* Cambridge University Press.

[3] Kushner, Boris A. (1973) *Lectures on constructive mathematical analysis.* Amer. Math. Soc. 1984.

ON CONWAY NUMBERS AND GENERALIZED REAL NUMBERS

Conway's Theory of Games and Numbers Constructively Reconstructed

Frank Rosemeier
FernUniversität Hagen
Postfach 940, D-58084 Hagen (Germany)
frank.rosemeier@fernuni-hagen.de

Keywords: Constructive mathematics, Conway games, Conway numbers.

Mathematical Subject Classification: 03F65 (03H15, 91A05)

Abstract: John Horton Conway presents in his book "On Numbers and Games" a general method to create a class of numbers containing all real numbers as well as every ordinal number. Using the logical law of excluded middle (LEM) he equips this class with the structure of a totally ordered field. This paper is a first step to investigate the contribution of Conway's theory to the foundations of Constructive Nonstandard Analysis. In his book Conway suggests defining real numbers as (Conway) cuts in the set of rational numbers. Following his ideas, a constructive notion of real numbers will be developed. Parallels to and differences from the concept of generalized real numbers recently published by Fred Richman [Indag. Mathem., N. S., 9 (4) 595-606 (1998)] will be outlined.

Introduction

In his book "On Numbers and Games" [2] John H. Conway develops a very general theory of numbers and games, frequently using the logical law of excluded middle (LEM). This paper aims to start a constructive investigation of this theory. Following his ideas, constructive notions for Conway games and Conway numbers will be developed and a constructive version of Conway's theory will be given. We shall mark any application of (LEM), constructively rejected omniscience or choice principles are avoided. Whether the author's aim to avoid even countable choice has been achieved may be judged by mathematicians with more

P. Schuster, U. Berger and H. Osswald (eds.),
Reuniting the Antipodes – Constructive and Nonstandard Views of the Continuum, 211–227.

experience in working without choice. (Fred Richman suggested to drop countable choice at the Antipodes-Symposion, cf. also [7] and [9].)

Conway games are defined in Section 1 and operations of addition and subtraction for such games are presented in Section 2 resp. 3. The relations of order and equality are shown to have the expected properties in Sections 4–5. Section 6 (resp. 7) deals with *(real) Conway numbers.* In Section 8 real Conway numbers are compared with generalized Dedekind reals (cf. [6] and [9]).

1. CONWAY GAMES

1.1 Motivation *Conway games* are played by two players (usually called *Left* and *Right*) moving alternately according to specific rules without chance moves and without hidden information. Such a game is characterized by the positions each of the two players can reach from any position with the next move. Thus, a Conway game x will be described by two sets L_x and R_x, the sets of *Left* resp. *Right options* (i. e. positions reachable by Left resp. Right from the starting position of x within one move). As every position P in a game x can be identified with the *shortened game* x_P (which is played according to the rules of x starting from position P) the sets L_x and R_x will be identified with sets of Conway games. Vice versa, whenever L and R are sets of Conway games, we can construct a new Conway game $\{L|R\}$, in which Left may move to any element of L whereas Right may move to any element of R. Having this in mind the following definition can be given.

1.2 Definition *(Conway games)*
For every set X let $\Gamma(X) := \mathcal{P}(X) \times \mathcal{P}(X)$ be the set of pairs of subsets of X. Define $G_0 := \Gamma(\emptyset),\ G_1 := \Gamma(G_0),\ G_2 := \Gamma(G_1),\ \ldots$

$$G_\omega := \Gamma\Big(\bigcup_{k=0}^{\infty} G_k\Big),\ G_{\omega+1} := \Gamma(G_\omega),\ \text{etc.}$$

i. e. define $G_{\alpha+1} := \Gamma(G_\alpha)$ for every ordinal α, and for any lim-ordinal λ define $G_\lambda := \Gamma\left(\bigcup\{\, G_\alpha : \alpha \text{ contained in } \lambda\,\}\right)$. (Containment is introduced in A.3 of [8], where the appendix presents a constructive notion of ordinal numbers which is compatible with constructive Conway theory.)

$\mathrm{Ug}_j := \bigcup\{\, G_\alpha : \alpha \in \mathrm{On}_j\,\}$ may be called *j-th (Conway) game class* ($j \in \mathbb{N}_0$), and elements of $\mathrm{Ug}_\star := \bigcup_{j=0}^{\infty} \mathrm{Ug}_j$ are *(Conway) games.* (Conway denotes by **Ug** his proper class of "unimpartial" games, i. e. games possibly favouring one of the players, cf. [2] p. 78. The ordinal number classes On_j corresponding to the ordinals ω_{j-1} are defined in A.2 of [8].)

1.3 Notation *(Left/Right Options)*
The *projections* $\mathrm{pr_L} : \mathrm{Ug}_\star \to \mathrm{Ug}_\star$, $(L, R) \mapsto L$ and $\mathrm{pr_R} : \mathrm{Ug}_\star \to \mathrm{Ug}_\star$, $(L, R) \mapsto R$ are used to define two sets of games for every game $x \in \mathrm{Ug}_\star$: $\mathrm{L}_x := \mathrm{pr_L}(x)$ and $\mathrm{R}_x := \mathrm{pr_R}(x)$, the set of *Left* resp. *Right options* in x. Two games are called *identical* if their sets of Left options and their sets of Right options coincide: $x \equiv y :\Longleftrightarrow \mathrm{L}_x = \mathrm{L}_y$ and $\mathrm{R}_x = \mathrm{R}_y$ (x and y have the *same form*). If $x \equiv (\mathrm{L}_x, \mathrm{R}_x)$ is a game, x^{L} will be a typical element of L_x (*typical Left option*) and x^{R} will be a typical element of R_x (*typical Right option*).

Let $\{x_1, \ldots, x_n | y_1, \ldots, y_m\}$ abbreviate $(\{x_1, \ldots, x_n\}, \{y_1, \ldots, y_m\})$. Instead of $y \equiv (\{z\}, \emptyset)$ we will write $y \equiv \{z|\}$ etc. Sometimes the expression $\{x^{\mathrm{L}} | x^{\mathrm{R}}\}$ will be used as notation for the game x.

1.4 Examples (The four simplest games)
(1) $\perp \equiv \{|\} \equiv (\emptyset, \emptyset)$, the *empty game*, in which both players are unable to move, is the only element of Ug_0.
(2) $1_{\mathrm{L}} \equiv \{\perp|\} \equiv (\{\perp\}, \emptyset)$, the *Left unit game*, in which Left has a move to $\perp$ while Right is unable to move, is an element of Ug_1;
(3) $1_{\mathrm{R}} \equiv \{|\perp\} \equiv (\emptyset, \{\perp\})$, the *Right unit game*, in which Right has a move to $\perp$ while Left is unable to move, is an element of Ug_1;
(4) $* \equiv \{\perp|\perp\} \equiv (\{\perp\}, \{\perp\})$, the *Nim unit game*, in which both players have a move to $\perp$, is an element of Ug_1. (Nim is described in [1].)

1.5 Convention *(Normal play convention)*
A player unable to move *loses*, the other player is the *winner*. (No game can go on forever, cf. the Descending Chain Condition in A.3 of [8].)

1.6 Definition *(Outcome classes)*
For any $x \in \mathrm{Ug}_\star$ define (a game theoretic interpretation is given in 1.7)
$x \geq 0 :\Longleftrightarrow \forall x^{\mathrm{R}} \in \mathrm{R}_x : x^{\mathrm{R}} \rhd 0$ (Left can win if Right starts),
$x \rhd 0 :\Longleftrightarrow \exists x^{\mathrm{L}} \in \mathrm{L}_x : x^{\mathrm{L}} \geq 0$ (Left can win if Left starts),
$x \leq 0 :\Longleftrightarrow \forall x^{\mathrm{L}} \in \mathrm{L}_x : x^{\mathrm{L}} \lhd 0$ (Right can win if Left starts),
$x \lhd 0 :\Longleftrightarrow \exists x^{\mathrm{R}} \in \mathrm{R}_x : x^{\mathrm{R}} \leq 0$ (Right can win if Right starts),
$x > 0 :\Longleftrightarrow x \geq 0$ and $x \rhd 0$ (x is *positive*, Left can win),
$x < 0 :\Longleftrightarrow x \leq 0$ and $x \lhd 0$ (x is *negative*, Right can win),
$x \parallel 0 :\Longleftrightarrow x \rhd 0$ and $x \lhd 0$ (x is *fuzzy*, the first player can win),
$x = 0 :\Longleftrightarrow x \geq 0$ and $x \leq 0$ (x is *zero*, the second player can win).
(Here 'can win' stands for 'has a winning strategy'. Readers who rely

on this intuitive concept may prefer to define the outcome classes by the expresssions in parentheses; they can arrive at the formal definition given here by considering 1.7. Other readers may use the formal definitions given here to make precise the concept of *winning strategy* using 1.7.)

1.7 Remark Left can win x in case Right moves first (i. e. $x \geq 0$) if all possible Right moves lead to games which Left can win, provided Left is allowed to make the first move there. Left can win x in case Left moves first (i. e. $x \rhd 0$) if there is a Left (winning) move leading to a game which Left can win, provided Right has to move first there. Interpret the outcome classes in favour of Right ($x \leq 0$, $x \lhd 0$) analoguously.

1.8 Examples The outcome classes for the games from 1.4:
(1) $\perp = 0$ as both players are unable to move in $\perp$;
(2) $1_{\mathrm{L}} > 0$ as Left wins (by moving to $\perp$ or since Right has no move);
(3) $1_{\mathrm{R}} < 0$ as Right wins (by moving to $\perp$ or since Left has no move);
(4) $* \parallel 0$ as the first player wins by moving to $\perp$.

1.9 Proposition
(1) *For all games* $x \in \mathrm{Ug}_\star$ *we have*
(i) $\neg(x \geq 0$ and $x \lhd 0)$, (ii) $\neg(x \leq 0$ and $x \rhd 0)$.
(2) *The logical* law of excluded middle,
(LEM) ψ or $\neg\psi$ for every proposition ψ,
is equivalent to each of the following statements:
(a) $x \geq 0$ or $x \lhd 0$ *for all* $x \in \mathrm{Ug}_\star$,
(b) $x \leq 0$ or $x \rhd 0$ *for all* $x \in \mathrm{Ug}_\star$.

Proof:
(1): (i) and (ii) are proved by *mutual game induction*, i. e. by *game induction* (transfinite induction on $\alpha \in \mathrm{On}_j$, $j \in \mathbb{N}_0$, $x \in G_\alpha$) for the conjunction of (i) and (ii); the induction basis will not be mentioned as there are no options of $\{|\}$, the only element of $G_0 = \Gamma(\emptyset)$.
Ind. Step: (i) Suppose we have $x \geq 0$ and $x \lhd 0$, hence $x^{\mathrm{R}} \rhd 0$ for all $x^{\mathrm{R}} \in \mathrm{R}_x$ and $x^{\mathrm{R}} \leq 0$ for some $x^{\mathrm{R}} \in \mathrm{R}_x$; but $x^{\mathrm{R}} \rhd 0$ and $x^{\mathrm{R}} \leq 0$ would contradict Ind. Hyp. (ii).
(ii) $x \leq 0$ and $x \rhd 0$ would yield similarly a contradiction to Ind. Hyp. (i).

(2): "(LEM) $\Longrightarrow$ (a), (b)" is also proved by mutual game induction:
Since $\neg(x \lhd 0) \Longrightarrow \neg(\exists\, x^{\mathrm{R}} \in \mathrm{R}_x : x^{\mathrm{R}} \le 0) \Longrightarrow \forall\, x^{\mathrm{R}} \in \mathrm{R}_x : \neg(x^{\mathrm{R}} \le 0)$
$\Longrightarrow \forall\, x^{\mathrm{R}} \in \mathrm{R}_x : x^{\mathrm{R}} \rhd 0$ [by Ind. Hyp. (b)] $\Longrightarrow x \ge 0$,
we can deduce (a) from (LEM) via ($x \lhd 0$ or $\neg(x \lhd 0)$); (b) is deduced similarly from (LEM) and Ind. Hyp. (a).
"(a) $\Longrightarrow$ (LEM)": Let $x_\psi \equiv (\emptyset, \mathrm{R}_\psi) \in \mathrm{Ug}_\star$, $\mathrm{R}_\psi := \{\, x \in \{\bot\} : \psi \,\}$, then
$x_\psi \ge 0 \Longleftrightarrow \forall\, y \in \mathrm{R}_\psi : y \rhd 0 \Longleftrightarrow \bot \notin \{\, x \in \{\bot\} : \psi \,\} \Longleftrightarrow \neg\psi$ and
$x_\psi \lhd 0 \Longleftrightarrow \exists\, y \in \mathrm{R}_\psi : y \le 0 \Longleftrightarrow \bot \in \{\, x \in \{\bot\} : \psi \,\} \Longleftrightarrow \psi$.
"(b) $\Longrightarrow$ (LEM)" is proved analogously. □

2. ADDITION

2.1 Motivation Two games $x, y \in \mathrm{Ug}_\star$ are played simultaneously by the *simultaneous play rule*: The player to move may choose from the allowed moves in exactly one of the components x or y leaving the other component unchanged. This leads to the following inductive definition.

2.2 Definition *(Addition of Conway games)*
For games $x, y \in \mathrm{Ug}_\star$ their *(disjunctive) sum* is $x{+}y \equiv (\mathrm{L}_{x+y}, \mathrm{R}_{x+y})$ with
$\mathrm{L}_{x+y} := (\mathrm{L}_x + y) \cup (x + \mathrm{L}_y) = \{\, x^{\mathrm{L}} + y : x^{\mathrm{L}} \in \mathrm{L}_x \,\} \cup \{\, x + y^{\mathrm{L}} : y^{\mathrm{L}} \in \mathrm{L}_y \,\}$,
$\mathrm{R}_{x+y} := (\mathrm{R}_x {+} y) \cup (x {+} \mathrm{R}_y) = \{\, x^{\mathrm{R}} + y : x^{\mathrm{R}} \in \mathrm{R}_x \,\} \cup \{\, x + y^{\mathrm{R}} : y^{\mathrm{R}} \in \mathrm{R}_y \,\}$;
we write concisely $x + y \equiv \{x^{\mathrm{L}} + y, x + y^{\mathrm{L}} | x^{\mathrm{R}} + y, x + y^{\mathrm{R}}\}$.

2.3 Remark For all $\alpha, \beta \in \mathrm{On}_j$, $j \in \mathbb{N}_0$, we find $\gamma \in \mathrm{On}_j$ with $x + y \in G_\gamma$ whenever $x \in G_\alpha$, $y \in G_\beta$; hence $\mathrm{add}_{\mathrm{Ug}_\star} : \mathrm{Ug}_\star \times \mathrm{Ug}_\star \to \mathrm{Ug}_\star$, $(x, y) \mapsto x + y$ satisfies $\mathrm{add}_{\mathrm{Ug}_\star}(x, y) \in \mathrm{Ug}_j$ for all $x, y \in \mathrm{Ug}_j$, $j \in \mathbb{N}_0$.

2.4 Examples
(1) $\bot + \bot \equiv \{|\} + \{|\} \equiv \{|\} \equiv \bot$;
(2) $1_{\mathrm{L}} + \bot \equiv \{\bot|\} + \{|\} \equiv \{\bot + \bot|\} \equiv 1_{\mathrm{L}}$ by (1);
(3) $\bot + 1_{\mathrm{R}} \equiv 1_{\mathrm{R}}$ can be seen similarly;
(4) $1_{\mathrm{L}} + 1_{\mathrm{R}} \equiv \{\bot + 1_{\mathrm{R}} | 1_{\mathrm{L}} + \bot\} \equiv \{1_{\mathrm{R}} | 1_{\mathrm{L}}\}$ by (2) and (3).

2.5 Proposition ($\mathrm{Ug}_\star$ monoid)
$\mathrm{Ug}_\star$ *is a* commutative monoid *with neutral element* $\bot \equiv \{|\}$:
For all games $x, y, z \in \mathrm{Ug}_\star$ *we have*
(1) $x + \bot \equiv x$,
(2) $x + y \equiv y + x$,
(3) $(x + y) + z \equiv x + (y + z)$.

Proof: The proofs are carried out by ordinary game inductions:
(1): $x + \bot \equiv \{x^{\mathrm{L}} + \bot | x^{\mathrm{R}} + \bot\} \equiv \{x^{\mathrm{L}} | x^{\mathrm{R}}\}_{[\text{Ind. Hyp.}]} \equiv x$.
(2): $y + x \equiv \{y^{\mathrm{L}} + x, y + x^{\mathrm{L}} | y^{\mathrm{R}} + x, y + x^{\mathrm{R}}\}$
$\equiv \{x + y^{\mathrm{L}}, x^{\mathrm{L}} + y | x + y^{\mathrm{R}}, x^{\mathrm{R}} + y\}_{[\text{Ind. Hyp.}]} \equiv x + y$.
(3): $(x + y) + z \equiv \{(x^{\mathrm{L}} + y) + z, (x + y^{\mathrm{L}}) + z, (x + y) + z^{\mathrm{L}} | \ldots\}$
$\equiv \{x^{\mathrm{L}} + (y + z), x + (y^{\mathrm{L}} + z), x + (y + z^{\mathrm{L}}) | \ldots\}_{[\text{Ind. Hyp.}]} \equiv x + (y + z)$. □

2.6 Lemma (Outcome classes and addition)
For all games $x, y \in \mathrm{Ug}_\star$ *the following statements hold.*
(1) $x \geq 0$ and $y \geq 0 \implies x + y \geq 0$,
(2) $x \geq 0$ and $y \rhd 0 \implies x + y \rhd 0$,
(3) $x + y \geq 0$ and $y \leq 0 \implies x \geq 0$,
(4) $x + y \geq 0$ and $y \lhd 0 \implies x \rhd 0$,
(5) $x + y \rhd 0$ and $y \leq 0 \implies x \rhd 0$.
(2.7 gives a game theoretic interpretation of some of these implications.)

Proof:
(1) and (2) are mutually proved by a straightforward game induction.
(3), (4) and (5) are also proved by mutual game induction (cf. 1.9):
(3): $x + y \geq 0 \implies \forall x^{\mathrm{R}} \in \mathrm{R}_x : x^{\mathrm{R}} + y \rhd 0$
$\implies \forall x^{\mathrm{R}} \in \mathrm{R}_x : x^{\mathrm{R}} \rhd 0$ [by Ind. Hyp. (5)].
(4): $x + y \geq 0 \implies \forall y^{\mathrm{R}} \in \mathrm{R}_y : x + y^{\mathrm{R}} \rhd 0$
$\implies x \rhd 0$ [by Ind. Hyp. (5)],
because $y \lhd 0 \implies \exists y^{\mathrm{R}} \in \mathrm{R}_y : y^{\mathrm{R}} \leq 0$.
(5): $x + y \rhd 0 \implies \exists x^{\mathrm{L}} \in \mathrm{L}_x : x^{\mathrm{L}} + y \geq 0$ or $\exists y^{\mathrm{L}} \in \mathrm{L}_y : x + y^{\mathrm{L}} \geq 0$;
first case: $\exists x^{\mathrm{L}} \in \mathrm{L}_x : x^{\mathrm{L}} + y \geq 0 \implies x^{\mathrm{L}} \geq 0$ [by Ind. Hyp. (3)];
second case: $\exists y^{\mathrm{L}} \in \mathrm{L}_y : x + y^{\mathrm{L}} \geq 0 \implies x \rhd 0$ [by Ind. Hyp. (4)],
because $y \leq 0 \implies \forall y^{\mathrm{L}} \in \mathrm{L}_y : y^{\mathrm{L}} \lhd 0$. □

2.7 Interpretation The *first implication* of 2.6 asserts that Left can win the sum if Right starts, provided Left can win each component. Indeed, Left can find a good reply to any move of Right because there is a good reply in any component, thus Left will win by choosing always the same component as Right and playing a winning move there. The *second implication* of 2.6 means that Left having the first move can win the sum $x+y$, provided Left can win one component x with Right moving first and the other component y having the first move. Indeed, Left may start with a move from $x + y$ to $x + y^{\mathrm{L}} \geq 0$ choosing a winning move $y^{\mathrm{L}} \geq 0$ in y. Interpret the remaining implications of 2.6 analogously.

3. SUBTRACTION

3.1 Motivation

The *antigame* $-x$ is played like the original game x in which the roles of Left and Right have been interchanged: The allowed moves for Left in $-x$ correspond to the Right moves in x and the allowed moves for Right in $-x$ correspond to the Left moves in x, where the roles have to be interchanged in the options too. The following definition formalizes this idea.

3.2 Definition *(Subtraction)*
For every game $x \in \mathrm{Ug}_\star$ its *antigame* is given by $-x \equiv (\mathrm{L}_{-x}, \mathrm{R}_{-x})$ with $\mathrm{L}_{-x} := -\mathrm{R}_x = \{\, -x^{\mathrm{R}} : x^{\mathrm{R}} \in \mathrm{R}_x \,\}$ and $\mathrm{R}_{-x} := -\mathrm{L}_x = \{\, -x^{\mathrm{L}} : x^{\mathrm{L}} \in \mathrm{L}_x \,\}$; we write concisely $-x \equiv \{-x^{\mathrm{R}} | -x^{\mathrm{L}}\}$. The *difference* of $x, y \in \mathrm{Ug}_\star$ is $x - y \equiv x + (-y) \equiv \{x^{\mathrm{L}} - y, x - y^{\mathrm{R}} | x^{\mathrm{R}} - y, x - y^{\mathrm{L}}\}$.

3.3 Remark We have $-x \in \mathrm{Ug}_j$ whenever $x \in \mathrm{Ug}_j$, $j \in \mathbb{N}_0$; using $\mathrm{add}_{\mathrm{Ug}_\star}$ from 2.3 we obtain $\mathrm{sub}_{\mathrm{Ug}_\star} : \mathrm{Ug}_\star \times \mathrm{Ug}_\star \to \mathrm{Ug}_\star$, $(x, y) \mapsto x - y$ with $\mathrm{sub}_{\mathrm{Ug}_\star}(x, y) \equiv \mathrm{add}_{\mathrm{Ug}_\star}(x, -y) \in \mathrm{Ug}_j$ whenever $x, y \in \mathrm{Ug}_j$, $j \in \mathbb{N}_0$.

3.4 Examples
(1) $-\bot \equiv -\{|\} \equiv \{|\} \equiv \bot$;
(2) $-1_{\mathrm{L}} \equiv -\{\bot|\} \equiv \{|-\bot\} \equiv 1_{\mathrm{R}}$ by (1);
(3) $-* \equiv -\{\bot|\bot\} \equiv \{-\bot|-\bot\} \equiv *$ by (1);
(4) $1_{\mathrm{L}} - 1_{\mathrm{L}} \equiv 1_{\mathrm{L}} + 1_{\mathrm{R}} \equiv \{1_{\mathrm{R}}|1_{\mathrm{L}}\}$ by (2) and 2.4 (4).

3.5 Note *For all games* $x \in \mathrm{Ug}_\star$ *we have*
(1) $x \leq 0 \iff -x \geq 0$,
(2) $x \lhd 0 \iff -x \rhd 0$,
(3) $x < 0 \iff -x > 0$.
((1) and (2) are proved by mutual game induction, then (3) follows.)

3.6 Proposition
For all games $x, y \in \mathrm{Ug}_\star$ the following statements hold.
(1) $-(-x) \equiv x$,
(2) $-(x+y) \equiv (-x)+(-y)$,
(3) $-(x-y) \equiv y-x$,
(4) $x-x=0$.
(Example 3.4 (4) shows that (4) cannot be replaced by $x-x \equiv \bot$.)

Proof:
The proofs of (1) and (2) are carried out by ordinary game inductions:
(1) $-(-x) \equiv -\left\{-x^{\mathrm{R}} \middle| -x^{\mathrm{L}}\right\} \equiv \left\{-(-x^{\mathrm{L}}) \middle| -(-x^{\mathrm{R}})\right\} \underset{\text{[Ind. Hyp.]}}{\equiv} x$.
(2) $-(x+y) \equiv \left\{-(x^{\mathrm{R}}+y), -(x+y^{\mathrm{R}}) \middle| -(x^{\mathrm{L}}+y), -(x+y^{\mathrm{L}})\right\}$
$\equiv \left\{(-x^{\mathrm{R}})+(-y), (-x)+(-y^{\mathrm{R}}) \middle| (-x^{\mathrm{L}})+(-y), (-x)+(-y^{\mathrm{L}})\right\}$ [Ind. Hyp.]
$\equiv (-x)+(-y)$.
(3) is a consequence of (1), (2) and 2.5 (2).
(4): We prove $x-x \leq 0$. (This, (3) and 3.5 (1) yields $x-x \geq 0$.)
We have $x^{\mathrm{L}} - x \lhd 0$ for all $x^{\mathrm{L}} \in \mathrm{L}_x$ as $x^{\mathrm{L}} - x^{\mathrm{L}} \leq 0$ by Ind. Hyp. and we have $x - x^{\mathrm{R}} \lhd 0$ for all $x^{\mathrm{R}} \in \mathrm{R}_x$ as $x^{\mathrm{R}} - x^{\mathrm{R}} \leq 0$ by Ind. Hyp. Hence $(x-x)^{\mathrm{L}} \lhd 0$ holds for all $(x-x)^{\mathrm{L}} \in \mathrm{L}_{x-x}$, i. e. $x - x \leq 0$. □

4. ORDER

4.1 Definition *(Order)*
For games $x, y \in \mathrm{Ug}_\star$ define
$x \geq y :\Longleftrightarrow x-y \geq 0$ (x is *at least as favourable* for Left as y),
$x \leq y :\Longleftrightarrow x-y \leq 0$ (x is *at most as favourable* for Left as y),
$x \rhd y :\Longleftrightarrow x-y \rhd 0$ (x is *partly more favourable* for Left than y),
$x \lhd y :\Longleftrightarrow x-y \lhd 0$ (x is *partly less favourable* for Left than y),
$x > y :\Longleftrightarrow x-y > 0$ (x is *more favourable* for Left than y),
$x < y :\Longleftrightarrow x-y < 0$ (x is *less favourable* for Left than y),
$x = y :\Longleftrightarrow x-y = 0$ (x and y are *equally favourable* for Left),
$x \parallel y :\Longleftrightarrow x-y \parallel 0$ (x and y are *incompatible*).

4.2 Hint *For all games $x, y \in \mathrm{Ug}_\star$ we have*
(1) $x \leq y \Longleftrightarrow y \geq x \Longleftrightarrow -x \geq -y$,
(2) $x \lhd y \Longleftrightarrow y \rhd x \Longleftrightarrow -x \rhd -y$,
(3) $x < y \Longleftrightarrow y > x \Longleftrightarrow -x > -y$.
(The proofs are straightforward with 3.5 and 3.6.)

4.3 Lemma (Characterization of order)
For all games $x, y \in \mathrm{Ug}_\star$ the following statements hold.

(1) $x \leq y \iff \forall\, x^{\mathrm{L}} \in \mathrm{L}_x : x^{\mathrm{L}} \lhd y$ and $\forall\, y^{\mathrm{R}} \in \mathrm{R}_y : x \lhd y^{\mathrm{R}}$,
(2) $x \lhd y \iff \exists\, x^{\mathrm{R}} \in \mathrm{R}_x : x^{\mathrm{R}} \leq y$ or $\exists\, y^{\mathrm{L}} \in \mathrm{L}_y : x \leq y^{\mathrm{L}}$,
(3) $x > y \iff x \geq y$ and $x \rhd y$,
(4) $x = y \iff x \geq y$ and $y \geq x$,
(5) $x \parallel y \iff x \rhd y$ and $y \rhd x$.

Proof:
(1): $x \leq y \iff \forall\, (x-y)^{\mathrm{L}} \in \mathrm{L}_{x-y} : (x-y)^{\mathrm{L}} \lhd 0$
$\iff \forall\, x^{\mathrm{L}} \in \mathrm{L}_x : x^{\mathrm{L}} - y \lhd 0$ and $\forall\, y^{\mathrm{R}} \in \mathrm{R}_y : x - y^{\mathrm{R}} \lhd 0$.
(2) is proved similarly, while (3), (4) and (5) are plain. □

4.4 Note (Properties of order)
For all games $x, y, z \in \mathrm{Ug}_\star$ the following statements hold.

(1) $x \geq x$,
(2) $x \geq y$ and $y \geq z \implies x \geq z$,
(3) $x \geq y$ and $y \rhd z \implies x \rhd z$,
(4) $x \geq y \iff x + z \geq y + z$,
(5) $x \rhd y \iff x + z \rhd y + z$,
(6) $x^{\mathrm{L}} \lhd x$ and $x \lhd x^{\mathrm{R}}$ whenever $x^{\mathrm{L}} \in \mathrm{L}_x$, $x^{\mathrm{R}} \in \mathrm{R}_x$.

(For (1) and (6) use 3.6 (4), for (2), (3), (4) and (5) use 2.6.)

4.5 Remark (Properties of strict order)
For all games $x, y, z \in \mathrm{Ug}_\star$ we have

(1) $\neg(x > x)$,
(2) $x > y \implies \neg(y > x)$,
(3) $x > y$ and $y > z \implies x > z$,
(4) $x > y \iff x + z > y + z$.

((1) and (2) are proved with 1.9 (1); (3) and (4) are consequences of 4.4.)

4.6 Result $\geq$ is a *preorder relation* (reflexive and transitive) on $\mathrm{Ug}_\star$ with associated equivalence relation $=$, and $>$ is a strict partial order relation on $\mathrm{Ug}_\star$.

5. EQUALITY

5.1 Note The equivalence relation $=$ (cf. 4.1 and 4.6) is *invariant with respect to translations and reflections*, i.e. for all $x, y, z \in \mathrm{Ug}_\star$

(1) $x = y \Longrightarrow x + z = y + z$,

(2) $x = y \Longrightarrow -x = -y$.

(The proofs are straightforward with 4.3 (4), 4.4 and 3.5.)

5.2 Observation Addition and subtraction preserve equality, and $\geq$, $\rhd$, $>$ as well as $\|$ allow substitution of equals, which is to say that for all $x_1, x_2, y_1, y_2, x, y, z \in \mathrm{Ug}_\star$ the following statements hold.

(1) $x_1 = x_2$ and $y_1 = y_2 \Longrightarrow x_1 + y_1 = x_2 + y_2$,

(2) $x_1 = x_2$ and $y_1 = y_2 \Longrightarrow x_1 - y_1 = x_2 - y_2$,

(3) $x = y$ and $y \,\varrho\, z \Longrightarrow x \,\varrho\, z$ whenever $\varrho \in \{\geq, \rhd, >, \|\}$,

(4) $x \,\varrho\, y$ and $y = z \Longrightarrow x \,\varrho\, z$ whenever $\varrho \in \{\geq, \rhd, >, \|\}$.

((1) and (2) are consequences of 5.1, the proofs of (3) and (4) for $\geq$ and $\rhd$ are plain with 4.4, then (3) and (4) for $>$ and $\|$ follow easily .)

5.3 Result $\mathbf{Ug}_\star$, i.e. $\mathrm{Ug}_\star$ modulo $=$, is a partially ordered group. (Here bold print symbolizes the employment of $=$ as equality.)

5.4 Lemma

Let $x, y \in \mathrm{Ug}_\star$ and $L', R' \subseteq G_\alpha$ with $\alpha \in \mathrm{On}_j$, $j \in \mathbb{N}_0$. Then

(1) $(\mathrm{L}_x \cup L', \mathrm{R}_x) = x$, *if* $L' \lhd x$ *(i. e. if $x' \lhd x$ for all $x' \in L'$);*

(2) $(\mathrm{L}_y, \mathrm{R}_y \cup R') = y$, *if* $y \lhd R'$ *(i. e. if $y \lhd y'$ for all $y' \in R'$);*

(3) $(\mathrm{L}_x \cup L', \mathrm{R}_x \cup R') = x$, *if* $L' \lhd x \lhd R'$
(i. e. if $x' \lhd x \lhd y'$ for all $x' \in L'$, $y' \in R'$).

Proof:

In (1) let $\tilde{x} \equiv (\mathrm{L}_x \cup L', \mathrm{R}_x)$ and prove $\tilde{x} \leq x$: We have $L' \lhd x$ by assumption and $\mathrm{L}_x \lhd x$ by 4.4 (6), hence $\mathrm{L}_{\tilde{x}} = (\mathrm{L}_x \cup L') \lhd x$. Moreover, by 4.4 (6), we have $\tilde{x} \lhd \mathrm{R}_{\tilde{x}} = \mathrm{R}_x$. ($x \leq \tilde{x}$ is proved similarly.)
For (2) apply (1) with $x = -y$, $L' = -R'$, for (3) apply (1) and (2). □

5.5 Notation

For every subset $A \subseteq X$ of a set X with preorder $\leq$ we call

$\uparrow\! A := \{ x \in X : \exists\, a \in A : a \leq x \}$ the *upwards closure* of A,

$\downarrow\! A := \{ x \in X : \exists\, a \in A : x \leq a \}$ the *downwards closure* of A.

5.6 Hint $\uparrow$ and $\downarrow$ are *closure operators*:

(1) $A \subseteq \uparrow A$ and $A \subseteq \downarrow A$,

(2) $\uparrow\uparrow A = \uparrow A$ and $\downarrow\downarrow A = \downarrow A$,

(3) $A \subseteq B \Longrightarrow \uparrow A \subseteq \uparrow B$ and $\downarrow A \subseteq \downarrow B$,

(4) $(A \subseteq \uparrow B \iff \uparrow A \subseteq \uparrow B)$ and $(A \subseteq \downarrow B \iff \downarrow A \subseteq \downarrow B)$.

(The proofs are plain; *closure spaces* are presented in [5].)

5.7 Proposition

For all games $x, y \in \mathrm{Ug}_\star$ the following statements hold.

(1) $\mathrm{L}_x \subseteq \downarrow\mathrm{L}_y$ and $\mathrm{R}_y \subseteq \uparrow\mathrm{R}_x \Longrightarrow x \leq y$,

(2) $\downarrow\mathrm{L}_x = \downarrow\mathrm{L}_y$ and $\uparrow\mathrm{R}_y = \uparrow\mathrm{R}_x \Longrightarrow x = y$.

(The implications are no equivalences; for instance, $x \equiv \{*|\} = \{|\} \equiv y$ by 5.4 as $* \equiv \{\perp|\perp\} \lhd \perp \equiv \{|\}$, but $\mathrm{L}_x = \{*\} \supsetneq \emptyset = \downarrow\mathrm{L}_y$.)

Proof:

(1): Because of $\mathrm{L}_x \subseteq \downarrow\mathrm{L}_y$ we have $\mathrm{L}_x \lhd y$ (i. e. $x^{\mathrm{L}} \lhd y$ for all $x^{\mathrm{L}} \in \mathrm{L}_x$) and because of $\mathrm{R}_y \subseteq \uparrow\mathrm{R}_x$ we have $x \lhd \mathrm{R}_y$ (i. e. $x \lhd y^{\mathrm{R}}$ for all $y^{\mathrm{R}} \in \mathrm{R}_y$).
(2) is a consequence of (1). □

5.8 Interpretation Proposition 5.7 (2) is interpreted as follows: *To omit any dominated option leaves the value of a game unchanged.*
(A *dominated* option of x is a Left option $x^{\mathrm{L}} \in \downarrow(\mathrm{L}_x \setminus \{x^{\mathrm{L}}\})$ or a Right option $x^{\mathrm{R}} \in \uparrow(\mathrm{R}_x \setminus \{x^{\mathrm{R}}\})$.)

6. CONWAY NUMBERS

6.1 Motivation The difference $x^{\mathrm{L}} - x$ (resp. $x - x^{\mathrm{R}}$) is the so-called *incentive* for a move from x to x^{L} (resp. from x to x^{R}), cf. [2] p. 207.
If $x^{\mathrm{L}} < x$ (resp. $x^{\mathrm{R}} > x$) does always hold, Left (resp. Right) will try to avoid moving in x because every move would be disadvantageous. (High/low values are advantageous for Left resp. Right.)
A game $z \in \mathrm{Ug}_\star$ with this *negative incentive property*, in which in addition all (Left and Right) options also have this same property, is called *Conway number*.

6.2 Definition *(Conway numbers)*
A *Conway number* is a Conway game $z \in \mathrm{Ug}_\star$ satisfying the following conditions:
(N1) $z^{\mathrm{L}} \lhd z^{\mathrm{R}}$ for all $z^{\mathrm{L}} \in \mathrm{L}_z$, $z^{\mathrm{R}} \in \mathrm{R}_z$,
(N2) all $z^{\mathrm{L}} \in \mathrm{L}_z$ and all $z^{\mathrm{R}} \in \mathrm{R}_z$ are Conway numbers.
Analogously to 1.2 define sets $N_\alpha \subset G_\alpha$ for every $\alpha \in \mathrm{On}_j$, $j \in \mathbb{N}_0$:
$N_0 := G_0$, $N_1 := \{\, z \in \Gamma(N_0) : (\mathrm{N1})\,\}$, $N_2 := \{\, z \in \Gamma(N_1) : (\mathrm{N1})\,\}$, ...
$N_\omega := \{\, z \in \Gamma\,(\bigcup_{k=0}^{\infty} G_k) : (\mathrm{N1})\,\}$, $N_{\omega+1} := \{\, z \in \Gamma(G_\omega) : (\mathrm{N1})\,\}$, etc.

$\mathrm{No}_j := \bigcup\{\, N_\alpha : \alpha \in \mathrm{On}_j \,\}$ may be called the *j-th Conway number class* ($j \in \mathbb{N}_0$), and $\mathrm{No}_\star := \bigcup_{j=0}^{\infty} \mathrm{No}_j$ is the set of all *Conway numbers.* (Conway denotes by **No** his proper class of numbers, cf. [2] p. 4.)

6.3 Examples $(M, \emptyset)$ and $(\emptyset, M)$ are Conway numbers for any set of Conway numbers $M \subset N_\alpha$ ($\alpha \in \mathrm{On}_j$, $j \in \mathbb{N}_0$). Especially $\perp \equiv \{|\}$, $1_{\mathrm{L}} \equiv \{\perp|\}$ and $1_{\mathrm{R}} \equiv \{|\perp\}$ are Conway numbers.

$* \equiv \{\perp|\perp\}$ is not a Conway number as $\perp \lhd \perp$ does not hold.

6.4 Remark (Properties of Conway numbers)
For all Conway numbers $z, z_1, z_2 \in \mathrm{No}_\star$ we have
(1) $z^{\mathrm{L}} < z < z^{\mathrm{R}}$ whenever $z^{\mathrm{L}} \in \mathrm{L}_z$, $z^{\mathrm{R}} \in \mathrm{R}_z$;
(2) $z_1 \lhd z_2 \iff z_1 < z_2$;
(3) $-z, z_1 + z_2 \in \mathrm{No}_\star$.
Therefore $\mathbf{No}_\star$, i.e. $\mathrm{No}_\star$ modulo $=$, is a subgroup of $\mathbf{Ug}_\star$.
((1) is proved using (N1) and 4.4 (3) by *Conway number induction*, i.e. by transfinite induction on $\alpha \in \mathrm{On}_j$, $j \in \mathbb{N}_0$, $z \in N_\alpha$; using 4.3 (2) and 4.4 we obtain (2) from (1); now (3) is proved with ((2) by game inductions.)

6.5 Lemma (Simplicity Lemma)
Let $x \in \mathrm{Ug}_\star$ be a game. Then $x = z$ for any number $z \in \mathrm{No}_\star$ with $\mathrm{L}_x \lhd z \lhd \mathrm{R}_x$ (i.e. with $x^{\mathrm{L}} \lhd z \lhd x^{\mathrm{R}}$ for all $x^{\mathrm{L}} \in \mathrm{L}_x$, $x^{\mathrm{R}} \in \mathrm{R}_x$), $\mathrm{L}_z \subseteq \downarrow\mathrm{L}_x$ and $\mathrm{R}_z \subseteq \uparrow\mathrm{R}_x$.
(If the assumptions hold, z is the *"simplest"* Conway number between L_x and R_x, because – as $z^{\mathrm{L}} \le x^{\mathrm{L}}$ and $x^{\mathrm{R}} \le z^{\mathrm{R}}$ hold for some $x^{\mathrm{L}} \in \mathrm{L}_x$ and $x^{\mathrm{R}} \in \mathrm{R}_x$ – it is impossible that always $x^{\mathrm{L}} \lhd z^{\mathrm{L}} \lhd x^{\mathrm{R}}$ or $x^{\mathrm{L}} \lhd z^{\mathrm{R}} \lhd x^{\mathrm{R}}$.)

Proof:
$z \le x$: Because of $\mathrm{L}_z \subseteq \downarrow\mathrm{L}_x$ we have $\mathrm{L}_z \lhd x$ (i.e. $z^{\mathrm{L}} \lhd x$ for all $z^{\mathrm{L}} \in \mathrm{L}_z$), and by assumption we have $z \lhd \mathrm{R}_x$. ($x \le z$ is proved similarly.) □

6.6 Auxiliary Theorem (Dyadic Conway numbers)
Set $D_k := 2^{-k}\mathbb{Z}$ *for all* $k \in \mathbb{N}_0$, *and let* $D_\infty := \bigcup_{k=0}^{\infty} D_k$ *denote the set of* dyadic rationals (*i. e. rationals* $\frac{m}{2^k}$ *with* $m \in \mathbb{Z}$, $k \in \mathbb{N}_0$).
(1) *There is a* unique map $c_\infty : D_\infty \to \mathrm{No}_\star$ *with image* $c_\infty[D_\infty] \subseteq \mathrm{No}_1$ *and*
(i) $c_\infty(0) \equiv \{|\} \equiv \bot$, *the neutral element of* $\mathrm{No}_\star$,
(ii) $c_\infty(n) \equiv \{c_\infty(n-1)|\}$ *for all* $n \in \mathbb{N}$,
(iii) $c_\infty(-n) \equiv \{|-c_\infty(n-1)\}$ for all $n \in \mathbb{N}$,
(iv) $c_\infty\left(\frac{2\ell+1}{2^k}\right) \equiv \left\{c_\infty\left(\frac{\ell}{2^{k-1}}\right) \middle| c_\infty\left(\frac{\ell+1}{2^{k-1}}\right)\right\}$ *for all* $k \in \mathbb{N}$, $\ell \in \mathbb{Z}$.
(2) c_∞ *is* reflection preserving *and* strictly increasing, *i. e. we have* $c_\infty(-s) \equiv -c_\infty(s)$ *and* $c_\infty(s) < c_\infty(t)$ *if* $s < t$ *and* $s, t \in D_\infty$.
(3) $\overline{c}_\infty := \overline{\mathrm{pr}} \circ c_\infty$ *is an* injective group homomorphism $\overline{c}_\infty : D_\infty \to \mathbf{No}_\star$, *where* $\overline{\mathrm{pr}} : \mathrm{No}_\star \to \mathbf{No}_\star$ *is the canonical projection* ($\mathbf{No}_\star$ as in 6.4).

Proof:
(1): First construct $c_0 : \mathbb{Z} \to \mathrm{No}_\star$ satisfying $c_0[\mathbb{Z}] \subset \mathrm{No}_1$ as well as (i), (ii) and (iii) with c_0 instead of c_∞, then extend $c_{k-1} : D_{k-1} \to \mathrm{No}_\star$ to $c_k : D_k \to \mathrm{No}_\star$ using (iv) with c_k instead of c_∞. Finally define $c_\infty(s)$ to be $c_k(s)$ if $s \in D_k$. Uniqueness is proved by inductions on n and on k. (2) is proved with (1) (i)–(iv), and (3) is proved with 6.5. □

6.7 Convention Dyadic rationals can be interpreted as Conway numbers by dint of 6.6. Numbers of the form $c_\infty(s)$ with $s \in D_\infty$ may be called *dyadic Conway numbers.* Occurrences of c_∞ will usually be suppressed, provided that no misunderstandings are to be expected: $0 \equiv \{|\}$, $1 \equiv \{0|\}$, $2 \equiv \{1|\}$, $3 \equiv \{2|\}$, ..., $\frac{1}{2} \equiv \{0|1\}$, $\frac{1}{4} \equiv \{0|\frac{1}{2}\}$, $\frac{3}{4} \equiv \{\frac{1}{2}|1\}$, $-1 \equiv \{|0\}$, $-2 \equiv \{|-1\}$, $-3 \equiv \{|-2\}$, ..., $-\frac{1}{2} \equiv \{-1|0\}$, $-\frac{1}{4} \equiv \{-\frac{1}{2}|0\}$ etc. (Writing 0 for $\bot$ is consistent with Definition 1.6, as $x \varrho \bot$ is equivalent to $x \varrho 0$ for every $\varrho \in \{\geq, \triangleright, \leq, \triangleleft, >, <, \|, =\}$; in 1.4 we have $1_{\mathrm{L}} \equiv 1$ and $1_{\mathrm{R}} \equiv -1$.)

7. REAL CONWAY NUMBERS

7.1 Definition *(Real Conway numbers)*
A Conway number $z \in \mathrm{No}_\star$ is called *real* if
(R1) $-n < z < n$ for some $n \in \mathbb{N}$,
(R2) for all $z^{\mathrm{L}} \in \mathrm{L}_z$ there is an $m \in \mathbb{N}$ with $z - z^{\mathrm{L}} > 2^{-m}$ and for all $z^{\mathrm{R}} \in \mathrm{R}_x$ there is an $m \in \mathbb{N}$ with $z^{\mathrm{R}} - z > 2^{-m}$.

We call a (real) Conway number z *located* if $c(s) < z$ or $z < c(t)$ holds whenever $s, t \in D_\infty$ with $s < t$. We set $\mathrm{No}_{\mathrm{real}} := \{ z \in \mathrm{No}_\star : z \text{ is real} \}$ and $\mathbf{R} := \overline{\mathrm{pr}}[\{ z \in \mathrm{No}_{\mathrm{real}} : z \text{ is located} \}]$ (with $\overline{\mathrm{pr}}$ from 6.6).

7.2 Example Dyadic Conway numbers (cf. 6.7) are located reals.

7.3 Remark (Properties of real Conway numbers)
(1) $\mathbf{R}$ is a subgroup of $\mathbf{No}_\star$.
(2) A real Conway number z is located if and only if for every $k \in \mathbb{N}$ there are $s, t \in D_\infty$ with $z - 2^{-k} < s < z < t < z + 2^{-k}$.
(3) For $z_1, z_2 \in \mathrm{No}_{\mathrm{real}}$ with $z_1 < z_2$ there is $m \in \mathbb{N}$ with $z_2 - z_1 > 2^{-m}$.

7.4 Auxiliary Theorem (Rational Conway numbers)
(1) $c : \mathbb{Q} \to \mathrm{No}_\star$, $q \mapsto (c_\infty[\{ s \in D_\infty : s < q \}], c_\infty[\{ s \in D_\infty : s > q \}])$ *satisfies* $c(s) = c_\infty(s)$ *for all* $s \in D_\infty$ (with c_∞ and D_∞ from 6.6).
(2) *All* rational *Conway numbers (elements of* $c[\mathbb{Q}]$*) are located reals.*
(3) c *is reflection preserving and strictly increasing, i. e. we have* $c(-q) \equiv -c(q)$ *and* $c(p) < c(q)$ *if* $p < q$ *and* $p, q \in \mathbb{Q}$.
(4) $\overline{c} := \overline{\mathrm{pr}} \circ c$ *is an* injective group homomorphism $\overline{c} : \mathbb{Q} \to \mathbf{No}_\star$ *with image* $\mathbf{Q} := \overline{c}[\mathbb{Q}] \subset \mathbf{R}$ (where $\overline{\mathrm{pr}} : \mathrm{No}_\star \to \mathbf{No}_\star$ is as in 6.6).
(5) *The following statements is equivalent to* (LEM) *from* 1.9:
(a) *Every* $z \in \mathbf{R}$ *can be approximated on both sides by rational numbers, i. e. for all* $z \in \mathbf{R}$ *and all* $k \in \mathbb{N}$ *there are* $r_1, r_2 \in \mathbf{Q}$ *with* $z - 2^{-k} < r_1 < z < r_2 < z + 2^{-k}$.
(b) *All Conway reals are located.*

Proof: (1),(2): $c(q)$ is a located real Conway number because c_∞ is strictly increasing, while $c(s) = c_\infty(s)$ can be seen with 5.4.
(3) is proved easily using the definition of c, and (4) is proved with 6.5.
(5): "(a) $\iff$ (b)" is proved with 7.3, "(LEM) $\Longrightarrow$ (b)" with 1.9 (2), and for "(b) $\Longrightarrow$ (LEM)" use the real Conway number $z_\psi \equiv -x_\psi$ with x_ψ as in the proof of 1.9. □

7.5 Definition *(Rational cuts)*
A *rational cut* is a pair (P, Q) of subsets $P, Q \subseteq \mathbb{Q}$ such that

(C1) P and Q are *downwards/upwards closed*: $\downarrow P = P$ and $\uparrow Q = Q$, i.e. $p < p' \in P \Longrightarrow p \in P$ resp. $q > q' \in Q \Longrightarrow q \in Q$;
(C2) P and Q are *disjoint*, thus $P < Q$, ($p < q$ for all $p \in P$, $q \in Q$;)
(C3) P and Q are *open*, i.e. for all $p \in P$ there is $p' \in P$ with $p' > p$ and for all $q \in Q$ there is $q' \in Q$ with $q' < q$.

A rational cut (P, Q) is called *bounded* if P and Q are non-empty, and (P, Q) is called *located* if $P \cup Q$ is dense in $\mathbb{Q}$ (equivalently $p \in P$ or $q \in Q$ whenever $p, q \in \mathbb{Q}$ with $p < q$).

7.6 Theorem (Real Conway numbers and rational cuts)
(1) *For all rational cuts (P, Q) the Conway number $\hat{c}(P, Q) \equiv (c[P], c[Q])$ satisfies* (R2)*; it is real (and located) if (P, Q) is bounded (and located).*
(2) *For every real Conway number z the rational cut $\check{c}(z) := (P_z, Q_z)$ with $P_z := \{\, p \in \mathbb{Q} : c(p) < z \,\}$ and $Q_z := \{\, q \in \mathbb{Q} : c(q) > z \,\}$ is located if z is located. Whenever (P, Q) is located we have $\check{c}(\hat{c}(P, Q)) = (P, Q)$.*
(3) *There is a bijection between the set* $\mathbf{R}$ *(defined in* 7.1*) and the set* $\mathbb{R}$ *of* Dedekind reals, *i. e. bounded and located rational cuts.*

Proof: (1) is proved straightforwardly.
(2): The assertions of the first sentence are easily verified; $s < \hat{c}(P, Q)$ can hold with a located rational cut (P, Q) only if $s \in P$.
(3): Using 7.3 (1) we have $\mathrm{L}_z \subseteq \downarrow c[P_z]$, $\mathrm{R}_z \subseteq \uparrow c[Q_z]$ for $z \in \mathbf{R}$, thus $z = \hat{c}(\check{c}(z))$ holds by the Simplicity Lemma 6.5, as $c[P_z] < z < c[Q_z]$. □

8. GENERALIZED REAL NUMBERS

Generalized Dedekind reals (introduced by Fred Richman in [6] and Peter Schuster in [9]) are just rational cuts (cf. 7.5). Addition for these cuts is defined by $(P, Q) + (P', Q') := (P + P', Q + Q')$ in contrast to 2.2, whereas reflection $-(P, Q) := (-Q, -P)$ corresponds to Definition 3.2. The partial order relation for cuts defined by $(P, Q) \leq (P', Q') :\Longleftrightarrow (P \subseteq P'$ and $Q \supseteq Q')$ is different from the one defined in 4.1, because all cuts of the form $(P, Q) = (\mathbb{Q} \setminus \uparrow\{p\}, \mathbb{Q} \setminus \downarrow\{q\})$, with $\uparrow$ and $\downarrow$ from 5.5, have $\hat{c}(P, Q) = 0$ if $p < 0 < q$.
In [6] multiplication is defined only for *weakly positive* cuts (i. e. for cuts with $(P, Q) \geq 0$). Conway's definition of *multiplication* (cf. [2] p. 19)

$$xy \equiv \{x^{\mathrm{L}}y + xy^{\mathrm{L}} - x^{\mathrm{L}}y^{\mathrm{L}}, x^{\mathrm{R}}y + xy^{\mathrm{R}} - x^{\mathrm{R}}y^{\mathrm{R}} \mid x^{\mathrm{L}}y + xy^{\mathrm{R}} - x^{\mathrm{L}}y^{\mathrm{R}}, x^{\mathrm{R}}y + xy^{\mathrm{L}} - x^{\mathrm{R}}y^{\mathrm{L}}\}$$

is possibly useful to define a product for arbitrary cuts.

Acknowledgments

I am grateful to Peter Schuster for drawing my attention to the Symposion "Reuniting the Antipodes" and to Conway's article [3], for several preprints of [9] as well as for encouraging me to give a talk on Conway numbers in Venice. Volkswagen foundation Hannover and FernUniversität Hagen supported my participation at the Symposion "Reuniting the Antipodes" in May 1999 at Venice International University.
Furthermore I appreciate the valuable instructions of Prof. Holger P. Petersson before my departure and after my return to FernUniversität Hagen, and the helpful discussions with participants of the "Oberseminar über Algebra und Topologie", where I presented some details of constructive Conway theory in June 1999. Moreover I want to thank Dipl. Math. Peter Brühne and HDoz. Goswin Große-Erdmann for proof-reading penultimate versions of this paper as well as Prof. Reinhard Börger for discussing some technical items. Their notes, suggestions and hints turned out to be very helpful during the revision.

The anonymous referee stimulated the revised version of this paper by insisting on precision concerning the employed notion of ordinal numbers, a topic I discussed in November 1999 with Per Martin-Löf in Munich. I am indebted to both of them for giving me a fresh impetus to improve the performance and to write [8], an extended version of my talk given in Hagen.

This paper is dedicated to my wife Monika Rosemeier and to our newborn daughter Lena Tabea. Thanks to their patience I could finish the revisions of it in time. Let us appreciate the indispensable value of playing for the development of human life.

References

[1] Elwyn R. Berlekamp, John H. Conway, Richard K. Guy: *Winning Ways*, Academic Press, London, GB, 1982, 3rd ed. 1985.

[2] John H. Conway: *On Numbers and Games*, Academic Press, London, GB, 1976.

[3] John H. Conway: *The surreals and the reals*, in [4] 93–103.

[4] Philip Ehrlich (ed.): *Real Numbers, Generalizations of the reals and Theories of Continua*, Kluwer, Dordrecht, NL, 1994.

[5] Norman M. Martin, Stephen Pollard: *Closure Spaces and Logic*, Kluwer, Dordrecht, NL, 1996.

[6] Fred Richman: *Generalized Real Numbers in Constructive Mathematics*, Indag. Mathem., N.S., **9** (4) (1998) 595–606.

[7] Fred Richman: *The fundamental theorem of algebra: a constructive development without choice*, Pacific J. Math., **196** (2000) 213–230.

[8] Frank Rosemeier: *A Constructive Approach to Conway's Theory of Games*, to appear in Seminarberichte aus dem Fachbereich Mathematik, FernUniversität Hagen (ISSN 0944-5838).

http:// www.fernuni-hagen.de/MATHEMATIK/ALGGEO/
Mitarbeiter/Rosemeier/publ.html

[9] Peter Schuster: *A Constructive Look at Generalised Cauchy Reals*, Math. Logic Quarterly **46** (2000) 125-134.

THE CONSTRUCTIVE CONTENT OF NONSTANDARD MEASURE EXISTENCE PROOFS—IS THERE ANY?

David A. Ross*
Department of Mathematics, University of Hawaii
Honolulu, HI 96822 USA
ross@math.hawaii.edu

Keywords: Loeb measure, constructive analysis, nonstandard analysis

Mathematical Subject Classification: 28E05, 03F60, 26E35

Abstract: Some examples are given which cast some light on such questions as: What (if any) constructive sense can be made out of typical nonstandard measure existence results? Can NSA give any insight into constructive measure theory?

"You want proof? I'll give you proof!"

— *caption,*
Sidney Harris cartoon
of two mathematicians fighting

Introduction

The differences – both methodological and sociological – between Constructive and Nonstandard Analysis will be familiar to everyone at this meeting. Though the latter is technically a part of classical methematics, it is often considered even less constructive than most of the rest of classical mathematics. One obvious reason is the essential role of the

*The author is grateful to the organizers for inviting him to this meeting, as well as to Nigel J. Cutland, who was his host at the University of Hull during the Spring 1999 academic semester.

P. Schuster, U. Berger and H. Osswald (eds.),
Reuniting the Antipodes – Constructive and Nonstandard Views of the Continuum, 229–239.

Axiom of Choice in its foundations; beyond that, nonstandard analysis is an opportunistic methodology, in that it will employ all techniques available to render a proof as quick and painless as possible. (The author has even utilized such atrocities as measurable cardinals; see, for example, [14].) Consequently, even classical mathematicians are often reluctant to integrate artifacts constructed by nonstandard methods (for example, Loeb measures and hulls of Banach spaces) into their mathematical ontology.

On the other hand, nonstandard arguments often have in practice a very constructive flavor, much more than do the corresponding standard proofs. Indeed, those who use nonstandard arguments often say of their proofs that they are "constructive modulo an ultrafilter." This is especially true in measure theory.

Work we've heard about at this conference (e.g., [12, 10]) suggests that many nonstandard arguments can be given an *actual* (and not just intuitive) constructive interpretation. However, the fragment of nonstandard analysis which has so far been amenable to such interpretation apparently does not include the so-called 'nonstandard hull' constructions (such as Loeb measures), which have been at the core of almost every interesting application of nonstandard analysis published during the last quarter century. The question raised in the title of this paper therefore remains quite open.

In this paper I address questions such as: What (if any) constructive sense can be made out of typical nonstandard measure existence results? Can NSA give any insight into constructive measure theory? In no case do I give any sort of formal answer; rather, I discuss a few concrete examples which have some relevance to the broad question. Although the provocative wording of this paper's title suggests a negative conclusion, in fact I only isolate one problem which threatens to become a general concern – namely, that an apparently inconsequential change in a nonstandard argument (for example, in the way a sample set is chosen) can radically affect the constructiveness (or even the veracity) of the ensuing argument.

For the remainder of the paper I assume an understanding of nonstandard analysis at the level of any of the many excellent texts which now exist, though I will work within the framework of 'Robinsonian' nonstandard analysis rather than the more syntactic approaches such as Nelson's Internal Set Theory. I will use the word "constructive" itself intuitively, *usually* in the sense of Bishop, sometimes just in my (flawed) interpretation of the sense of Bishop.

1. NONSTANDARD MEASURE THEORY

1.1. Outline

The reader is referred to [15] for a development of nonstandard measure theory; for completeness and context, I begin by recalling the steps in a general procedure for obtaining a measure using the Loeb construction: (1) **"Construct" an internal measure** $(\Omega, \mathcal{A}, P)$**.** Often Ω is hyperfinite, which means that we can use transfer of finitary combinatorial arguments both to guide the construction and to analyze the resulting measure. (2) **Convert to a standard finitely additive measure** $(\Omega, \mathcal{A}, {}^\circ P)$**.** Except in the most trivial cases, the measure $(\Omega, \mathcal{A}, {}^\circ P)$ will be both external (a source of trouble when mathematicians who use non-Robinsonian flavors of nonstandard analysis try to reproduce the Loeb measure) and non-σ-additive (in fact, $\mathcal{A}$ will generally not be a σ-algebra). (3) **Extend to a standard measure** $(\Omega, \mathcal{A}_L, P_L)$**.** This extension can be realized by a trivial application of the Caratheodory Extension Theorem; however, the extension can be obtained directly by an elementary and intuitive argument; see [15]. (4) **"Push down" to a standard space** X**.** This last step is neither necessary nor even desirable if it is *not* important that the final measure live on some predetermined standard space - this is in fact the situation for many of the most important applications of the methodology, such as stochastic analysis. However, even in such cases one often has constructed - alongside the space $(\Omega, \mathcal{A}, {}^\circ P)$ - an internal random variable or process, which would itself then need to be pushed down. This step is mainly required when we are working in the presence of some fixed topological space X, $\Omega \subseteq^* X$, and we are trying to construct a a final measure on X. Usually the standard part map will be measurable, and so we can obtain a measure on X as the image under this map of $(\Omega, \mathcal{A}_L, P_L)$. (The resulting measure on X will be a kind of weak standard part of $(\Omega, \mathcal{A}_L, P_L)$.)

1.2. Lebesgue measure

One of the easiest measures to produce in this way is Lebesgue measure on $[0, 1]$; the following specific argument is often attributed to Fisher.

Begin by letting $(\Omega, \mathcal{A}, P)$ be internal uniform counting measure on $\Omega = \{0, \delta t, 2\delta t, \ldots, 1 - \delta t\}$, where H is an infinite integer and $\delta t = 1/H$. We assume that P is normalized to 1; that is, for any internal subset A of ${}^*[0, 1]$, $P(A) = card(A \cap \Omega)/H$.

Following the above procedure, we obtain an external measure P_L on Ω, which we then push down to a measure λ on $[0,1]$ using standard part map. It is an easy exercise to show that intervals get right measure,

It is not difficult to find at least *some* constructive content in this simple proof. Note first that if f is a continuous function on $[0,1]$, then $\int_0^1 f dt \approx \sum_k {}^*f(k\delta t)$. For any such f, as *finite* H gets large then the sum approaches the integral. Given a modulus of continuity for f, one can certainly bound the rate at which this convergence takes place. It follows that this measure can be recaptured constructively by computable finitary approximations.

In other words, this nonstandard argument can be constructively interpreted as a *shorthand for a weak limit argument.*

It is important to note that part of the reason that this measure *does* have such an interpretation is that the sample Ω is distributed in a constructive way. This distribution has some peculiarities; for example, the hyperirrationals form a P-nullset. While this might not seem like a very important point, the manner in which the hyperfinite sample space Ω is distributed appears to be central for the question of whether the resulting proof admits a constructive interpretation. The next example reveals that this is precisely the difficulty one encounters when extending this Lebesgue measure construction to more general invariant measures.

1.3. Haar measure

Something akin to the method from the last section can be used to build measures on locally compact groups (Haar measure):

Theorem 1 *Let G be a locally compact Hausdorff group; then there is a nontrivial Borel measure m on G such that $mE = m(g^{-1}E)$ for all $g \in G$ and Borel E*

A nonstandard proof begins by distributing sample points in ${}^*G$ 'just like' we did for Lebesgue measure. There is an immediate problem, namely that intervals no longer exist either to indicate where the points should be scattered or to verify that the resulting measure is invariant. However, a trick borrowed from the theory of dynamical systems makes everything work.

In particular, let K_0 be an internally compact set which contains all nearstandard points. Let u_e be an infinitesimal neighborhood of the identity (that is, $e \in u_e \subseteq monad(e)$, where e is the identity of G).

Observe that $\{u_g : g \in^* G\}$ (where $u_g := gu_e$) is an internal open cover of K_0; there is then an internally finite subcover $\{u_g : g \in \Omega\}$. We may assume that Ω has the minimal internal cardinality among all such subcovers.

As with Lebesgue measure, let P be an internal uniform measure on Ω, weighted by a suitable normalizing factor. Apply the Loeb construction, then push P_L down to a standard measure m on G.

It is not hard to verify that that m is G-invariant; details (in the slightly simpler case where G is compact) appear in [13] and [15].

Suppose we try to find the constructive content of this argument as we did with Lebesgue measure. We immediately face two new problems not present in that case: (1) How do we choose K_0 and u_e? *Existence* of K_0 and u_e in the nonstandard argument follows from sufficient saturation. For the finitary/limiting analogue, one might replace u_e with a sequence of neighborhoods which is a local base at e, and similarly substitute for K_0 a sequence of compact sets with union G; it is not difficult to frame natural conditions on G under which such sequences can be given constructively. A more perplexing problem is (2) How do we find the minimal subcover? In fact, the difficulty in finding such subcovers in general is what leads to the awkward constructive formulation of "compact" that one finds in, say, [2]. I suspect that this aspect of the nonstandard argument will not be ameneble to constructive interpretation.

In the case of Lebesgue measure, the decision of how to distribute the sample Ω was dictated not by any minimality considerations, but rather by the geometry of the underlying space. As it happens, for sufficiently small $\delta' > \delta$ this sample *does* correspond to a minimal cover by open intervals of width δ'. It is natural to speculate that for groups G with simple enough structure, such structure can be used in an analogous way to distribute a sample. The work of Gordon [7] on the approximation of topological goups by hyperfinite groups certainly has some bearing here, and carries with it its own caveats, as not all reasonable groups can be approximated in this sense.

We will return to the question of how to distribute hyperfinite samples in a another context in Section 2.2, where it again seems to be the main sticking point in the constructive interpretation of a nonstandard argument.

1.4. Ergodic Theorem

While versions of Birkhoff's Ergodic Theorem have already been given constructive treatments (e.g. [3]), T. Kamae [8] has given a lovely non-standard proof which is of interest here. Recall the theorem's statement:

Theorem 2 *Let $(X, \mathcal{B}, m)$ a probability space, T a measurable transformation of X, and $f \in L_1(m)$; then*

$$\lim_{n\to\infty} \frac{1}{n}\sum_{k=0}^{n-1} f(T^k(x))$$

exists for almost all x.

Kamae's proof has two stages:

(1) Show that it suffices to assume that $(X, \mathcal{B}, m)$ is a uniform Loeb probability measure on $X = \{0, 1, 2, \ldots, H-1\}$ for some infinite integer H, and $T(x) = x + 1 \mod H$.

(2) Show that

$$\int \overline{f} dm \leq \int \underline{f} dm\,,$$

where

$$\overline{f} = \overline{\lim}_{n\to\infty} \frac{1}{n}\sum_{k=0}^{n-1} f(T^k(x)) \quad \text{and} \quad \underline{f} = \underline{\lim}_{n\to\infty} \frac{1}{n}\sum_{k=0}^{n-1} f(T^k(x))\,.$$

Although the first stage employs some very powerful machinery (e.g. the Von Neumann/Maharam structure theory for finite measures), it is of less interest for the purposes of this paper than the second. To prove the inequality, for any given cycle $x, Tx, T^2x, \ldots$ and $\epsilon > 0$ the limit $\overline{f}$ is approximated within ϵ by the partial sums on finite cycle blocks $T^k x, \ldots, T^{k+M(k,x,\epsilon)}x$. As $M(x, k, \epsilon)$ is an internal function of x which takes standard values, one can apply overspill to obtain a uniform, standard bound $M_\infty(\epsilon)$ on the values $M(x, k, \epsilon)$. The inequality then follows by considering sufficiently large partial sums, where the error in the inequality is bounded by a constant multiple of $M_\infty(\epsilon)/M$, with M the number of terms considered in the partial sums.

The essential use of nonstandard analysis in this proof appears to be the use of overspill. However, Kamae's argument here has become a common technique in *standard* probability in proving arguments about sums of random variables; see, for example, [11]. The uniform finite bound is replaced in the standard arguments by a similar bound on a set of large measure, and the strong control over the convergence allows the argument to succeed. While the standard papers where this has been used are not overtly constructive, it is quite clear from inspection that the proofs often have a constructive kernel, again a consequence of the control over the convergence.

It is possible that the evolution of the Kamae technique illustrates the real future for the extraction of constructive content from Loeb measure proofs: rather than a formal interpretation (in the sense of [12]), or an interpretation as a weak limit (as in Section 1.2), the nonstandard space serves as a kind of a testbed where a constructive combinatorial argument can be tried; if it works there, the key features of the argument might then be adapted to the case of a general space.

1.5. Other measure-theoretic results

It is worth noting that other well-known *standard* results in measure-theory have so far resisted nonstandard proof. Among these are three of special interest: the Radon-Nikodym Theorem, the Lyapunov Theorem[1], and the Vitaly-Hahn-Sacks Theorem [2]. These three have one notable feature in common, namely that easy discrete analogues exist (e.g., see [9] for Peter Loeb's pretty discrete version of the Lyapunov Theorem) Moreover, the failure of the 'obvious' nonstandard proof stems from the difficulty in pushing an object constructed nonstandardly down to a standard one. It seems plausible to me that these arguments *might* be sufficient for constructive analogues of the standard results, and in any event an understanding of why these nonstandard proofs fail will likely be relevant to the questions being considered here.

2. CONSTRUCTIVE ARGUMENTS

2.1. A result in constructive measure theory

The remainder of this paper is concerned not so much with examining nonstandard arguments with a constructive eye, but rather with examining constructive arguments with a nonstandard eye.

---

[1]If $\mu_1, \ldots, \mu_n$ are finite atomless measures on a space X, and $r \in [0,1]$, then there is a measurable A such that $\mu_i(A) = r\mu_i(X)$ for all $i \leq n$.

[2]Let $\{\mu_n\}_{n\in\mathbb{N}}$ be a sequence of probability measures on a common measurable space, and suppose $\mu(A) := \lim_{n\to\infty} \mu_n(A)$ exists for all measurable A. Then μ is a measure.

In this section, I give a nonstandard proof for an existence result which is central to the constructive treatment of integration, and for which the usual proofs ([4, 6, 5][3]) are far from obvious. This proof uses only very simple nonstandard tools, so should be amenable to the techniques of Palmgren and Martin-Löf. However, it avoids altogether one tool from the standard proof (a constructive partition of unity) which is almost certainly required to make the proof genuinely constructive. It would therefore be interesting to see what Palmgren's machinery would make of this proof.

If X is a Hausdorff topological space, write $C_K(X)$ for the vector space of real continuous functions on X with compact support, and $C_K^+(X)$ for the nonnegative elements of $C_K(X)$. The result, as proved in [4, 6], is the following:

Theorem 3 *Suppose that X is a locally compact metric space, that T is a nonnegative linear functional on $C_K(X)$, that $\{f_n\}_{n\in\mathbb{N}}$ is a sequence from $C_K^+(X)$, and that $\sum_{n=1}^{\infty} Tf_n < Tf_0$. Then there exists an $x \in X$ with $\sum_{n=1}^{\infty} f_n(x) < f_0(x)$*

For brevity I will prove this result under the simplifying assumption that X is compact, so that $C_K(X) = C(X)$, etc. Nonstandard proofs of the full result appear in [16][4].

For $m \in \mathbb{N}$ put

$$\phi_m(x) = \sum_{n=1}^{m} f_n(x) - f_0(x) \quad \text{and} \quad \alpha_m = \sum_{n=1}^{m} Tf_n - Tf_0 = T\phi_m\,.$$

Put also

$$\alpha = \sum_{n=1}^{\infty} Tf_n - Tf_0 = \lim_{m\to\infty} \alpha_m = \sup_{m\in\mathbb{N}} \alpha_m$$

(convergence is assumed to exist by hypothesis), and

$$\phi(x) = \sum_{n=1}^{\infty} f_n(x) - f_0(x) = \lim_{m\to\infty} \phi_m(x) = \sup_{m\in\mathbb{N}} \phi_m(x)$$

---

[3]Per Martin-Löf has kindly informed me that this result was also known to and used by Brouwer.

[4]**Remark:** The last section of [16] looks again at the standard constructive proof of this theorem, and shows that nonstandard arguments can be used to considerably simplify the steps of that proof. This raises the interesting philosophical question: if nonstandard methods are used to verify the steps in a constructive existence proof, to what extent is the resulting argument constructive? A similar question is hinted at in the second paragraph of Section 2.2 of the present paper.

(where the sum/limit might converge, or diverge to $+\infty$).

Proof. Without loss of generality $T(1) = 1$. Fix $M \in {}^*\mathbb{N} \setminus \mathbb{N}$, note ${}^*T\phi_M = \alpha_M \approx \alpha < 0$, so for some (standard) $\epsilon > 0$, ${}^*T\phi_M < -\epsilon$, so $T(\phi_M + \epsilon) < 0$. By nonnegativity of T there is some $x_0 \in {}^*X$ with $\phi_M(x_0) + \epsilon < 0$. Put $x =^\circ x_0$. For any standard $m > 0$, $\phi_m(x) \approx {}^*\phi_m(x_0) \leq \phi_M(x_0) < -\epsilon$. Taking the supremum over all standard m, $\phi(x) \leq -\epsilon < 0$, as desired. ⊣

2.2. Lower Semicontinuity

The main argument of the last proof is very similar to the nonstandard proof that if $\{h_i\}_{i \in I} \subseteq C_K(X)$ and $h = \sup_i h_i$ then h is lower semicontinuous on X. (Recall that $h : X \to \mathbb{R}$ is *lower semicontinuous*, or LSC, if $h^{-1}(r, \infty)$ is open in X for every $r \in \mathbb{R}$; equivalently, provided whenever $x = {}^\circ x_0, h(x_0) \gtrsim h(x)$.) Indeed, suppose $x =^\circ x_0$, and let $r \in (-\infty, \infty]$ standard with $r > {}^*h(x_0)$. For any standard $i \in I, h_i(x) \approx {}^*h_i(x_0) \leq {}^*h(x_0) < r$; taking the supremum over I, $h(x) \leq r$. Since r is arbitrary, $h(x) \lesssim h(x_0)$.

Every LSC function h attains a minimum on a compact set K. This is of course false constructively, since it even fails for continuous functions. Even the weaker fact that $h(K)$ has a greatest lower bound – which is true constructively when h is continuous – isn't constructively true for general LSC h. Otherwise, *any* proof of Theorem 3, even the nonconstructive one we gave above, could be coupled with an algorithm for finding the GLB for $\phi(K)$ in order to construct the x for the theorem. (This asumes that such a GLB algorithm would carry with it – as does the corresponding algorithm for continuous h – an algorithm for finding $x \in K$ with $\phi(x)$ arbitrarily close to this GLB.)

Interestingly, the nonstandard proof that a LSC function h attains its minimum on a compact K is no more difficult than when h is continuous. Recall the proof in the continuous case: let $\hat{K}$ be any hyperfinite subset of ${}^*K$ such that $\mathrm{st}(\hat{K}) = K$, let x_0 minimize ${}^*h$ on $\hat{K}$, and let $x =^\circ x_0$. If $y \in K$, then by choice of $\hat{K}$ there is a $y_0 \in \hat{K}$ with $y_0 \approx y$, and $h(y) \approx h(y_0) \geq^* h(x_0) \gtrsim h(x)$, so h attains its minimum at x.

If h is only LSC, then we only have $h(y) \leq h(y_0)$ instead of $h(y) \approx h(y_0)$, and this inequality is in the wrong direction for the argument to work! This is an essential problem: take $X = [0, 1]$, let $h(x) = 1$ for

$x \neq \pi/4$, $h(x) = 0$ otherwise. If $\hat{K}$ is the set Ω from Section 1.2, then ${}^*h$ is identically 0 on $\hat{K}$ so does not attain its maximum there[5].

Suppose however that we assume (as we can by saturation) that $\hat{K}$ contains every standard $x \in K$; then we may take $y_0 = y$, and the argument works.

Note that this uses, in an *essential* way, inclusion of K in $\hat{K}$. Even in the case of a metric K, where the nonstandard proof for continuous h has an immediate obvious translation to the constructive proof that $h(K)$ has a GLB, it is very difficult to imagine any reasonable constructive interpretation of the corresponding LSC argument.

3. FINAL REMARK

To a working nonstandard analyst, there is no obvious conceptual difference in the way in which Ω is chosen in Sections 1.2 and 1.3; the same is true for $\hat{K}$ in Section 2.2. However, this subtle difference seems to be the locus of nonconstructivity for some of the results under consideration. In particular, any attempt to formalize the process of finding constructive content for nonstandard measure existence arguments will need to understand this difference, and perhaps more fully explain the role it plays.

In this paper I have only been concerned with results provable in 'large' nonstandard models, those endowed with 'sufficient' saturation. However, constructivists might find the work of Michael Benedikt (e.g., [1]) of some interest; he considers Loeb measures in 'small' nonstandard models, and these measures are sufficiently tied to standard sequences that they might be more amenable to constructive interpretation. Also of some interest might some of the attempts to capture Loeb measures within systems such as IST by coding external sets (as Michael Reeken has discussed elsewhere at the conference).

References

[1] Benedikt, M., Nonstandard analysis in selective universes, Developments in nonstandard mathematics (Aveiro, 1994), Pitman Res. Notes Math. Ser. **336** (ed. Cutland, N.J., Neves, V., Oliveira , F. and Sousa-Pinto,J.), Longman, Harlow, 1995

[5] Of course, this function h does not make sense as a constructive object.

[2] Bishop, E.A., Foundations of constructive analysis, McGraw-Hill, New York (1967)

[3] Bishop, E.A, A constructive ergodic theorem, *J. Math. Mech*, **17**, (1968)

[4] Bishop, E.A. and Cheng, H., Constructive measure theory, *Amer. Math. Soc. Memoir* **116** (1972)

[5] Bridges, D.S., Constructive functional analysis, Pitman (1979)

[6] Chan, Y.K., A short proof of an existence theorem in constructive measure theory, *Proc. Amer. Math. Soc.* **48** (1975)

[7] Gordon, E. I., Nonstandard methods in commutative harmonic analysis, Translations of Mathematical Monographs **164**, American Mathematical Society, Providence, RI, (1997)

[8] Kamae, T., A simple proof of the ergodic theorem using nonstandard analysis, *Israel J. Math* **42** (1982)

[9] Loeb, P.A., A combinatorial analog of Lyapunov's theorem for infinitesimally generated atomic vector measures, *Proc. Amer. Math. Soc.* **39** (1973)

[10] Martin-Löf, P., Mathematics of infinity, COLOG-88 (Tallinn, 1988), Lecture Notes in Comput. Sci. **417** (ed. Martin-Löf, P. and Mints, G.), Springer, Berlin (1990)

[11] Keane, M.S., The essence of the law of large numbers, Algorithms, fractals, and dynamics (ed. Takahashi, Y .) Plenum, New York (1995)

[12] Palmgren, E., Developments in constructive nonstandard analysis, *Bull. Symbolic Logic* **4** (1998)

[13] Ross, D.A., Measures invariant under local homeomorphisms, *Proc. Amer. Math. Soc.* **102** (1988)

[14] Ross, D.A., Unions of Loeb nullsets, *Proc. Amer. Math. Soc.*, **124** (1996)

[15] Ross, D.A., Loeb measure and probability, Nonstandard Analysis: Theory and Applications (ed. Arkeryd, L., Cutland, N., Henson, C.W., and Wolff, M.), NATO Adv. Sci. Inst. Ser. C Math. Phys. Sci., 493, Kluwer Acad. Publ., Dordrecht (1997)

[16] Ross, D.A., A nonstandard proof of a lemma from constructive measure theory, to appear.

KRUSKAL'S TREE THEOREM IN A CONSTRUCTIVE THEORY OF INDUCTIVE DEFINITIONS

Monika Seisenberger*
Mathematisches Institut, Ludwig-Maximilians-Universität München
Theresienstraße 39, D-80333 München, Germany
seisenb@rz.mathematik.uni-muenchen.de

Keywords: Kruskal's theorem, finite trees, well quasi orders

Mathematical Subject Classification: 03F65

Abstract: We give a constructive proof of Kruskal's tree theorem by using an inductive characterization of well quasi orders.

1. INTRODUCTION

This paper is about a famous theorem in infinitary combinatorics, Kruskal's tree theorem, in a context of constructive mathematics. Usually, Kruskal's theorem is formulated in terms of well quasi orders:

> If $(A, \leq)$ is a well quasi order, then so is $T(A)$, the set of finite trees with labels in A, together with the embeddability relation.

Here a set A with a reflexive and transitive relation $\leq$ is a well quasi order if for every infinite sequence $(a_i)_{i<\omega}$ there exist indices $i<j<\omega$ such that $a_i \leq a_j$. A tree is embeddable into another one if there exists a one to one map on them respecting infima of nodes such that the label of each node is less or equal to that of its image.

Kruskal's theorem [Kru60] has a very elegant classical proof due to Nash-Williams [Nas63] using a minimal-bad-sequence-argument. This proof is often referred to in term rewriting, the main field of application where Kruskal's theorem implies termination of simplification orders.

*Research supported by the PhD Program "Logic in Computer Science", LMU München.

P. Schuster, U. Berger and H. Osswald (eds.),
Reuniting the Antipodes – Constructive and Nonstandard Views of the Continuum, 241–255.

To obtain a constructive proof we reformulate Kruskal's theorem. The main idea is to use an inductive definition of the notion of a well quasi order: call a finite sequence $[a_1, \ldots, a_n]$ bad if $a_i \not\leq a_j$ for all $i < j \leq n$. Intuitively, being a well quasi order means that no bad sequence can be badly extended infinitely often. In other words: the extension relation between finite bad sequences (denoted $\ll$) is wellfounded, where by 'wellfounded' we mean that the induction principle for this relation holds. Technically this is expressed by the statement that all bad sequences are in the accessible part of the relation $\ll$. In fact, this already is the case if the empty sequence is in the accessible part, i.e $\mathsf{acc}_{\ll}[]$. Hence, the inductive formulation of Kruskal's theorem is

$$\mathsf{acc}_{\ll_A}[] \to \mathsf{acc}_{\ll_{T(A)}}[].$$

Constructively the inductive characterization of a well quasi order implies the classical one. The converse is an instance of Brouwer's axiom of bar induction. However, since we prove the stronger inductive formulation, in our approach we don't make use of Brouwer's axiom.

The idea of reformulating Kruskal's theorem in order to get a constructive proof can also be found in the proofs of Schmidt [Sch79], Rathjen and Weiermann [RW93], and Hasegawa [Has94] using (ordinal) notation systems. Schmidt gave a proof of Kruskal's theorem by determining the maximal order type of the set of labeled trees (i.e. the supremum of the order types of all well orders the quasi order of trees can be extended to). Rathjen and Weiermann [RW93] reorganized this proof such that its constructive content became more visible and thereby solved the problem of determining the proof theoretic strength of Kruskal's theorem (see [Sim85] or [Gal91]). Hasegawa's proof [Has94] is similar to that of [RW93]. He uses a different well ordering but doesn't carry out a constructive well ordering proof. Our proof follows the combinatorial idea of [RW93]. However, using the inductive characterization we do not need to introduce ordinal notations, nor do we have to refer to other results. Although almost completely formalized, this proof is short and intuitive and, as inductive definitions are available in most theorem provers, could serve as an interactive proof script.

The inductive characterization of a well quasi order was introduced by Martin-Löf [Mar70]. Coquand [Coq93] used it for a short proof of Higman's lemma (on a two element alphabet), a specialization of Kruskal's theorem to sequences. Using essentially the same method, Richman and Stolzenberg [RS93] proved Higman's lemma for decidable quasi orders. A proof of Higman's lemma not requiring decidability was given by

Fridlender [Fri97], inspired by ideas of Veldman. Our proof of Kruskal's theorem requires decidable relations, an analogous result for arbitrary relations would be highly desirable.

Many other theorems typically known to hold classically can be reformulated by means of inductive definitions thus admitting constructive proofs. Recent applications in infinitary combinatorics and algebra were given by Coquand, Persson [CP99], [Per99] and Fridlender [Fri99].

2. WELL QUASI ORDERS

In this section we introduce the formal inductive definition of a well quasi order and prove two simple properties of it.

Conventions. In the following let A be a decidable quasi order, i.e. a set with a binary, decidable, reflexive, and transitive[1] relation $\leq_A$. $A^* := \{[a_1, \ldots, a_n] \colon n \in \mathbb{N},\, a_i \in A\}$. We use

a, b	for elements of A (letters)
as, bs	for elements of A^* (words, i.e. finite sequences of letters)

Let $as = [a_1, \ldots, a_n]$. We write $as * a$ for the sequence $[a_1, \ldots, a_n, a]$ and $as@bs$ for the concatenation of two sequences as and bs. $[]$ is the empty sequence. If $i \leq$ length as, the i-th element of as is denoted by $(as)_i$.

Definition. We call a finite sequence $[a_1, \ldots, a_n]$ good if there exists $i < j \leq n$ such that $a_i \leq_A a_j$, otherwise bad. Let $\mathrm{Bad}(A)$ be the set of all finite bad sequences in A. Let $as = [a_1, \ldots, a_n] \in \mathrm{Bad}(A)$. We set

$$A_{as} := \{b \in A \colon \forall i \leq n.\, a_i \not\leq_A b\},$$

$\leq_{A_{as}}$ is the relation $\leq_A$ restricted to the set A_{as}. Moreover, we define a relation $\ll_A \subseteq \mathrm{Bad}(A) \times \mathrm{Bad}(A)$ by

$$bs \ll_A as :\leftrightarrow bs = as * a \text{ for some } a \in A_{as}.$$

The accessible part (also called the wellfounded part) of the relation $\ll_A$ is inductively given by the rule

$$\frac{\forall bs \ll_A as \quad \mathsf{acc}_{\ll_A} bs}{\mathsf{acc}_{\ll_A} as}.$$

[1]For historical reasons we have formulated Kruskal's theorem for quasi orders. Therefore we assume $\leq_A$ to be transitive, although transitivity is not used in the proof.

According to the definition of $\mathsf{acc}_{\ll_A}$ the following induction principle holds for any formula $\phi(as)$:

$$\frac{\forall as.\, \forall bs \ll_A as\ \phi(bs) \rightarrow \phi(as)}{\forall as.\, \mathsf{acc}_{\ll_A} as \rightarrow \phi(as)}.$$

Analogously, the acc-notion will be used for other relations. Now $(A, \leq_A)$ is a well quasi order iff $\ll_A$ is wellfounded on $\mathrm{Bad}(A)$, i.e. if all bad sequences are in the accessible part of the relation $\ll_A$.

The remark below shows that being a well quasi order (as defined in the introduction) is equivalent to $\mathsf{acc}_{\ll_A}[]$.

Remark. $\mathsf{acc}_{\ll_A}[] \iff A$ is a well quasi order.

Proof. "$\Longrightarrow$". We show more generally

$$\forall as.\, \mathsf{acc}_{\ll_A} as \rightarrow \forall (a_i)_{i<\omega}, \forall n.\, [a_1, \ldots, a_n] = as \rightarrow \exists i < j.\, a_i \leq_A a_j$$

by induction on $\mathsf{acc}_{\ll_A}$. The statement will follow by $as = []$. Let $as \in \mathrm{Bad}(A)$ and assume the induction hypothesis:

$$\forall bs \ll_A as, \forall (a_i)_{i<\omega}, \forall n.\, [a_1, \ldots, a_n] = bs \rightarrow \exists i < j.\, a_i \leq_A a_j.$$

Now, let an infinite sequence $(a_i)_{i<\omega}$ and n such that $[a_1, \ldots, a_n] = as$ be given. Case 1: there exists an $i \leq n$ such that $a_i \leq_A a_{n+1}$. Then we are done. Case 2: $\forall i \leq n\ a_i \not\leq_A a_{n+1}$, i.e. $as * a_{n+1} \ll_A as$. By induction hypothesis we obtain i and j such that $a_i \leq_A a_j$ and we have also proved our assertion.
"$\Longleftarrow$". This direction is an instance of Brouwer's axiom of bar induction which is intuitionistically considered acceptable. We also give a classical proof: assume that $\mathsf{acc}_{\ll_A}[]$ does not hold. Then, by the definition of $\mathsf{acc}_{\ll_A}$ there is an element $a_1 \in A$ such that $\neg\mathsf{acc}_{\ll_A}[a_1]$. By iteration (using the Axiom of Dependent Choice) we get a sequence $(a_i)_{i<\omega}$, such that for all n holds: $\neg\mathsf{acc}_{\ll_A}[a_1, \ldots, a_n]$ and $\forall i < j \leq n\ a_i \not\leq_A a_j$, contradicting A being a well quasi order. ◇

Lemma 1. $(\forall a \in A\, \mathsf{acc}_{\ll_{A_{[a]}}}[]) \rightarrow \mathsf{acc}_{\ll_A}[]$.

Proof. Assume $\forall a \in A\, \mathsf{acc}_{\ll_{A_{[a]}}}[]$. By definition of $\mathsf{acc}_{\ll_A}$ we need to show $\forall a \in A\, \mathsf{acc}_{\ll_A}[a]$. Let $a \in A$. By assumption we have $\mathsf{acc}_{\ll_{A_{[a]}}}[]$.

Then $\mathsf{acc}_{\ll_A}[a]$ follows from the following more general assertion, if we set $as = [a]$ and $bs = []$:

$$\forall as \in \mathrm{Bad}(A), \forall bs \in \mathrm{Bad}(A_{as}).\, \mathsf{acc}_{\ll_{A_{as}}} bs \to \mathsf{acc}_{\ll_A} as@bs.$$

$\mathsf{Ind}(\mathsf{acc}_{\ll_{A_{as}}})$. Fix as, bs and assume $\mathsf{ih}\colon \forall bs' \ll_{A_{as}} bs\, \mathsf{acc}_{\ll_A} as@bs'$. Again by definition of $\mathsf{acc}_{\ll_A}$ it suffices to show $\mathsf{acc}_{\ll_A} as@bs{*}c$ for an arbitrary $c \in A_{as@bs}$ and since $bs{*}c \ll_{A_{as}} bs$ we are done by induction hypothesis. $\diamond$

Definition. A quasi embedding from $(A, \leq_A)$ to $(B, \leq_B)$ is an injective map $f\colon A \to B$, that doesn' t create any new $\leq$-relations; i.e. for all $a_1, a_2 \in A$:

$$f(a_1) \leq_B f(a_2) \to a_1 \leq_A a_2.$$

Lemma 2. Let $f\colon A \to B$ be a quasi embedding. Then it holds

$$\mathsf{acc}_{\ll_B}[] \to \mathsf{acc}_{\ll_A}[].$$

Proof. We show $\forall bs.\, \mathsf{acc}_{\ll_B} bs \to \forall as.\, f(as) = bs \to \mathsf{acc}_{\ll_A} as$, where $f(as)$ means that f is applied component wise to as, i.e. $f([a_1, \ldots, a_n]) = [f(a_1), \ldots, f(a_n)]$. Then $\mathsf{acc}_{\ll_B}[] \to \mathsf{acc}_{\ll_A}[]$ follows by $as = bs = []$.
$\mathsf{Ind}(\mathsf{acc}_{\ll_B})$. Fix bs and assume $\mathsf{ih}\colon \forall bs' \ll_B bs, \forall as. f(as) = bs' \to \mathsf{acc}_{\ll_A} as$. Let as s.t. $f(as) = bs$. We have to show $\mathsf{acc}_{\ll_A} as$, i.e. $\forall a \in A_{as}\, \mathsf{acc}_{\ll_A} as{*}a$. $a \in A_{as}$ satisfies $(as)_i \not\leq_A a$ for all $i \leq$ length as and, since f is a quasi embedding, it follows $f((as)_i) \not\leq_B f(a)$, i.e. $(bs)_i \not\leq_B f(a)$. Hence $bs * f(a) \ll_B bs$, and by ih, applied to $bs * f(a)$ and $as * a$ we obtain $\mathsf{acc}_{\ll_A} as * a$. $\diamond$

3. KRUSKAL'S TREE THEOREM

Now, we are going to state and prove the inductive formulation of Kruskal's theorem. We start with a formal definition of $(T(A), \leq_{T(A)})$ and then introduce an additional structure on trees due to [Sch79] which will be essential for the combinatorial idea of the proof.

Definition. The set $T(A)$ of finite trees with labels in A is inductively defined by

$$\frac{a \in A \qquad t_1, \ldots, t_n \in T(A)}{a\, t_1 \ldots t_n \in T(A)},$$

where $a\,t_1 \ldots t_n$, $0 \leq n$, is the tree consisting of a root with label a and the subtrees $t_1, \ldots, t_n$. Moreover, we define the embeddability relation $\leq_{T(A)}$ inductively by the following rules:

$$\frac{t \leq_{T(A)} u_j,\ \text{for some } j \leq m}{t \leq_{T(A)} a\,u_1 \ldots u_m} \qquad \frac{a \leq_A b \quad [t_1, \ldots, t_n] \leq_{T(A)^*} [u_1, \ldots, u_m]}{a\,t_1 \ldots t_n \leq_{T(A)} b\,u_1 \ldots u_m}.$$

Here $\leq_{T(A)^*}$ refers to the Higman embedding. $\leq_{A^*}$ on A^* is inductively defined by

$$\frac{}{[] \leq_{A*} []} \qquad \frac{as \leq_{A^*} bs}{as \leq_{A^*} bs{*}b} \qquad \frac{as \leq_{A^*} bs \quad a \leq_A b}{as{*}a \leq_{A^*} bs{*}b}.$$

Definition. Given quasi orders $(A_1, \leq_{A_1}), \ldots, (A_n, \leq_{A_n})$ we define disjoint union $\dot{\cup}\{A_i : i \leq n\}$ and cartesian product $\times\{A_i : i \leq n\}$ via

$$a \leq_{\dot{\cup} A_i} a' \leftrightarrow a, a' \in A_i \wedge a \leq_{A_i} a' \text{ for some } i \leq n,$$
$$(a_1, \ldots, a_n) \leq_{\dot{\times} A_i} (a'_1, \ldots, a'_n) \leftrightarrow a_i \leq_{A_i} a'_i \text{ for all } i \leq n$$

Definition. Let $(A_0, \leq_{A_0}), \ldots, (A_n, \leq_{A_n})$ and $0 < \alpha_0 < \cdots < \alpha_n \leq \omega$ be given. By $T\binom{A_n \ \ldots \ A_0}{\alpha_n \ \ldots \ \alpha_0}$ we denote the set of all trees with labels in $\dot{\cup}\{A_i : i \leq n\}$ such that every node with a label in A_i has less than α_i immediate successors. The embeddability relation on $T\binom{A_n \ \ldots \ A_0}{\alpha_n \ \ldots \ \alpha_0}$ is the restriction of $\leq_{T(\cup\{A_i : i \leq n\})}$ on $T\binom{A_n \ \ldots \ A_0}{\alpha_n \ \ldots \ \alpha_0}$.

Kruskal's tree theorem.

$$\mathsf{acc}_{\ll_A}[] \rightarrow \mathsf{acc}_{\ll_{T\binom{A}{\omega}}}[].$$

We first give a rough idea of the proof, in order to motivate further definitions. To show $\mathsf{acc}_{\ll_{T\binom{A}{\omega}}}[]$, because of lemma 1, it suffices to show $\mathsf{acc}_{\ll_{T\binom{A}{\omega}[t]}}[]$ for an arbitrary tree t. If there is a quasi embedding of $T\binom{A}{\omega}_{[t]}$ into an appropriate tree set $T\binom{B\ A_{[a]}}{n\ \ \omega}$ and if $\mathsf{acc}_{\ll_{T\binom{B\ A_{[a]}}{n\ \ \omega}}}[]$ holds we are done by lemma 2. Now, $\binom{B\ A_{[a]}}{n\ \ \omega}$ is in a certain sense lexicographically smaller than $\binom{A}{\omega}$. So we will be finished by induction on this order once we have defined it. However, such a definition would require quantification over sets. Since in this definition we only need sets B of a certain structure, we can restrict ourselves to elements of the set of cartesian compositions of A, a set of names denoted $\mathsf{Cart}(A)$, thus avoiding second order quantification. We define a relation $<$ on $\mathsf{Cart}(A)$ such that for all cartesian compositions $\mathcal{X}$ $\mathsf{acc}_{<_{\mathsf{Cart}(A)}}\mathcal{X}$ implies $\mathsf{acc}_{\ll_X}[]$. In a second step we define the relation $<_{\mathsf{Lex}}$ and show that all cartesian compositions $\mathcal{X}$ are in the accessible part of $<_{\mathsf{Cart}(A)}$, i.e. $\mathsf{acc}_{<_{\mathsf{Cart}(A)}}\mathcal{X}$ holds. The definition of $\mathsf{Cart}(A)$ as well as the notion $\mathcal{X}^x$ and lemma 3 are due to [RW93]. In [Has94] the set $\mathsf{Cart}(A)$ corresponds, with a small restriction, to the class of algebras.

Definition (Cartesian compositions). We define the set $\mathsf{Cart}(A)$ inductively by the following rules:

1 If as is a bad sequence in A^*, then $\mathcal{A}_{as} \in \mathsf{Cart}(A)$ is a name for A_{as}.

2 If $\mathcal{X}_1, \mathcal{X}_2 \in \mathsf{Cart}(A)$ are names for X_1, X_2,
then $\mathcal{X}_1 \dot\cup \mathcal{X}_2 \in \mathsf{Cart}(A)$ is a name for $X_1 \dot\cup X_2$.

3 If $\mathcal{X}_1, \mathcal{X}_2 \in \mathsf{Cart}(A)$ are names for X_1, X_2,
then $\mathcal{X}_1 \times \mathcal{X}_2 \in \mathsf{Cart}(A)$ is a name for $X_1 \times X_2$.

4 If $\mathcal{X} \in \mathsf{Cart}(A)$ is a name for X,
then $\mathcal{X}^* \in \mathsf{Cart}(A)$ is a name for X^*.

5 Let $\mathcal{X}_0, \ldots, \mathcal{X}_m \in \mathsf{Cart}(A)$ be names for quasi orders $X_0, \ldots, X_m$ and let $0 < \gamma_0 < \cdots < \gamma_m \leq \omega$. Then $\mathcal{T}\begin{pmatrix} \mathcal{X}_m & \ldots & \mathcal{X}_0 \\ \gamma_m & \ldots & \gamma_0 \end{pmatrix}$ is a name for $T\begin{pmatrix} X_m & \ldots & X_0 \\ \gamma_m & \ldots & \gamma_0 \end{pmatrix}$.

Remark. Every name $\mathcal{X} \in \mathsf{Cart}(A)$ obviously denotes a set X. Note that $\mathcal{A}_{[]}$ is a name for A. Occasionally we will use the binary operations $\dot\cup$ and $\times$ for finitely many arguments, including the one and zero case. For these cases the definition is to be extended in the obvious way.

Definition (Recursive Definition of $\mathcal{X}^x$ for given $\mathcal{X} \in \mathsf{Cart}(A)$ and $x \in X$).

1 $(\mathcal{A}_{as})^a := \mathcal{A}_{as*a}$.

2 $(\mathcal{X}_1 \dot{\cup} \mathcal{X}_2)^x := \begin{cases} \mathcal{X}_1{}^x \dot{\cup} \mathcal{X}_2, & \text{if } x \in X_1, \\ \mathcal{X}_1 \dot{\cup} \mathcal{X}_2{}^x, & \text{if } x \in X_2. \end{cases}$

3 $(\mathcal{X}_1 \times \mathcal{X}_2)^{(x_1,x_2)} := (\mathcal{X}_1{}^{x_1} \times \mathcal{X}_2) \dot{\cup} (\mathcal{X}_1 \times \mathcal{X}_2{}^{x_2})$.

4 $(\mathcal{X}^*)^{[x_1,\ldots,x_n]} := \dot{\bigcup}\{(\mathcal{X}^{x_1})^* \times \mathcal{X} \times (\mathcal{X}^{x_2})^* \times \cdots \times \mathcal{X} \times (\mathcal{X}^{x_j})^* : j \leq n\}$.

5 Let $t \in T\binom{X_m \ \ldots \ X_0}{\gamma_m \ \ldots \ \gamma_0}$ be given. Let t consist of the root $x \in \mathcal{X}_i$ and the subtrees $t_1, \ldots, t_n$. We may assume that $\mathcal{T}\binom{\mathcal{X}_m \ \ldots \ \mathcal{X}_0}{\gamma_m \ \ldots \ \gamma_0}^{t_j}$ for all $0 \leq j \leq n$ is already defined. Set

$$\mathcal{L} := \mathcal{X}_i \times \dot{\bigcup}\{(\mathcal{T}(\ldots)^{t_1})^* \times \cdots \times (\mathcal{T}(\ldots)^{t_j})^* : j \leq n\}.$$

If $n = 0$, i.e. if t only consists of a root, then

$$\mathcal{T}\binom{\mathcal{X}_m \ \ldots \ \mathcal{X}_0}{\gamma_m \ \ldots \ \gamma_0}^t := \mathcal{T}\binom{\mathcal{X}_m \ \ldots \ \mathcal{X}_i^x \ \ldots \ \mathcal{X}_0}{\gamma_m \ \ldots \ \gamma_i \ \ldots \ \gamma_0}.$$

If $n = \gamma_k$ for $k < i$, we define

$$\mathcal{T}\binom{\mathcal{X}_m \ \ldots \ \mathcal{X}_0}{\gamma_m \ \ldots \ \gamma_0}^t := \mathcal{T}\binom{\mathcal{X}_m \ \ldots \ \mathcal{X}_i^x \ \ldots \ \mathcal{X}_k \dot{\cup} \mathcal{X}_i \dot{\cup} \mathcal{L} \ \ldots \ \mathcal{X}_0}{\gamma_m \ \ldots \ \gamma_i \ \ldots \ \gamma_k \ \ldots \ \gamma_0},$$

If $n \neq \gamma_k$ for all $k < i$, then we define

$$\mathcal{T}\binom{\mathcal{X}_m \ \ldots \ \mathcal{X}_0}{\gamma_m \ \ldots \ \gamma_0}^t := \mathcal{T}\binom{\mathcal{X}_m \ \ldots \ \mathcal{X}_i^x \ \ldots \ \mathcal{X}_i \dot{\cup} \mathcal{L} \ \ldots \ \mathcal{X}_0}{\gamma_m \ \ldots \ \gamma_i \ \ldots \ n \ \ldots \ \gamma_0},$$

where the column with $\mathcal{X}_i \dot{\cup} \mathcal{L}$ and n has to be filled in at the appropriate place.

Definition. On $\mathsf{Cart}(A)$ we define a relation $<_{\mathsf{Cart}(A)}$ via

$$\mathcal{Y} <_{\mathsf{Cart}(A)} \mathcal{X} \leftrightarrow \exists x \in X \ \mathcal{Y} = \mathcal{X}^x.$$

Lemma 3. Let $\mathcal{X} \in \mathsf{Cart}(A), x \in X$. Then there exists a quasi embedding $e \colon X_{[x]} \to X^x$.

Proof. $\mathsf{Ind}(\mathsf{Cart}(A))$.

1 Let $a \in A_{as}$. Then, because of $(A_{as})_{[a]} \subseteq (A_{as})^a$, the identity is a quasi embedding from $(A_{as})_{[a]}$ to $(A_{as})^a$.

2 Let $x \in X_1 \dot\cup X_2$. W.l.o.g. $x \in X_1$. By **ih** there is a quasi embedding $e_{X_1,x}\colon X_{1[x]} \to {X_1}^x$. We define

$$\begin{aligned} e\colon\quad & (X_1 \dot\cup X_2)_{[x]} \to {X_1}^x \dot\cup X_2 \\ & y \mapsto \begin{cases} e_{X_1,x}(y), & \text{if } y \in X_{1[x]}, \\ y, & \text{if } y \in X_2. \end{cases} \end{aligned}$$

3 Let $(x_1, x_2) \in X_1 \times X_2$. By **ih**, we already have quasi embeddings $e_{X_i,x_i}\colon X_{i[x_i]} \to {X_i}^{x_i}$ for $x_i \in X_i$, $i \in \{1,2\}$. Set

$$\begin{aligned} e\colon\quad & (X_1 \times X_2)_{[(x_1,x_2)]} \to ({X_1}^{x_1} \times X_2) \,\dot\cup\, (X_1 \times {X_2}^{x_2}) \\ & (y_1, y_2) \mapsto \begin{cases} (e_{X_1,x_1}(y_1), y_2), & \text{if } y_1 \in X_{1[x_1]}, \\ (y_1, e_{X_2,x_2}(y_2)), & \text{otherwise.} \end{cases} \end{aligned}$$

4 Let $[x_1, \dots, x_n] \in X^*$. By **ih** we already have quasi embeddings $e_{X,x_i}\colon X_{[x_i]} \to X^{x_i}$, for all $1 \le i \le n$. We look for a quasi embedding

$$e\colon {X^*}_{[[x_1,\dots,x_n]]} \to \dot\bigcup\{(X^{x_1})^* \times X \times (X^{x_2})^* \times X \times \cdots \times (X^{x_j})^* : j \le n\}.$$

If $n = 0$ then the quasi embedding is the empty quasi embedding. Otherwise, let $[y_1, \dots, y_m] \in {X^*}_{[[x_1,\dots,x_n]]}$. Since $[x_1, \dots, x_n] \not\le_{X^*} [y_1, \dots, y_m]$ there exists $0 \le l < n$ such that $[x_1, \dots, x_l] \le_{X^*} [y_1, \dots, y_m]$ holds, but $[x_1, \dots, x_{l+1}] \le_{X^*} [y_1, \dots, y_m]$ doesn't. We choose $j_1 < \dots < j_l$ minimal such that $x_i \le y_{j_i}$, for all $i \le l$. Then we have $[y_1, \dots, y_{j_1-1}] \in (X_{[x_1]})^*, \dots, [y_{j_l+1}, \dots, y_m] \in (X_{[x_{l+1}]})^*$, and by induction hypothesis we may define

$$\begin{aligned} ws_1 &:= [e_{X,x_1}(y_1), \dots, e_{X,x_1}(y_{j_1-1})] \in (X^{x_1})^* \\ &\dots \\ ws_{l+1} &:= [e_{X,x_{l+1}}(y_{j_l+1}), \dots, e_{X,x_{l+1}}(y_m)] \in (X^{x_{l+1}})^* \end{aligned}$$

and set $e([y_1, \dots, y_m]) := (ws_1, y_{j_1}, ws_2, y_{j_2}, \dots, y_{j_l}, ws_{l+1})$.

5 Let $t \in T\binom{X_m\ \dots\ X_0}{\gamma_m\ \dots\ \gamma_0}$ be given. **Ind**(structure of t). Let us regard first the case t consisting only of a root with a label $x \in X_i$. Let $u \in T\binom{X_m\ \dots\ X_0}{\gamma_m\ \dots\ \gamma_0}_{[t]}$. Then for every node in u with a label $y \in X_i$ it holds $x \not\le_{X_i} y$, i.e. u lies in $T\begin{pmatrix} X_m & \dots & X_{i[x]} & \dots & X_0 \\ \gamma_m & \dots & \gamma_i & \dots & \gamma_0 \end{pmatrix}$. A quasi

embedding into $T\begin{pmatrix} X_m \dots X_i{}^x \dots X_0 \\ \gamma_m \dots \gamma_i \dots \gamma_0 \end{pmatrix}$ can easily be constructed by **ih**, applied to to X_i and x.

Now, assume t consists of a root $x \in X_i$ and the immediate sub trees $t_1, \dots, t_n$ with $n = \gamma_k < \gamma_i$ (The case $n \neq \gamma_k$, for all $k < i$ runs analogously). By induction hypothesis, for all $j \leq n$ there exist quasi embeddings

$$e_{t_j} : T\begin{pmatrix} X_m \dots X_0 \\ \gamma_m \dots \gamma_0 \end{pmatrix}_{[t_j]} \to T\begin{pmatrix} X_m \dots X_0 \\ \gamma_m \dots \gamma_0 \end{pmatrix}^{t_j}.$$

We look for a quasi embedding

$$e\colon T\begin{pmatrix} X_m \dots X_0 \\ \gamma_m \dots \gamma_0 \end{pmatrix}_{[t]} \to T\begin{pmatrix} X_m \dots X_i{}^x \dots X_k\dot{\cup}X_i\dot{\cup}L \dots X_0 \\ \gamma_m \dots \gamma_i \dots \gamma_k \dots \gamma_0 \end{pmatrix}.$$

Let $u \in T\begin{pmatrix} X_m \dots X_0 \\ \gamma_m \dots \gamma_0 \end{pmatrix}_{[t]}$ be a tree with root y and immediate subtrees $u_1, \dots, u_r$. We may assume that the quasi embeddings on the subtrees are already defined, and have to map the root label y suitably to a label in $X_0, \dots, X_k\dot{\cup}X_i\dot{\cup}L, \dots, X_i{}^x, \dots$, or X_m such that the condition concerning the number of immediate subtrees is fulfilled.

$t \not\leq_{T(\dots)} u$ only can hold for one of the following reasons:

- i $y \notin X_i$. Then map y to itself respectively in the case $y \in X_k$ to y in the first component of $X_k\dot{\cup}X_i\dot{\cup}L$. We set $e(y\, u_1 \dots u_r) := y\, e(u_1) \dots e(u_r)$.
- ii $y \in X_i$, but $x \not\leq y$. Hence $y \in X_{i[x]}$ and by induction hypothesis we have $e_{X_i,x}(y) \in X_i{}^x$. We set $e(y\, u_1 \dots u_r) := e_{X_i,x}(y)\, e(u_1) \dots e(u_r)$.
- iii $y \in X_i$ and $x \leq y$, but y has less than γ_k successors. Then anyhow we may map y to itself, if we regard y as a label in the second component of $X_k\dot{\cup}X_i\dot{\cup}L$. We set $e(y\, u_1 \dots u_r) := y\, e(u_1) \dots e(u_r)$.
- iv $y \in X_i$, $x \leq y$ and y has more than γ_k immediate successors, but $[t_1, \dots, t_{\gamma_k}] \not\leq_{T(\dots)^*} [u_1, \dots, u_r]$. Then there exists an l, $0 \leq l < \gamma_k$ such that $[t_1, \dots, t_l] \leq_{T(\dots)^*} [u_1, \dots, u_r]$ and $[t_1, \dots, t_{l+1}] \not\leq_{T(\dots)^*} [u_1, \dots, u_r]$. We will be mapping u to a tree with $l < \gamma_k$ subtrees and a root label in L,

where this label contains the images of the remaining subtrees. More formally: Choose $j_1 < \cdots < j_l$ minimal such that $t_1 \leq_{T(\ldots)} u_{j_1}, \ldots, t_l \leq_{T(\ldots)} u_{j_l}$. Then it holds

$$[u_1, \ldots, u_{j_1-1}] \in T\begin{pmatrix} X_m & \ldots & X_0 \\ \gamma_m & \ldots & \gamma_0 \end{pmatrix}^*_{[t_1]}$$
$$\ldots$$
$$[u_{j_l+1}, \ldots, u_r] \in T\begin{pmatrix} X_m & \ldots & X_0 \\ \gamma_m & \ldots & \gamma_0 \end{pmatrix}^*_{[t_{l+1}]}$$

and by ih it follows

$$ts_1 := [e_{t_1}(u_1), \ldots, e_{t_1}(u_{j_1-1})] \in T\begin{pmatrix} X_m & \ldots & X_0 \\ \gamma_m & \ldots & \gamma_0 \end{pmatrix}^{t_1 *}$$
$$\ldots$$
$$ts_{l+1} := [e_{t_{l+1}}(u_{j_l+1}), \ldots, e_{t_{l+1}}(u_r)] \in T\begin{pmatrix} X_m & \ldots & X_0 \\ \gamma_m & \ldots & \gamma_0 \end{pmatrix}^{t_{l+1} *}$$

Finally we put

$$e(y\, u_1 \ldots u_r) := (\, y,\ (ts_1, \ldots, ts_{l+1}))\, e(u_{j_1}) \ldots e(u_{j_l}).$$

It is left to the reader to check that e is actually a quasi embedding. $\diamond$

Lemma 4. $\forall \mathcal{X} \in \mathsf{Cart}(A).\, \mathsf{acc}_{<_{\mathsf{Cart}(A)}} \mathcal{X} \to \mathsf{acc}_{\ll_X} [\,]$.

Proof. $\mathsf{Ind}(\mathsf{acc}_{<_{\mathsf{Cart}(A)}})$. Let $\mathcal{X} \in \mathsf{Cart}(A)$. Assume $\mathsf{ih}\colon \forall \mathcal{Y} <_{\mathsf{Cart}(A)} \mathcal{X}\, \mathsf{acc}_{\ll_Y}[\,]$. Because of lemma 1 it suffices to show $\forall x \in X\, \mathsf{acc}_{\ll_{X_{[x]}}}[\,]$. Let $x \in X$. By lemma 3 there exists a quasi embedding $e\colon X_{[x]} \to X^x$. Therefore, using lemma 2 we only have to show $\mathsf{acc}_{\ll_{X^x}}[\,]$. But this follows by ih. $\diamond$

Definition. On the set Lex:=

$$\left\{ \begin{pmatrix} \mathcal{X}_n & \ldots & \mathcal{X}_1 \\ \gamma_n & \ldots & \gamma_1 \end{pmatrix},\ \mathsf{acc}_{<_{\mathsf{Cart}(A)}} \mathcal{X}_i,\ i \leq n,\ 0 \leq n < \omega,\ 0 < \gamma_1 < \cdots < \gamma_n \leq \omega \right\}$$

we define a relation $<_{\mathsf{Lex}}$: $\begin{pmatrix} \mathcal{Y}_m \dots \mathcal{Y}_1 \\ \delta_m \dots \delta_1 \end{pmatrix} <_{\mathsf{Lex}} \begin{pmatrix} \mathcal{X}_n \dots \mathcal{X}_1 \\ \gamma_n \dots \gamma_1 \end{pmatrix} \iff$

$$m, n > 0 \text{ and}$$
$$\delta_m = \gamma_n \ \wedge \ \mathcal{Y}_m <_{\mathsf{Cart}(A)} \mathcal{X}_n \text{ or}$$
$$\delta_m = \gamma_n \ \wedge \ \mathcal{Y}_m = \mathcal{X}_n \ \wedge \ \begin{pmatrix} \mathcal{Y}_{m-1} \dots \mathcal{Y}_1 \\ \delta_{m-1} \dots \delta_1 \end{pmatrix} <_{\mathsf{Lex}} \begin{pmatrix} \mathcal{X}_{n-1} \dots \mathcal{X}_1 \\ \gamma_{n-1} \dots \gamma_1 \end{pmatrix}.$$

Lemma 5. Assume $\mathsf{acc}_{<_{\mathsf{Cart}(A)}} \mathcal{X}_0, \dots, \mathsf{acc}_{<_{\mathsf{Cart}(A)}} \mathcal{X}_n$ and let $0 < \gamma_0 < \dots < \gamma_n \leq \omega$. Then it follows $\mathsf{acc}_{<_{\mathsf{Lex}}} \begin{pmatrix} \mathcal{X}_n \dots \mathcal{X}_0 \\ \gamma_n \dots \gamma_0 \end{pmatrix}$.

Proof. We show

$$\forall \gamma \leq \omega.\ \forall \mathcal{X}.\ \mathsf{acc}_{<_{\mathsf{Cart}(A)}} \mathcal{X} \to \forall \mathbb{X}.\ \mathsf{acc}_{<_{\mathsf{Lex}}} \mathbb{X} \to$$
$$\forall n, \mathcal{X}_1, \dots, \mathcal{X}_n, \gamma_1, \dots, \gamma_n.\ \gamma_1 < \dots < \gamma_n < \gamma \wedge \mathbb{X} = \begin{pmatrix} \mathcal{X}_n \dots \mathcal{X}_1 \\ \gamma_n \dots \gamma_1 \end{pmatrix} \to$$
$$\mathsf{acc}_{<_{\mathsf{Lex}}} \begin{pmatrix} \mathcal{X}\ \mathcal{X}_n \dots \mathcal{X}_1 \\ \gamma\ \gamma_n \dots \gamma_1 \end{pmatrix}$$

by $\mathsf{Ind}_1(\gamma)$, $\mathsf{Ind}_2(\mathsf{acc}_{<_{\mathsf{Cart}(A)}})$ and $\mathsf{Ind}_3(\mathsf{acc}_{<_{\mathsf{Lex}}})$. We need to prove $\mathsf{acc}_{<_{\mathsf{Lex}}} \mathbb{Y}$ for all $\mathbb{Y} <_{\mathsf{Lex}} \begin{pmatrix} \mathcal{X}\ \mathcal{X}_n \dots \mathcal{X}_1 \\ \gamma\ \gamma_n \dots \gamma_1 \end{pmatrix}$.

i Let $\mathbb{Y} = \begin{pmatrix} \mathcal{X}^x\ \mathcal{Y}_m \dots \mathcal{Y}_1 \\ \gamma\ \delta_m \dots \delta_1 \end{pmatrix}$ such that $x \in X$, $\delta_1 < \dots < \delta_m < \gamma$ and $\forall i \leq m\ \mathsf{acc}_{<_{\mathsf{Cart}(A)}} \mathcal{Y}_i$. By an m-fold application of ih_1 we obtain $\mathsf{acc}_{<_{\mathsf{Lex}}} \begin{pmatrix} \mathcal{Y}_m \dots \mathcal{Y}_1 \\ \delta_k \dots \delta_1 \end{pmatrix}$ and by ih_2 $\mathsf{acc}_{<_{\mathsf{Lex}}} \begin{pmatrix} \mathcal{X}^x\ \mathcal{Y}_m \dots \mathcal{Y}_1 \\ \gamma\ \delta_m \dots \delta_1 \end{pmatrix}$.

ii Let $\mathbb{Y} = \begin{pmatrix} \mathcal{X}\ \mathcal{Y}_m \dots \mathcal{Y}_1 \\ \gamma\ \delta_m \dots \delta_1 \end{pmatrix}$ such that $\begin{pmatrix} \mathcal{Y}_m \dots \mathcal{Y}_1 \\ \delta_m \dots \delta_1 \end{pmatrix} <_{\mathsf{Lex}} \begin{pmatrix} \mathcal{X}_n \dots \mathcal{X}_1 \\ \gamma_n \dots \gamma_1 \end{pmatrix}$. Then $\mathsf{acc}_{<_{\mathsf{Lex}}} \begin{pmatrix} \mathcal{X}\ \mathcal{Y}_m \dots \mathcal{Y}_1 \\ \gamma\ \delta_m \dots \delta_1 \end{pmatrix}$ immediately follows by ih_3.

The lemma follows from $\mathsf{acc}_{<_{\mathsf{Lex}}}()$ by $n+1$ applications of this assertion.

◇

Lemma 6. Assume $\mathsf{acc}_{\ll_A}[]$. Then $\forall \mathcal{X}. \mathcal{X} \in \mathsf{Cart}(A) \to \mathsf{acc}_{<_{\mathsf{Cart}(A)}} \mathcal{X}$.

Proof. $\mathsf{Ind}(\mathsf{Cart}(A))$.

1 We show $\forall as \in \mathrm{Bad}(A).\, \mathsf{acc}_{\ll_A} as \to \mathsf{acc}_{<_{\mathsf{Cart}(A)}} \mathcal{A}_{as}$. $\mathsf{Ind}(\mathsf{acc}_{\ll_A})$. Let $as \in \mathrm{Bad}(A)$ and assume ih: $\forall bs \ll_A as\, \mathsf{acc}_{<_{\mathsf{Cart}(A)}} \mathcal{A}_{bs}$. We need to prove $\forall a \in A_{as}\, \mathsf{acc}_{<_{\mathsf{Cart}(A)}} (\mathcal{A}_{as})^a$. But this follows by ih because of $(\mathcal{A}_{as})^a = \mathcal{A}_{as*a}$ and $as*a \ll_A as$.

2 Let $\mathcal{X}_1$ and $\mathcal{X}_2$ s.t. $\mathsf{acc}_{<_{\mathsf{Cart}(A)}} \mathcal{X}_1$ and $\mathsf{acc}_{<_{\mathsf{Cart}(A)}} \mathcal{X}_2$. In order to show $\mathsf{acc}_{<_{\mathsf{Cart}(A)}} \mathcal{X}_1 \dot\cup \mathcal{X}_2$ use induction on $\mathsf{acc}_{<_{\mathsf{Cart}(A)}} \mathcal{X}_1$ and $\mathsf{acc}_{<_{\mathsf{Cart}(A)}} \mathcal{X}_2$. Let $x \in \mathcal{X}_1 \dot\cup \mathcal{X}_2$, w.l.o.g. $x \in X_1$, then the induction hypothesis implies $\mathsf{acc}_{<_{\mathsf{Cart}(A)}} \mathcal{X}_1{}^{x_1} \dot\cup \mathcal{X}_2$.

3 Let $\mathcal{X}_1$ and $\mathcal{X}_2$ s.t. $\mathsf{acc}_{<_{\mathsf{Cart}(A)}} \mathcal{X}_1$ and $\mathsf{acc}_{<_{\mathsf{Cart}(A)}} \mathcal{X}_2$. Now we show $\mathsf{acc}_{<_{\mathsf{Cart}(A)}} \mathcal{X}_1 \times \mathcal{X}_2$ by induction on $\mathsf{acc}_{<_{\mathsf{Cart}(A)}} \mathcal{X}_1$ and $\mathsf{acc}_{<_{\mathsf{Cart}(A)}} \mathcal{X}_2$. We have

$$\mathsf{ih}_1 : \forall x_1 \in X_1, \forall \mathcal{X}_2.\, \mathsf{acc}_{<_{\mathsf{Cart}(A)}} \mathcal{X}_2 \to \mathsf{acc}_{<_{\mathsf{Cart}(A)}} \mathcal{X}_1^{x_1} \times \mathcal{X}_2$$
$$\mathsf{ih}_2 : \forall x_2 \in X_2\, \mathsf{acc}_{<_{\mathsf{Cart}(A)}} \mathcal{X}_1 \times \mathcal{X}_2^{x_2}$$

Now, let $(x_1, x_2) \in \mathcal{X}_1 \times \mathcal{X}_2$. $\mathsf{acc}_{<_{\mathsf{Cart}(A)}} (\mathcal{X}_1^{x_1} \times \mathcal{X}_2) \dot\cup (\mathcal{X}_1 \times \mathcal{X}_2^{x_2})$ follows by ih_1, applied to x_1 and $\mathcal{X}_2$, ih_2 and 2.

4 Let $\mathcal{X}$ such that $\mathsf{acc}_{<_{\mathsf{Cart}(A)}} \mathcal{X}$ be given. We prove $\mathsf{acc}_{<_{\mathsf{Cart}(A)}} \mathcal{X}^*$ by induction on $\mathsf{acc}_{<_{\mathsf{Cart}(A)}} \mathcal{X}$. Assume ih: $\forall x \in \mathcal{X}\, \mathsf{acc}_{<_{\mathsf{Cart}(A)}} (\mathcal{X}^x)^*$. We have to show $\mathsf{acc}_{<_{\mathsf{Cart}(A)}} \mathcal{X}^*$, i.e. $\forall w \in X^*\, \mathsf{acc}_{<_{\mathsf{Cart}(A)}} (\mathcal{X}^*)^w$. Let $w = [x_1, \ldots, x_n]$. Now $\mathsf{acc}_{<_{\mathsf{Cart}(A)}} \dot\bigcup\{(\mathcal{X}^{x_1})^* \times \mathcal{X} \times \cdots \times \mathcal{X} \times (\mathcal{X}^{x_j})^* : j \leq n\}$ follows by ih, 2., and 3.

5 Let $\mathcal{X}_0, \ldots, \mathcal{X}_m$ such that $\mathsf{acc}_{<_{\mathsf{Cart}(A)}} \mathcal{X}_i$ for all $i \leq m$ and $0 < \gamma_0 < \cdots < \gamma_m \leq \omega$. Since $\mathsf{acc}_{<_{\mathsf{Cart}(A)}} \mathcal{X}_0, \ldots, \mathsf{acc}_{<_{\mathsf{Cart}(A)}} \mathcal{X}_m$ implies $\mathsf{acc}_{<_{\mathsf{Lex}}} \begin{pmatrix} \mathcal{X}_m & \cdots & \mathcal{X}_0 \\ \gamma_m & \cdots & \gamma_0 \end{pmatrix}$ by lemma 5, we get $\mathsf{acc}_{<_{\mathsf{Cart}(A)}} \mathcal{T}\begin{pmatrix} \mathcal{X}_m & \cdots & \mathcal{X}_0 \\ \gamma_m & \cdots & \gamma_0 \end{pmatrix}$ when we have shown more generally

$$\forall \mathbb{X} \neq ().\, \mathsf{acc}_{<_{\mathsf{Lex}}} \mathbb{X} \to \mathsf{acc}_{<_{\mathsf{Cart}(A)}} \mathcal{T}(\mathbb{X}).$$

$\mathsf{Ind}_1(\mathsf{acc}_{<_{\mathsf{Lex}}})$. Fix $\mathbb{X}$ and assume $\mathsf{ih}_1 : \forall \mathbb{Y} <_{\mathsf{Lex}} \mathbb{X}.\, \mathsf{acc}_{<_{\mathsf{Cart}(A)}} \mathcal{T}(\mathbb{Y})$. Suppose $\mathbb{X}$ to be of a form $\left(\begin{smallmatrix} \mathcal{X}_r & \cdots & \mathcal{X}_0 \\ \gamma_r & \cdots & \gamma_0 \end{smallmatrix}\right)$ show $\mathsf{acc}_{<_{\mathsf{Cart}(A)}} \mathcal{T}(\mathbb{X})^t$ for an arbitrary $t \in T(\mathbb{X})$ by

Ind_2 (structure of t). Assume first that t is a branch with a label $x \in X_i$. Since $\begin{pmatrix} \mathcal{X}_r \dots \mathcal{X}_i^x \dots \mathcal{X}_0 \\ \gamma_r \dots \gamma_i \dots \gamma_0 \end{pmatrix} <_{\mathsf{Lex}} \mathbb{X}$ we may conclude $\mathsf{acc}_{<_{\mathsf{Cart}(A)}} \mathcal{T}\begin{pmatrix} \mathcal{X}_r \dots \mathcal{X}_i^x \dots \mathcal{X}_0 \\ \gamma_r \dots \gamma_i \dots \gamma_0 \end{pmatrix}$ by ih_1.

Now, assume that t consists of the root $x \in \mathcal{X}_i$ and the subtrees $t_1, \dots, t_n$, $n = \gamma_k < \gamma_i$. (The case $n \neq \gamma_k$ for all $k < i$ is analogous.) By ih_2 it follows $\mathsf{acc}_{<_{\mathsf{Cart}(A)}} \mathcal{T}\begin{pmatrix} \mathcal{X}_r \dots \mathcal{X}_0 \\ \gamma_r \dots \gamma_0 \end{pmatrix}^{t_j}$, for all $j \leq n$. Because of 2., 3., and 4. we are able to conclude $\mathsf{acc}_{<_{\mathsf{Cart}(A)}} (\mathcal{X}_i \times \dot{\cup}\{(\mathcal{T}(\mathbb{X})^{t_1})^* \times \dots \times (\mathcal{T}(\mathbb{X})^{t_j})^* : j \leq n\})$. Using the abbreviation $\mathcal{L} = \mathcal{X}_i \times \dot{\cup}\{(\mathcal{T}(\mathbb{X})^{t_1})^* \times \dots \times (\mathcal{T}(\mathbb{X})^{t_j})^* : j \leq n\}$ we get $\mathsf{acc}_{<_{\mathsf{Cart}(A)}} \mathcal{L}$ and $\mathsf{acc}_{<_{\mathsf{Cart}(A)}} (\mathcal{X}_k \dot{\cup} \mathcal{X}_i \dot{\cup} \mathcal{L})$ by a further application of 2. Therefore we have

$$\mathcal{T}\begin{pmatrix} \mathcal{X}_r \dots \mathcal{X}_i^x \dots \mathcal{X}_k \dot{\cup} \mathcal{X}_i \dot{\cup} \mathcal{L} \dots \mathcal{X}_0 \\ \gamma_r \dots \gamma_i \dots \gamma_k \dots \gamma_0 \end{pmatrix} <_{\mathsf{Lex}} \mathcal{T}\begin{pmatrix} \mathcal{X}_r \dots \mathcal{X}_i \dots \mathcal{X}_k \dots \mathcal{X}_0 \\ \gamma_r \dots \gamma_i \dots \gamma_k \dots \gamma_0 \end{pmatrix}$$

and by ih_1 we obtain $\mathsf{acc}_{<_{\mathsf{Cart}(A)}} \mathcal{T}\begin{pmatrix} \mathcal{X}_r \dots \mathcal{X}_i^x \dots \mathcal{X}_k \dot{\cup} \mathcal{X}_i \dot{\cup} \mathcal{L} \dots \mathcal{X}_0 \\ \gamma_r \dots \gamma_i \dots \gamma_k \dots \gamma_0 \end{pmatrix}$. $\diamond$

Proof of Kruskal's theorem. We want to show $\mathsf{acc}_{\ll_A} [] \to \mathsf{acc}_{\ll_{\mathcal{T}(\substack{A \\ \omega})}} []$. Assume $\mathsf{acc}_{\ll_A} []$. For all $\mathcal{X} \in \mathsf{Cart}(A)$ we have $\mathsf{acc}_{<_{\mathsf{Cart}(A)}} \mathcal{X}$ by lemma 6 and therefore $\mathsf{acc}_{\ll_{\mathcal{X}}} []$ by lemma 4. This holds especially for $\mathcal{X} = \mathcal{T}(\substack{A_{[]} \\ \omega})$ and since $A_{[]} = A$ we may conclude $\mathsf{acc}_{\ll_{\mathcal{T}(\substack{A \\ \omega})}} []$. $\diamond$

Acknowledgments

I would like to thank Ulrich Berger, Thierry Coquand, Daniel Fridlender, Wim Veldman, and Andreas Weiermann for helpful discussions and valuable comments.

References

[Coq93] Coquand, T. (1993). A proof of Higman's lemma by structural induction. Unpublished manuscript.

[CP99] Coquand, T. and Persson, H. (1999). Gröbner Bases in Type Theory. In Altenkirch, T., et. al., editors, *Types for Proofs and Programs*, Springer LNCS 1657:33–46.

[Fri97] Fridlender, D. (1997). *Higman's Lemma in Type Theory.* PhD thesis, Chalmers University of Technology and University of Göteburg, Sweden.

[Fri99] Fridlender, D. (1999). An Interpretation of the Fan Theorem in Type Theory. In Altenkirch, T., et. al., editors, *Types for Proofs and Programs*, Springer LNCS 1657:93–105.

[Gal91] Gallier, J. H. (1991). What's so special about Kruskal's Theorem and the ordinal Γ_0? A survey of some results in proof theory. APAL 53:199–260.

[Has94] Hasegawa, R. (1994). Well–Ordering of Algebra and Kruskal's Theorem. In Jones, N. D., et. al., editors, *Logic, Language and Computation.* Springer LNCS 792:133–172.

[Kru60] Kruskal, J. B. (1960). Well–quasi–orderings, the tree theorem and Vaszonyi's conjecture. *Trans. Am. Math. Soc.*, 95:210–255.

[Mar70] Martin-Löf, P. (1970). *Notes on Constructive Mathematics.* Almqvist & Wiksell, Stockholm.

[Nas63] Nash-Williams, C. S. J. A. (1963). On well–quasi–ordering finite trees. *Proc. Cambridge Phil. Soc.*, 59:833–835.

[Per99] Persson, H. (1999). *Type Theory and the Integrated Logic of Programs.* PhD thesis, Chalmers University of Technology and University of Göteburg, Sweden.

[RW93] Rathjen, M. and Weiermann, A. (1993). Proof–theoretic investigations on Kruskal's theorem. APAL 60:49–88.

[RS93] Richman, F. and Stolzenberg, G. (1993). Well quasi–ordered sets. *Advances in Math.*, 97:145–153.

[Sch79] Schmidt, D. (1979). Well–orderings and their maximal order types. Habilitationsschrift, Mathematisches Institut der Universität Heidelberg.

[Sim85] Simpson, S. G. (1985). Nonprovability of certain combinatorial properties of finite trees. In Harrington, L., et. al., editors, *Harvey Friedman's Research on the Foundations of Mathematics*, pages 87–117. North–Holland, Amsterdam.

REAL NUMBERS AND FUNCTIONS EXHIBITED IN DIALOGUES

Rudolf Taschner
Technical University of Vienna
Wiedner Hauptstrasse 8-10, A 1040 Vienna, Austria
rtasch@pop.tuwien.ac.at

Keywords: Real numbers, real functions

Mathematical Subject Classification: 03F55, 03F60, 03F65, 11A63, 26A15

Abstract: It was the late Paul Lorenzen who rediscovered the ancient Socratic method of putting forward arguments and proofs by dialogues, and who applied it to definitions and theorems of mathematics. In the light generated by such dialogues the differences between the fundamental assumptions of classical mathematics, of recursive constructivism, of Bishop's constructive mathematics, and of intuitionistic mathematics become apparent, especially in connection with various definitions of real numbers, of sequences, and functions. Even attempts to introduce infinitesimals seem to fit into this dialogical frame.

Introduction

An essential problem that not only each of the diverging directions of constructive mathematics, but also the usual formal mathematics are up against, was raised by Nicolas Goodman[1]: He wrote about Bishop's constructivism — and the same would likewise apply to other foundational systems — that it consists without question of excellent mathematics, but: is it founded equally well philosophically? This question is far from esoteric. It challenges the mathematician to reflect on the soundness of the basis of his activities, and it refers to the task of safeguarding the status of mathematics within society.

[1]Nicolas D. Goodman: Reflections on Bishop's Philosophy of Mathematics, in: *Constructive Mathematics*, Lecture Notes 873, (135–145) Springer, Heidelberg, 1981.

P. Schuster, U. Berger and H. Osswald (eds.),
Reuniting the Antipodes – Constructive and Nonstandard Views of the Continuum, 257–269.

The best method to guarantee a fair deliberation of the diverse foundational concepts was invented almost 50 years ago by Paul Lorenzen[2]: it is the imaginary dispute of two intellectual combatants, of the so-called *proponent* and the so-called *opponent.*

1. DIALOGUES ABOUT REAL NUMBERS

To give an idea of this dispute, we try to consider the many different facets in using the notion "real number" in dialogues. At first we start with a vague, but generally accepted concept of real numbers: it is possible to grasp them by finite decimal numbers with a degree of precision that can be defined at will. In a slightly more accurate setting this means: given the real number a, it is possible to denominate for each natural number n a decimal number

$$[a]_n = z + 0.z_1z_2z_3\ldots z_n = z + \frac{z_1}{10} + \frac{z_2}{100} + \frac{z_3}{1000} + \ldots + \frac{z_n}{10^n}$$

with exactly n decimal places, with an integer z and with digits z_1, z_2, z_3, ..., z_n, such that for all natural numbers n and m the inequality

$$|[a]_n - [a]_m| \leq \frac{1}{10^n} + \frac{1}{10^m}$$

holds. This definition makes the naive idea of an "infinite decimal number"

$$a = z.z_1z_2z_3\ldots z_n\ldots$$

with $[a]_n$ being the "n-th approximate value" more precise. For instance, one can write

$$\begin{aligned}[\pi]_3 &= 3.141\\ [\pi]_{13} &= 3.1415926535897\\ [\pi]_{30} &= 3.141592653589793238462643383279\end{aligned}$$

with $\pi = 4\arctan 1$ being the arc of two right angles. Far more interesting examples are Brouwer's *oscillating numbers* b which have as approximate values

$$[b]_n = \frac{w_1}{10} + \frac{w_2}{100} + \frac{w_3}{1000} + \ldots + \frac{w_n}{10^n}$$

[2]cf. Paul Lorenzen: *Formale Logik*, Berlin, 1958. Paul Lorenzen: *Metamathematik*, Mannheim, 1962.

wherein the numbers w_1, w_2, w_3, ..., w_n independently take values -1, 0, or 1, and it is presupposed that these values will never change in the course of calculating $[b]_{n+1}$, $[b]_{n+2}$, $[b]_{n+3}$, ..., $[b]_{n+k}$, ... One has to bear in mind that *only* in the case of consistently nonnegative w_1, w_2, w_3, ..., w_n do the digits z_1, z_2, z_3, ..., z_n of $[b]_n$ coincide with these w_1, w_2, w_3, ..., w_n respectively. This remark already indicates that Brouwer's oscillating numbers erode the naive identification of real numbers with simple "infinite decimal numbers".

Two real numbers a and b are defined to be *equal* if and only if the inequalities

$$|[a]_n - [b]_n| \leq \frac{2}{10^n}$$

hold for each natural number n. Only this *definition* ensures that the infinite decimal number $0.999\ldots9\ldots$ is equal to 1.

Yet the explanation of real numbers and their equality given so far *does not* fix the appropriate use of this concept in the dialogue between proponent and opponent. These two partners first have to reach an agreement as to the *logic* they are going to use — the proponent in putting forward and defending propositions and the opponent in casting doubt on all the proponent's assertions that are not yet proven.

If both the proponent and the opponent agree to use the logic of *formal mathematics*, they thereby accept the concept that by setting the real number a both the integral part z as well as *all* decimal places z_1, z_2, z_3, ..., z_n, ... are given *at once*, such that for all n the n-th approximate value reads $[a]_n = z + 0.z_1z_2z_3\ldots z_n$. Beyond that, the *whole* infinite sequence $z, z_1, z_2, \ldots, z_n, \ldots$ is at disposal. Only this precondition permits the proponent to defend the assertion that each oscillating number would be either equal to zero or different from zero. He banishes the doubts of the opponent by demanding his partner to set an arbitrary oscillating number b. Along with b the whole sequence of integers w_1, w_2, w_3, ..., w_n, ... between -1 and $+1$ is announced by the opponent such that for all natural numbers n the relation

$$[b]_n = \frac{w_1}{10} + \frac{w_2}{100} + \frac{w_3}{1000} + \ldots + \frac{w_n}{10^n}$$

and in the limit the representation of b as infinite series

$$b = \sum_{n=1}^{\infty} \frac{w_n}{10^n}$$

holds. Therefore the proponent is able to distinguish between the following cases:

In the *first case*, all $w_1, w_2, w_3, \ldots, w_n, \ldots$ are equal to zero, therefore $b = 0$.

Otherwise the proponent denominates N to be the smallest natural number with $w_N \neq 0$. In this *second case*, one has

$$|b| \geq \frac{1}{10^N} - \sum_{k=1}^{\infty} \frac{1}{10^{N+k}} = \frac{1}{10^N}\left(1 - \frac{1}{9}\right) > 0$$

with the consequence $b \neq 0$.

It is evident that constructivists will reject the proponent's arguments, yet they will do so for different reasons:

In *recursive constructivism*, the indication of a real number a proceeds with the indication of the *recursive function* that produces for any n the integral part z and the finite sequence of decimal digits $z_1, z_2, z_3, \ldots, z_n$ of $[a]_n$ — always under the assumption $|[a]_n - [a]_m| \leq 10^{-n} + 10^{-m}$. The assertion of the proponent that each oscillating number would be either equal to zero or different from zero should be understood as if the proponent claimed to be able to distinguish between these possibilities by means of recursive methods. This would imply the proponent's power to solve the halting problem — an obvious absurdity.

In *intuitionism*, the indication of a real number a by one partner of the dialogue proceeds with the mere indication of the integral part z and of the n digits $z_1, z_2, z_3, \ldots, z_n$ of $[a]_n$ if the other partner of the dialogue wants to know this n-th approximate value. (This, of course, is combined with the assurance $|[a]_n - [a]_m| \leq 10^{-n} + 10^{-m}$ for any n and m.) The "method" — whatever this vague notion might mean — by which the presenting partner obtains these numbers, is no topic of the dialogue at all. The assertion of the proponent that each oscillating number would be either equal zero or different from zero, is rejected by the opponent by setting an oscillating number b: for all natural numbers n that the proponent submits, the opponent announces $[b]_n = 0.000\ldots0$ as n-th approximate value. But this certainly does not imply that *all* numbers $w_1, w_2, w_3, \ldots, w_n, \ldots$ of b would be zero. Thus the proponent cannot win this dispute.

It should be noted that from the position of *formal mathematics* the set of real numbers created by *recursive constructivism* is rather *small*: it consists only of countably many numbers although the recursive constructivist is unable to enumerate them all by recursive methods. On the other hand the continuum created by *intuitionism* is, in a certain sense, *larger* than the set of real numbers in formal mathematics: the oscillating numbers which may or may not be equal zero are abandoned in formal mathematics.

In *Bishop's constructivism*, each of the above-mentioned concepts is incorporated. Bishop is able to carry out this program by *breaking the symmetry* of the logical basis in the dialogue between proponent and opponent: On the one hand, the proponent is only allowed to use real numbers that are validated by recursive methods. Therefore the proponent has only those arguments at his disposal that are accepted by the recursive constructivist, and all the more by the formal mathematician and the intuitionist. On the other hand, the opponent is allowed to use real numbers in the sense of Brouwer's intuitionism. This results in the pool of numbers suitable for counter-arguments being so comprehensive that the recursive constructivist as well as the formal mathematician and the intuitionist can identify themselves with the defenses of the proponent.

The directions of foundational systems discussed here represent by no means a complete list. For instance, one can study a scenario in which either the proponent, or the opponent, or indeed both have an *infinitesimal* number Ω at their disposal, greater than all "standard" natural numbers. It could be of some interest which types of non-standard systems will develop, depending on agreements between proponent and opponent in using the infinitesimal Ω. A compelling argument, however, *against* the introduction of infinitesimals, is Occam's razor of which Ω is a sheer paradigmatic sacrifice. This is the reason why infinitesimals are only mentioned in passing, and will not be treated below.

2. DIALOGUES ABOUT REAL FUNCTIONS

To start with a universally accepted notion of *function*, the definition of a real number should first be generalized to that of a *sequence* $(c_1, c_2, c_3, \dots, c_k, \dots)$ of real numbers: It is possible to denominate for each natural number k and each natural number n a decimal number

$$[c_k]_n = z + 0.z_1 z_2 z_3 \dots z_n = z + \frac{z_1}{10} + \frac{z_2}{100} + \frac{z_3}{1000} + \dots + \frac{z_n}{10^n}$$

with exactly n decimal places, with an integer z and with digits z_1, z_2, z_3, ... , z_n, such that for all natural numbers k, n, and m the inequality

$$|[c_k]_n - [c_k]_m| \leq \frac{1}{10^n} + \frac{1}{10^m}$$

holds. This reflects precisely what one would expect of a sequence: each natural number k is associated with a real number c_k.

A *real function* f is defined as a pair of two sequences $(a_1, a_2, a_3, \ldots, a_k, \ldots)$ and $(b_1, b_2, b_3, \ldots, b_k, \ldots)$, suggestively written as double-sequence

$$f = (a_1 \longmapsto b_1, a_2 \longmapsto b_2, a_3 \longmapsto b_3, \ldots, a_k \longmapsto b_k, \ldots).$$

If one replaces the arrows by commas, there is, at first sight, no difference between the notions "function" and "sequence". This difference only consists of the supplementary restriction on a function to be *extensional.* By this we mean that different b_k and b_l cause correspondingly different a_k and a_l.

Given two functions

$$f = (a_1 \longmapsto b_1, a_2 \longmapsto b_2, a_3 \longmapsto b_3, \ldots, a_k \longmapsto b_k, \ldots)$$

and

$$g = (c_1 \longmapsto d_1, c_2 \longmapsto d_2, c_3 \longmapsto d_3, \ldots, c_k \longmapsto d_k, \ldots)$$

the function g is called an *extension* of the function f if each of the numbers a_k coincides with at least one of the numbers c_l, and if the corresponding equality $b_k = d_l$ follows from the equality $a_k = c_l$. The *equality* $f = g$ of these two functions consists of the fact that each of them is an extension of the other.

A very simple example of a function is the *reciprocal function* r defined on the sequence of all rational numbers $\pm n/m$ different from zero where n and m designate relatively prime natural numbers: r interchanges the numerator with the denominator,

$$r = \left(\ldots, \frac{n}{m} \longmapsto \frac{m}{n}, \ldots, \frac{-n}{m} \longmapsto \frac{-m}{n}, \ldots\right).$$

A more sophisticated example is the function s defined on the same sequence, now additionally including zero, but associating the rational number with the Egyptian fraction with the same denominator:

$$s = \left(\ldots, \frac{n}{m} \longmapsto \frac{1}{m}, \ldots, 0 \longmapsto 1, \ldots, \frac{-n}{m} \longmapsto \frac{1}{m}, \ldots\right).$$

It is important, in this connection, that the continuity of a function

$$f = (a_1 \longmapsto b_1, a_2 \longmapsto b_2, a_3 \longmapsto b_3, \ldots, a_k \longmapsto b_k, \ldots)$$

can be defined at all accumulation points of the sequence $(a_1, a_2, a_3, \ldots, a_k, \ldots)$ — they need not be members of the sequence itself[3]. The reciprocal function r for instance is continuous on the whole continuum of points apart from zero and the above defined function s turns out to be continuous at all points apart from each rational number, attaining there the value zero, whereas the function s is discontinuous at all points of the sequence by which it was defined.

It is clear that, within the context of *recursive constructivism*, this definition of a function — although rather clumsy from the viewpoint of the recursive mathematician — is sufficient, provided that all occurring sequences are given by recursive rules. To stress this point: If the proponent or the opponent introduces a sequence $(c_1, c_2, c_3, \ldots, c_k, \ldots)$ — especially a function — this goes along with the fixing of a recursive function F which produces for natural arguments k and n the decimal number $F(k, n) = [c_k]_n$ with exact n decimal places. It is evident that under this agreement the scope of functions f defined above accords with that of recursive functions.

It is further clear that, within the context of *formal mathematics*, this definition of a function is far from being sufficient, because only functions defined on countable sets are singled out. The easiest way to overcome this difficulty while at the same time maintaining the former definition would be to replace the inadequate sequences $(a_1, a_2, a_3, \ldots, a_k, \ldots)$ and $(b_1, b_2, b_3, \ldots, b_k, \ldots)$ by *nets* $(\ldots, a_\kappa, \ldots)$ and $(\ldots, b_\kappa, \ldots)$. In these nets the "counting"-parameter κ runs over an index-set which can be finite, countably infinite, and even *uncountable*. To avoid this awkward construction, it is easier to identify the index-set with the domain of the function, $a_\kappa = \kappa$, and to interpret $b_\kappa = f(\kappa)$ as corresponding value of the function. (Occam's razor forbids speculations to what extent non-standard systems of natural numbers could serve as sufficiently extensive index-sets.)

In the context of *intuitionism*, the definition of functions as double-sequences must be specified for the appropriate use in dialogues: The indication of a sequence $(c_1, c_2, c_3, \ldots, c_k, \ldots)$ proceeds with the indication of an *arbitrary*, but always a *finite* number of its members, and each of these members is indicated with its integral part and sections of its decimal digits which are of an arbitrary, but finite length. In other words: In conjunction with the announcement of a function f the pre-

[3]cf. Rudolf Taschner: *Lehrgang der konstruktiven Mathematik 1*, Wien, 1991.

senting partner of the dialogue expresses his readiness to announce as many pairs of real numbers

$$a_1 \longmapsto b_1, a_2 \longmapsto b_2, a_3 \longmapsto b_3, \dots , a_k \longmapsto b_k$$

as his antagonist requires, and each of these real numbers a_1, a_2, a_3, $\dots$, a_k, b_1, b_2, b_3, $\dots$, b_k only have to be presented by their approximate values with accuracies being agreed upon in advance between the partners of the dialogue. By taking the abundance of the intuitionistic continuum into consideration, the proponent can successfully defend the assertion that functions defined on an interval have to be *continuous*, and even *uniformly continuous* if the interval turns out to be compact.

We sketch the dialogue about this statement: The opponent doubts its validity and the proponent answers by challenging the opponent to present a function f defined on the unit interval $[0; 1]$. If one chooses for instance $(a_1, a_2, a_3, \dots , a_k, \dots)$ to be the sequence enumerating all decimal numbers of the form $0.z_1z_2z_3 \dots z_n$, it particularly follows that the opponent is obliged to present $(b_1, b_2, b_3, \dots , b_k, \dots)$ as according sequence such that

$$f = (a_1 \longmapsto b_1, a_2 \longmapsto b_2, a_3 \longmapsto b_3, \dots , a_k \longmapsto b_k, \dots) .$$

But far more is involved in the opponent's assertion that f be defined on the whole interval $[0; 1]$: The presentation of f has to move along with the explanation of the opponent, how to "compute" the real number $b = f(a)$ for *any* arbitrary real number a with approximate values $[a]_n$ of the form $0.z_1z_2z_3 \dots z_n$. To be more exact: For any natural number m which the partners of the dialogue have agreed to choose, there must be a device to calculate $[b]_m = [f(a)]_m$ from the arbitrary real number a submitted by the proponent. If the opponent wants to know data about a, the proponent is now in a winning position: The only data he is obliged to present, are finitely many approximate values of a. Therefore the opponent has to admit: There must be a number N such that for all real numbers c with the property $[c]_n = [a]_n$ for all $n = 1, 2, 3, \dots , N$ the "computation" of $[f(c)]_m$ is the same as that of $[f(a)]_m$, producing therefore the identical number $[f(c)]_m = [f(a)]_m = [b]_m$.

This insight is the key that immediately opens the door to the proof that f is continuous. Beyond that, the finite sequences $([a]_1, [a]_2, \dots , [a]_N)$ constitute a so-called *fan* because there are only finitely many direct successors $([a]_1, [a]_2, \dots , [a]_N, [a]_{N+1})$ of $([a]_1, [a]_2, \dots , [a]_N)$, and because only the numbers 0.0, 0.1, 0.2, $\dots$, 0.9

serve as $[a]_1$. This observation finally leads to the perception that f is uniformly continuous.

Whereas in formal or in recursive mathematics, continuity of functions plays an incidental role, in intuitionistic mathematics this property turns out to be the essential one of functions defined on "more" than sequences.

From the position of *formal mathematics*, the set of functions created by recursive constructivism is rather *small*. The set of functions created by intuitionism on the one hand encloses the former, but on the other it shrinks to a *tiny* one, if intervals are brought into discussion as domains. This contrasts with the gigantic profusion of functions existing in formal mathematics.

It is the aim of *Bishop's constructivism* to incorporate all of the above-mentioned positions. Again Bishop succeeds by *breaking the symmetry* of the logical basis in the dialogue between proponent and opponent: The proponent is only allowed to use recursive functions. Therefore the proponent has only those functions at hand that are accepted by the recursive constructivist, and therefore also by the formal mathematician and the intuitionist. On the other hand, the opponent is allowed to play with arbitrary functions in the sense of intuitionism or of *formal* mathematics. Only this enables Bishop's proponent to formulate definitions and to defend theorems which are accepted by the recursive constructivist, the intuitionist, and the formal mathematician as well — an admirable achievement in view of the divergence of these conflicting disciplines.

3. THE COMPARISON OF DIALOGUES

Now we rephrase Nicolas Goodman's question in the following way: Granting that doing mathematics in Bishop's manner involves difficulties which other foundational systems can avoid: what are ultimately the advantages of Bishop's constructivism? Of course, the aim of reconciliation is placed in the limelight, but this goal has nothing to do with a well-founded philosophy of mathematics. On the other hand, the deliberate rejection of philosophical considerations in connection with mathematics inevitably leads to the pragmatic attitude of Bishop and his followers which is designed to benefit mathematics as such. But we will show by one example that even this may not be the case:

It is easy to explain the Riemann-integrability within the unit interval $[0;1]$ of bounded functions defined on sequences that are dense in $[0;1]$. For instance the function s defined above, and more general: all functions

s^q with

$$s^q = \left(\ldots, \frac{n}{m} \longmapsto \frac{1}{m^q}, \ldots, 0 \longmapsto 1, \ldots, \frac{-n}{m} \longmapsto \frac{1}{m^q}, \ldots\right)$$

are Riemann-integrable for any positive q with integral zero. The constructive proof is easy: Given an arbitrary positive ε, the natural number M is constructed so large that $1/M^q$ is less than $\varepsilon/2$. Now one fixes $\delta_q = \varepsilon/2M^2$ to be an upper bound of the length $x_p - x_{p-1}$ of the intervals $[x_{p-1}; x_p]$ in the partition

$$0 = x_0 < x_1 < x_2 < \ldots < x_{P-1} < x_P = 1.$$

We now choose arbitrary rational intermediate points

$$0 \leq t_1 < t_2 < \ldots < t_P \leq 1$$

which obey for all $p = 1, 2, \ldots, P$ the condition $x_{p-1} \leq t_p \leq x_p$. As there are fewer than M^2 intervals $[x_{p-1}; x_p]$ containing a rational number n/m with denominator $m < M$, and because for all the other intervals $[x_{p-1}; x_p]$ the inequality $s^q(t_p) < \varepsilon/2$ must hold, one gains

$$0 \leq \sum_{p=1}^{P} s^q(t_p)(x_p - x_{p-1}) < M^2 \cdot 1 \cdot \delta_q + \frac{\varepsilon}{2} \cdot 1 = \varepsilon$$

which confirms the integrability of s^q with integral zero.

We therefore conclude that

$$\lim_{q \to 0+0} \int_0^1 s^q = 0,$$

on the other hand we obtain the function

$$s^0 := \lim_{q \to 0+0} s^q = \left(\ldots, \frac{n}{m} \longmapsto 1, \ldots, 0 \longmapsto 1, \ldots, \frac{-n}{m} \longmapsto 1, \ldots\right)$$

which is integrable too, but possesses 1 as value of the integral. This result deviates substantially from Lebesgue's theorem about dominated convergence.

The reason for this deviation is clear from the viewpoint of the *formal mathematician*: In his eyes, s^0 stands for the Dirichlet-function, attaining the value 1 on rational and the value 0 on irrational numbers. This function is of course not Riemann-integrable but possesses a Lebesgue-integral with value zero, wholly in accordance with the dominated convergence theorem. One can even support the formalist's argument by the following straightforward proof which uses a slight generalization of Riemann-integrals, formulated by Ralph Henstock[4] about 40 years ago: A function f defined on $[0; 1]$ is *integrable*, fixing the real number

$$\int_0^1 f$$

as its integral, if there exists for any positive ε a *function* δ defined on $[0; 1]$ with *positive* values such that for any partition

$$0 = x_0 < x_1 < x_2 < \ldots < x_{P-1} < x_P = 1$$

of $[0; 1]$ and any choice of intermediate points

$$0 \le t_1 < t_2 < \ldots < t_P \le 1$$

which obey for all $p = 1, 2, \ldots, P$ the two conditions $x_{p-1} \le t_p \le x_p$ and $x_p - x_{p-1} \le \delta(t_p)$, the inequality

$$\left| \sum_{p=1}^{P} f(t_p)(x_p - x_{p-1}) - \int_0^1 f \right| < \varepsilon$$

holds. Let for instance the sequence $(u_1, u_2, u_3, \ldots, u_k, \ldots)$ be an enumeration of the rationals between 0 and 1 and define the function δ by the formula $\delta(t) = \varepsilon/2^k$ in the case $t = u_k$, and $\delta(t) = 1$ otherwise. It is then very easy to check

$$\left| \sum_{p=1}^{P} s^q(t_p)(x_p - x_{p-1}) \right| < \varepsilon \qquad \text{therefore} \qquad \int_0^1 s^q = 0$$

for any $q > 0$, and the fact that δ does not depend on the choice of q especially proves

$$\left| \sum_{p=1}^{P} s^0(t_p)(x_p - x_{p-1}) \right| < \varepsilon \qquad \text{therefore} \qquad \int_0^1 s^0 = 0$$

[4]Ralph Henstock: *Theory of Integration* , London, 1963.

with s^0 interpreted as Dirichlet-function — according to the dominated convergence theorem.

But this argument of the formalist is by no means convincing in the eyes of the intuitionist: The reason is that the above defined function δ is totally untenable within the intuitionistic theory. Any function defined on the whole interval $[0;1]$ must be uniformly continuous there, and if the function is positive, it fixes a positive infimum. And indeed, in the constructive proof of the integrability of s^q, the number δ_q obviously depended on q, and this forbids the limiting process $q \to 0$ to be interchanged with the integral. Within intuitionism the generalization in the definition of Riemann-integrals formulated by Henstock does not extend the domain of Riemann-integrable functions in the ordinary sense — whereas in formal mathematics all Lebesgue-integrable and all Denjoy-integrable functions are affected by this generalization.

Bishop's theory reconciles these two diverging concepts. But in his system it is impossible to maintain the full strength of the dominated convergence theorem or the realization that there are no Lebesgue- or Denjoy-integrable functions except the Riemann-integrable ones. By this, he forfeits both the results peculiar to formal mathematics and insights of intuitionistic mathematics. To put it metaphorically: It is hard to understand how the proponent in the dialogue is able to follow the opponent's arguments and vice-versa. Bishop's mathematics turns out to be an intellectual effort in a no-man's-land — rather similar to formal mathematics which from the constructivist's viewpoint is an intellectual effort in the utopia of omniscience about infinite sequences.

To the extent that the partners accept the same logical tenets the *fair* constructive theories are recursive constructivism and intuitionism. To balance the first against the second reminds one of Weyl's paper on the foundational crisis of mathematics[5] written almost 80 years ago. One reason that weighs heavily in favor of intuitionism is geometry: The reciprocal function r for instance was introduced above by a simple recursive rule. But there is a totally different, however likewise immediate definition of this function in the Cartesian coordinate-system using a right triangle, and of course supposing the continuity of this function. The idea of moving the vertices $(x,0)$ and $(-y,0)$ of the triangle along the abscissa while fixing the vertex of the right angle at the point $(0,1)$

[5] Hermann Weyl: Über die neue Grundlagenkrise der Mathematik, *Mathematische Zeitschrift* **10** (1921), 39-79.

and thus obtaining $y = 1/x = r(x)$, seems far away from the concept of the recursive constructivist that the continuum of the real axes would "in reality" be a system of distinct atoms. But this remark is only an impetus for the motion in the hermeneutic circle which surrounds mathematical activities and tries to trace how the partners in a mathematical dialogue are able to understand each other.

Acknowledgments

I am indebted to Peter M. Schuster who banished all my scruples and encouraged me to write this article, and to Otmar Binder who helped me enormously with his impressive expert knowledge of the English language.

ON THE QUANTITATIVE STRUCTURE OF Δ_2^0

Sebastiaan A. Terwijn*

Mathematisches Institut, Ludwig-Maximilians-Universität München
Theresienstraße 39, D-80333 München, Germany
terwijn@rz.mathematik.uni-muenchen.de

Keywords: Computable measure theory, Turing degrees, completeness

Mathematical Subject Classification: 03D15, 03D30, 28E15

Abstract: We analyze the quantitative structure of Δ_2^0. Among other things, we prove that a set is Turing complete if and only if its lower cone is non-negligible, and that the sets of r.e.-degree form a small subset of Δ_2^0.

1. INTRODUCTION

We study an effective measure theory suited for the study of Δ_2^0, the second level of the arithmetical hierarchy (alternatively, the sets computable relative to the halting problem K). This work may be seen as part of the constructivist tradition in mathematics as documented in [6]. The framework for effectivizing measure theory that we employ uses martingales. Martingales were first applied to the study of random sequences by J. Ville [22]. Recursive martingales were studied in Schnorr [19], and became popular in complexity theory in more recent years through the work of Lutz [14, 15]. Lutz used recursive and more efficient martingales to analyze the quantitative structure of complexity classes like EXP, the exponential time computable sets (see [1] for a survey). Recursive martingales define a notion of null set which is suited for the study of the recursive sets [15]. By relativization, K-recursive martingales can be used to analyze the quantitative structure of Δ_2^0. In particular, we can study classes defined by Turing reducibility (which is

*Research supported by a Marie Curie fellowship of the European Union under grant no. ERB-FMBI-CT98-3248.

P. Schuster, U. Berger and H. Osswald (eds.),
Reuniting the Antipodes – Constructive and Nonstandard Views of the Continuum, 271–283.

trivial on the recursive sets) in a quantitative way. This is what we do below. We first prove that the lower cones of incomplete sets are small, and use this to prove that the sets of r.e.-degree form a small subset of Δ_2^0. In Section 5 we will make some remarks on other definitions of effective measure. In particular, we consider the stronger notion of recursive null set introduced by Schnorr [19] which is directly based on the intuitionistic notion of null set. (This notion was also used in Freidzon [7].) We show that the results of Sections 3 and 4 also hold for this stronger measure. Finally, in Section 6 we motivate the question whether the low sets have measure zero in Δ_2.

2. MEASURE IN Δ_2

We proceed by giving the relevant definitions and by fixing notation. Since we will only consider arithmetical sets in this paper we will write Δ_2 instead of Δ_2^0.

Our recursion theoretic notation follows Soare [20]. We identify a subset of ω (the natural numbers) with its characteristic string. So 2^ω is the power set of ω. $2^{<\omega}$ is the set of finite binary strings, and λ is the empty string. $A\restriction n$ denotes the initial segment of the set A of length n. K denotes the standard Turing complete r.e. set.

We denote the upper cone of a set A by $A^{\leq_T} = \{B : A \leq_T B\}$ and the lower cone by ${}^{\leq_T}A = \{B : B \leq_T A\}$.

Definition 1 A function $d : 2^{<\omega} \to \mathbb{Q}^+$ is a *martingale* if for every $\sigma \in 2^{<\omega}$, d satisfies the averaging condition

$$2d(\sigma) = d(\sigma 0) + d(\sigma 1). \tag{23.1}$$

A martingale d *succeeds on* a set A if

$$\limsup_{n\to\infty} d(A\restriction n) = \infty.$$

We say that d succeeds on, or *covers*, a class $\mathcal{A} \subseteq 2^\omega$ if d succeeds on every $A \in \mathcal{A}$. The *success set* $S[d]$ of d is the class of all sets on which d succeeds. A class $\mathcal{A}$ has Δ_2-measure 0, denoted $\mu_{\Delta_2}(\mathcal{A}) = 0$, if there is a martingale $d \in \Delta_2$ such that $\mathcal{A} \subseteq S[d]$. $\mathcal{A}$ has *measure 0 in* Δ_2 if $\mathcal{A} \cap \Delta_2$ has Δ_2-measure 0. $\mathcal{A}$ has *measure 1 in* Δ_2 if the complement of $\mathcal{A}$ has measure 0 in Δ_2.

The above definition is robust in several respects. For example, for our purposes it does not matter whether in (23.1) we use "=" instead of "$\leq$". Also, normally one would start to define martingales to be real-valued functions and then proceed to study computability issues by using computable approximations. Again, for the resulting measure μ_{Δ_2} this does not make any difference (see e.g. [15, 21]).

The following basic facts are the relativized versions of the same facts for recursive measure. Part (i) shows the consistency of the definition of "measure 1 in Δ_2." Proofs can be found e.g. in [15, 21].

Theorem 2.1 (Lutz)

(i) *Δ_2 does not have Δ_2-measure zero.*

(ii) *For every $A \in \Delta_2$ the singleton $\{A\}$ has Δ_2-measure zero.*

(iii) *μ_{Δ_2} is closed under Δ_2-unions (and is in particular finitely additive) i.e. if $\mathcal{A}_i \subseteq 2^\omega$, $i \in \omega$, is a sequence of classes and d_i is a sequence of martingales uniformly computable in K such that d_i succeeds on $\mathcal{A}_i$ for every i, then $\bigcup_i \mathcal{A}_i$ has Δ_2-measure zero.*

As an example of (iii) we may note that the r.e. sets have Δ_2-measure 0 since they are uniformly computable in K. We will later use the following result, saying that upper cones are small.

Theorem 2.2 (Lutz and Terwijn) *For every nonrecursive set $A \in \Delta_2$ the upper cone*

$$A^{\leq_T} = \{B : A \leq_T B\}$$

has Δ_2-measure zero.

Proof. This is an effectivization of Sacks's theorem that $A^{\leq_T}$ has measure 0 for every nonrecursive set. A proof is in Terwijn [21, Thm 6.2.1]. □

Corollary 2 *The Turing-incomplete sets in Δ_2 do not have measure 0 in Δ_2.*

Proof. By Theorem 2.2 the degree of K has measure 0 in Δ_2, whereas by Theorem 2.1 Δ_2 itself does not have Δ_2-measure 0. The corollary follows from the additivity of Δ_2-measure. □

3. WEAK COMPLETENESS

Informally, a set is weakly complete for some class if its lower cone is not of measure zero in that class. We can consider various classes and reducibilities for which this definition makes sense. For our purposes let us make the following definition.

Definition 3 A set $A \in \Delta_2$ is *weakly complete* if ${}^{\leq_T}A$ does not have measure 0 in Δ_2.

The halting set K is weakly complete since Δ_2 is not of measure zero in itself by Theorem 2.1.

First note that if we define a set to be weakly complete if its lower cone has measure 1 in Δ_2,[1] then we immediately have that every weakly complete set is complete. This can be seen using the following argument of Regan et al. [17]:

Theorem 3.1 *If $\mathcal{C} \subseteq \Delta_2$ is closed under symmetric difference $\triangle$ and has measure 1 in Δ_2 then $\mathcal{C} = \Delta_2$.*

Proof. Suppose $\mathcal{C} \neq \Delta_2$, say $A \in \Delta_2 \setminus \mathcal{C}$. It is easy to see that also $A \triangle \mathcal{C} = \{A \triangle C : C \in \mathcal{C}\}$ has measure 1 in Δ_2. But $\mathcal{C}$ and $A \triangle \mathcal{C}$ are disjunct, because if $A \triangle C \in \mathcal{C}$ then also $A = (A \triangle C) \triangle C \in \mathcal{C}$, hence we have reached a contradiction. □

So if A is a set such that almost every set in Δ_2 reduces to it, then A is Turing complete because ${}^{\leq_T}A$ is closed under $\triangle$. The following theorem shows that this is true already when the lower cone of A is not small in Δ_2. So there are no nontrivial examples of weakly complete sets in Δ_2. This still leaves the possibility that some set in Δ_2 (necessarily a Turing complete one) would have a big lower cone with respect to some strong reducibility, like m- or tt- or wtt-reducibility. Since Δ_2 has no complete sets with respect to the strong reducibilities we could call such sets proper weakly complete with respect to these reducibilities. However, this cannot happen since for every $A \in \Delta_2$ its wtt-lower cone $\{B : B \leq_{wtt} A\}$ has measure zero in Δ_2 [21, Theorem 4.4.4].

N.B. In complexity theory it is known that the sets that are weakly p-m-complete for EXP, the class of exponential time computable sets, have measure 1 in EXP, whereas the p-m-complete sets have measure 0 (see the survey [1] for references). The question whether there are p-T-incomplete weakly T-complete sets in EXP is open.

N.B. Ambos-Spies and Terwijn [21, Chapter 3] defined a measure tailored for the recursively enumerable sets, and studied the resulting

[1] Such sets are called *almost complete* in complexity theory. They were studied recently in Ambos-Spies et al. [3]

weak completeness notions. They proved that in the class of r.e. sets the weak completeness notions for truth-table, weak truth-table, and Turing reducibility coincide with the ordinary completeness notions, whereas those for many-one and bounded truth-table reducibility differ from their ordinary counterparts.

Theorem 3.2 *Every weakly complete set is Turing complete.*

Proof. Suppose that $A <_T K$. We define uniformly in K for every $e \in \omega$ a martingale d_e such that

$$R_e: \quad \{e\}^A \text{ total and 0,1-valued} \Longrightarrow \{e\}^A \in S[d_e].$$

By Theorem 2.1 (iii) this suffices to prove the theorem. Let $f \leq_T K$ be a function that is not dominated by any function $g <_T K$, for example, $f(x) = \mu s(K_s \restriction x = K \restriction x)$ (the smallest s such that all the $y \in K$ smaller than x are enumerated into K within s steps).

We now define d_e in stages s. On stage s we define d_e on all strings $\sigma \in 2^{<\omega}$ of length s. The value $d_e(\sigma)$ will depend only on $|\sigma|$. (Such martingales were called 'oblivious' in Ambos-Spies et al. [2].)

Stage $s = 0$. Define $d_e(\lambda) = 1$.

Stage $s + 1$. Given $d_e(\sigma)$, $|\sigma| = s$, use the oracle K to search for a string $\tau \sqsubset A$ with $|\tau| \leq f(s)$ such that $\{e\}^{\tau}_{|\tau|}(s)\downarrow$. If such τ does not exist, or if $\{e\}^{\tau}_{|\tau|}(s)\downarrow \notin \{0,1\}$, we do not make a bet, namely we let $d_e(\sigma i) = d_e(\sigma)$ for $i \in \{0,1\}$. If τ exists and $\{e\}^{\tau}_{|\tau|}(s)\downarrow = i \in \{0,1\}$ we define $d_e(\sigma i) = 2d_e(\sigma)$, that is, we bet all our money on $\{e\}^A(|\sigma|) = i$. This concludes the definition of d_e.

It is clear that d_e is defined on all strings for every e, uniformly in K. We check that R_e is satisfied. Suppose that $\{e\}^A$ is total and computes a set. Then the function

$$f_e(n) = \mu t(\exists \tau \sqsubset A\,[\,|\tau| = t \wedge \{e\}^{\tau}_t(n)\downarrow])$$

is an A-computable function. By choice of f we have that there are infinitely many s such that $f(s) \geq f_e(s)$. But this means that in the definition of d_e infinitely often the string τ is found and a bet is placed successfully. (Note that we never make a wrong bet.) Hence $\{e\}^A \in S[d_e]$. □

Corollary 4 *For every $\emptyset <_T A <_T K$ the sets incomparable with A have measure one in Δ_2.*

Proof. The sets incomparable with A are the complement of the union ${}^{\leq_T}A \cup A^{\leq_T}$ of the lower and the upper cone of A. The former has measure 0 in Δ_2 by Theorem 3.2 and the latter by Theorem 2.2. □

4. SMALLNESS OF THE R.E.-DEGREES

After Theorem 2.1 we observed that the r.e. sets form a small subset of Δ_2. We now apply Theorem 2.2 to obtain

Theorem 4.1 *The class of sets that have r.e.-degree has measure 0 in* Δ_2.

Proof. Let W_e be the e-th r.e. set. For every W_e we define a martingale d_e as in the proof of Theorem 3.2 such that

$$W_e <_T K \implies d_e \text{ succeeds on } {}^{\leq_T}W_e.$$

Note that the definition of d_e does not depend on the assumption $W_e <_T K$, only the success of d_e does. So we have a recursive sequence of Δ_2-martingales d_e, and Theorem 2.1 (iii) gives that $\bigcup\{{}^{\leq_T}W_e : W_e <_T K\}$ has Δ_2-measure 0. Finally, Theorem 2.2 and the finite additivity of μ_{Δ_2} give that

$$\{A : \exists e(A \equiv_T W_e)\} \subseteq \{A : \exists e(A \leq_T W_e <_T K)\} \cup \{A : K \leq_T A\}$$

has measure 0 in Δ_2. □

We could have proved this theorem in a slightly different way, namely by defining for every r.e. set W_e two martingales d_e^0 and d_e^1, the first succeeding on ${}^{\leq_T}W_e$ if $W_e <_T K$, and the second succeeding on $W_e{}^{\leq_T}$ if $\emptyset <_T W_e$. This needs, however, that in the proof of Theorem 2.2 the definition of the martingale is independent of the hypothesis $\emptyset <_T A$, and only the success of the martingale depends on this. As this is indeed the case, this second proof now gives that

$$\{A : \exists e(A \leq_T W_e <_T K)\} \cup \{A : \exists e(\emptyset <_T W_e \leq_T A)\}$$

has Δ_2-measure 0. This yields

Corollary 5 *Almost every set (in the sense of* Δ_2*-measure) in* Δ_2 *is Turing-incomparable to every r.e. set, except the recursive and the complete ones.*

The existence of sets as in Corollary 5 can be proved directly (Kleene-Post style arguments [8, 20]). Corollary 5 can be seen as a probabilistic version of this result, just as the main result of Kučera and Terwijn [11] can be seen as a probabilistic solution to Post's problem.

5. MODULATED MEASURE

In this section we make some remarks about other measures and strengthen the results of the previous sections. Instead of Δ_2-measure one could use a weaker notion of null set like Σ_1-measure (as defined in Martin-Löf [16][2]) or recursive measure to study Δ_2. These two examples, however, are unnatural for this purpose because Δ_2 contains elements that are Martin-Löf random (Σ_1-random), i.e. elements A such that the Σ_1-measure of $\{A\}$ is not zero.[3] This trivially makes results like Theorem 2.2 and Theorem 3.2 untrue for these measures, and is an indication that they are to weak for the study of Δ_2. Instead we consider another definition of measure, which is more in the spirit of constructive mathematics.

Schnorr [19] pointed out a nonconstructive aspect of Martin-Löfs definition of measure [16], and he introduced the following stronger notion of null set, based on the intuitionistic notion of null set in Brouwer [4]. The definition given here is not his original one, but one that is equivalent to it [19, Satz 9.7].

Definition 6 A class $\mathcal{A}$ has *modulated recursive measure 0* if and only if there is a recursive martingale d and a nondecreasing unbounded recursive function h such that $\mathcal{A} \subseteq S_h[d]$, where

$$S_h[d] = \{X : \limsup_{n\to\infty} \frac{d(X\restriction n)}{h(n)} \geq 1\}.$$

The pair (d, h) is also called a *total recursive sequential test.*

Like the recursive measure, we can relativize this definition to K. Thus we obtain the notion of *modulated Δ_2-measure 0* (mod-Δ_2-measure 0 for short), which is defined exactly as modulated recursive measure

[2]One can define Σ_1-measure by replacing in Definition 1 Δ_2 by Σ_1. (A function f is in Σ_1 if the set $\{\langle x, y\rangle : y \leq f(x)\}$ is r.e.)

[3]Martin-Löf random sets play an important role in the area of Kolmogorov complexity, see Li and Vitányi [13].

0 except that in the definition we replace recursive everywhere by K-recursive. Note that indeed we have that no set in Δ_2 is mod-Δ_2-random. (Every single $A \in \Delta_2$ can be covered by a martingale that computes A, thus growing with rate $h(n) = 2^n$.)

Let us summarize the various notions of null set that we have discussed:

(i) modulated recursive null

(ii) recursive null

(iii) Σ_1-null

Relativized to K these give

(iv) modulated Δ_2-null

(v) Δ_2-null

(vi) Σ_2-null

From the definitions it is immediate that (i)$\Rightarrow$(ii)$\Rightarrow$(iii). (ii)$\not\Rightarrow$(i) was proved by Wang [23] and (iii)$\not\Rightarrow$(ii) follows since the recursive sets have Σ_1-measure zero. The strict implications (iv)$\Rightarrow$(v)$\Rightarrow$(vi) follow from relativizing this. Finally, (iii)$\Rightarrow$(iv) follows from the analysis in Schnorr [19],[4] and (iv)$\not\Rightarrow$(iii) because, as noted above, there are $A \in \Delta_2$ such that $\{A\}$ is not Σ_1-null, whereas for every $A \in \Delta_2$ we have that $\{A\}$ is mod-Δ_2-null. So (i)$-$(vi) is indeed a sequence of notions of measure 0 increasing in strength. As discussed above, (iii) is too weak for the study of Δ_2 and (vi) is too strong, since Δ_2 has Σ_2-measure 0. The results from the previous sections used (v). In this section we will prove that these results also hold for (iv). This yields stronger statements, since Terwijn [21] proved that mod-Δ_2-measure is strictly weaker (there are less null sets) on Δ_2 than Δ_2-measure.[5] We will use the following notion.

[4] Satz 9.5 in [19] says that every null set defined by a total recursive sequential test has modulated recursive measure 0. Since every Σ_1-null set is covered by a total K-recursive sequential test the claim follows by relativization.

[5] Theorem 6.4.2 of [21] says that modulated recursive measure is strictly weaker than recursive measure even when restricted to REC. This answered a question of Lutz. The proof of this fact relativizes. That there are (necessarily nonrecursive) sets that are not in any modulated recursive null set but that can be covered by a recursive martingale was proved earlier by Wang [23].

Definition 7 For functions f and g, we say that g *densely dominates* f if

$$\liminf_{n\to\infty} \frac{\|\{i \le n : g(i) \ge f(i)\}\|}{n} \ge \frac{1}{2}. \tag{23.2}$$

Lemma 8 *There is a function $f \le_T K$ such that no function $g <_T K$ densely dominates f.*

Proof. Let $h \le_T K$ be a function not dominated by any $g <_T K$ (cf. the proof of Theorem 3.2). Define $f(x) = h(\lfloor \log x \rfloor)$. If g satisfies (23.2) then for almost every e there is a natural number $x \in [2^e, 2^{e+1})$ such that $g(x) \ge f(x)$. But then $g'(e) = \max\{g(x) : x \in [2^e, 2^{e+1})\}$ dominates h, contradiction. □

We will also need effective unions:

Lemma 9 *If $\mathcal{A}_i \subseteq 2^\omega$, $i \in \omega$, is a sequence of classes and (d_i, h_i) is a sequence of sequential tests uniformly computable in K such that $\mathcal{A}_i \subseteq S_{h_i}[d_i]$ for every i, then $\bigcup_i \mathcal{A}_i$ has mod-Δ_2-measure zero.*

Proof. We will not prove this in detail. The proof is almost the same as that of Theorem 2.1 (iii). The only extra ingredient that one needs is that for a uniformly computable sequence h_i as above one can define a computable monotone unbounded h growing slower than any of the h_i, which is easy to see. □

Theorem 5.1 *The lower cone of every $A <_T K$ has mod-Δ_2-measure zero.*

Proof. We build on the proof of Theorem 3.2. There we argued that if $\{e\}^A$ is total, then infinitely often we doubled the value of the martingale. Now if we use the f from Lemma 8 we can do this at more than half of the stages rather than just infinitely often. Namely, for f_e as in the proof of Theorem 3.2, we now have that infinitely often

$$\|\{i \le n : f(i) > f_e(i)\}\| \ge \frac{1}{2}n.$$

So if $B = \{e\}^A$ is total then, for $h(n) = 2^{\frac{1}{2}n}$, we have

$$\limsup_{n\to\infty} \frac{d(B\lceil n)}{h(n)} \ge 1,$$

that is, $B \in S_h[d]$. Now we can apply Lemma 9 to finish the proof. (In fact we need less than Lemma 9, since all the h_i's are the same here.) $\square$

An inspection of the proof of Theorem 2.2 in [21] shows that in fact we have

Theorem 5.2 *For every nonrecursive $A \in \Delta_2$ the upper cone $A^{\leq_T}$ has mod-Δ_2-measure 0.*

Now the stronger versions of Theorem 4.1 and Corollary 5 follow in the same way as these results follow from the proofs of Theorems 2.2 and 3.2:

Theorem 5.3 *The sets of r.e. degree have mod-Δ_2-measure 0.*

Corollary 10 *Almost every set (in the sense of modulated Δ_2-measure) in Δ_2 is Turing-incomparable to every r.e. set, except the recursive and the complete ones.*

We end this section by remarking that the previous results cannot be strengthened from mod-Δ_2-measure to Σ_1-measure. For Theorems 5.2 and 5.3 and Corollary 10 this is immediate by the existence of Σ_1-random sets X with $X \equiv_T K$. For Theorem 5.1 this follows from the existence of low Σ_1-random sets X, i.e. X with $K^X \equiv_T K$. The failure of Corollary 10 for Σ_1-measure is also illustrated by the result of Kučera [10] that there exist a nonrecursive r.e. set A and a low Σ_1^A-random set X such that $A \leq_T X$. That the converse of this, namely that an incomplete r.e. set bounds a Σ_1-random set, cannot happen was proved by Kučera [9].

6. QUESTIONS

As an example of the well-known incompatibility of measure and category one can prove that the 1-generic sets have Δ_2-measure zero. Much more interestingly, one can prove that the sets that have 1-generic *degree* have measure zero in Δ_2. This follows from the idea in the proof of Kurtz [12, Theorem 4.2]. This gives in fact that the downward closure of the 1-generic sets has Σ_1-measure zero. Demuth and Kučera [5] proved the slightly more general statement that no diagonally nonrecursive function is recursive in a 1-generic set.

Theorem 6.1 (Kurtz [12], Demuth and Kučera [5]) *The class of sets that have 1-generic degree has Σ_1-measure zero. Equivalently, no Martin-Löf random set has 1-generic degree.*

Proof. The proof given here is a rather straightforward effectivization of the proof of Theorem 4.2 in Kurtz [12]. For every e,$i \in \omega$ define the Σ_1^0-class $\mathcal{S}_i^e$ as follows. Let $\{\sigma_j\}_{j\in\omega}$ be a recursive enumeration of $2^{<\omega}$. Let τ_j be the least τ extending σ_j such that $\Phi_e(\tau)$ is defined on at least $i+j+1$ arguments. (As usual Φ_e denotes the e-th Turing reduction.) Define $\mathcal{S}_i^e$ to be the open set defined by the range of the partial recursive function $\sigma_j \mapsto \tau_j$.

First we note that

$$\mu(\Phi_e(\mathcal{S}_i^e)) \leq \sum_{j\in\omega} \mu(\Phi_e(\tau_j)) \leq \sum_{i\in\omega} 2^{(i+j+1)} = 2^{-i}. \tag{23.3}$$

Second, since the classes $\mathcal{S}_i^e$ are uniformly r.e. in e and i, it follows that the classes $\Phi_e(\mathcal{S}_i^e)$ are uniformly r.e. in e and i. So by (23.3), for every e the class $\bigcap_{i\in\omega} \Phi_e(\mathcal{S}_i^e)$ is a Σ_1-null set, and by a standard sum argument we have that $\bigcup_{e\in\omega}\bigcap_{i\in\omega} \Phi_e(\mathcal{S}_i^e)$ is Σ_1-null.

Now if G is 1-generic and G does not meet $\mathcal{S}_i^e$, then by 1-genericity there is an initial segment $\sigma \sqsubseteq G$ such that no $\tau \sqsupseteq \sigma$ is in $\mathcal{S}_i^e$. Hence for every $\tau \sqsupseteq \sigma$, $\Phi_e(\tau)$ is never defined on at least $i+1$ arguments. Therefore if $\Phi_e(G)$ is total then G meets every $\mathcal{S}_i^e$. So we have

$$\begin{aligned} \{A : A \leq_T G \text{ for some 1-generic } G\} &= \\ \bigcup_{e\in\omega}\{A : A = \Phi_e(G) \text{ for some 1-generic } G\} &\subseteq \\ \bigcup_{e\in\omega}\{A : A = \Phi_e(G) \text{ for some } G \text{ that meets } \mathcal{S}_i^e \text{ for every } i\} &\subseteq \\ \bigcup_{e\in\omega}\bigcap_{i\in\omega} \Phi_e(\mathcal{S}_i^e), & \end{aligned}$$

and since this last class has Σ_1-measure 0 the theorem follows. □

Since all sets in Δ_2 that have 1-generic degree are low (A is low whenever it holds that $K^A \leq_T K$) one can ask whether the low sets have measure zero in Δ_2. One would expect this to be the case, since the low sets behave in many respects like the recursive sets, which have measure zero in Δ_2.

Question 11 *Do the low sets have measure zero in Δ_2?*

Still more general, we have:

Question 12 *Does every jump class $\{B : B' \equiv_T S\}$, $S \in [\emptyset', \emptyset'']$ have measure zero in Δ_2?*

Acknowledgments

I would like to thank Klaus Ambos-Spies for helpful discussions and for his support.

References

[1] K. Ambos-Spies and E. Mayordomo, *Resource-bounded measure and randomness*, in: A. Sorbi (ed.), Complexity, logic, and recursion theory, 1–47, Dekker, New York, 1997.

[2] K. Ambos-Spies, E. Mayordomo, Y. Wang, X. Zheng, *Resource-bounded balanced genericity, stochasticity and weak randomness*, Proc. 13th Symp. on Theoret. Aspects of Comp. Sci., Lect. Notes in Comp. Sci. 1046 (1996) 63–74, Springer-Verlag.

[3] K. Ambos-Spies, W. Merkle, J. Reimann, and S. A. Terwijn, *Almost complete sets.* Proc. 17th Symp. on Theoret. Aspects of Comp. Sci., Lect. Notes in Comp. Sci. 1770 (2000) 419–430, Springer-Verlag.

[4] L. E. J. Brouwer, *Begründung der Mengenlehre unabhängig vom logischen Satz vom ausgeschlossenen Dritten*, Zweiter Teil, Verhandelingen Nederlandse Akademie van Wetenschappen, afd. Natuurk. Sect. 12,7 (1919).

[5] O. Demuth and A. Kučera, *Remarks on 1-genericity, semigenericity and related concepts*, Comment. Math. Univ. Carolinae, 28 (1987), 85–94.

[6] Yu. L. Ershov, S. S. Goncharov, A. Nerode, and J. B. Remmel (eds.), *Handbook of Recursive Mathematics*, Vol. 1 and 2, Elsevier, 1998.

[7] R. I. Freidzon, *Families of recursive predicates of measure zero*, J. Soviet Math. 6 (1976), 449–455. Translated from Zap. Naučn. Sem. Leningrad. Otdel. Mat. Inst. Steklov. (LOMI) 32, (1972) 121–128.

[8] S. C. Kleene and E. L. Post, *The upper semi-lattice of degrees of recursive unsolvability*, Ann. of Math. (2) 59 (1954) 379–407.

[9] A. Kučera, *Measure, Π^0_1-classes and complete extensions of PA*, in: H.-D. Ebbinghaus, G. H. Müller, and G. E. Sacks, eds., Recursion theory week 1984, Lect. Notes in Math. 1141 (1985) 245–259.

[10] A. Kučera, *On relative randomness*, Annals of Pure and Applied Logic 63 (1993) 61–67.

[11] A. Kučera and S. A. Terwijn, *Lowness for the class of random sets*, J. Symbolic Logic 64(4) (1999) 1396–1402.

[12] S. A. Kurtz, *Randomness and genericity in the degrees of unsolvability*, PhD Thesis, University of Illinois, 1981.

[13] M. Li and P. Vitányi, *An introduction to Kolmogorov complexity and its applications*, 2nd edition, Springer-Verlag, 1997.

[14] J. H. Lutz, *Category and measure in complexity classes*, SIAM J. Comput. 19 (1990) 1100–1131.

[15] J. H. Lutz, *Almost everywhere high nonuniform complexity*, J. Comput. System Sci. 44 (1992) 220–258.

[16] P. Martin-Löf, *The definition of random sequences*, Information and Control 9 (1966) 602–619.

[17] K. W. Regan, D. Sivakumar, and J.-Y. Cai, *Pseudorandom Generators, Measure Theory, and Natural Proofs*, 36th Annual Symp. on Foundations of Comp. Sci. (FOCS'95), 26–35, IEEE Computer Society Press, 1995.

[18] G. E. Sacks, *Degrees of unsolvability*, Annals of Mathematics Studies 55, Princeton University Press, 1963.

[19] C.-P. Schnorr, *Zufälligkeit und Wahrscheinlichkeit*, Lect. Notes in Math. 218, Springer-Verlag, 1971.

[20] R. I. Soare, *Recursively enumerable sets and degrees*, Springer-Verlag, 1987.

[21] S. A. Terwijn, *Computability and Measure*, PhD Thesis, ILLC/University of Amsterdam, 1998.

[22] J. Ville, *Étude critique de la notion de collectif*, Gauthier-Villars, 1939.

[23] Y. Wang, *Randomness and complexity*, PhD Thesis, Universität Heidelberg, 1996.

UNDERSTANDING AND USING BROUWER'S CONTINUITY PRINCIPLE

Wim Veldman
Subfaculteit Wiskunde, Katholieke Universiteit Nijmegen
Toernooiveld 1, 6525 ED Nijmegen, The Netherlands
veldman@sci.kun.nl

In memory of Johan J. de Iongh (1915-1999)

Trying to learn to use words, and every attempt
Is a wholly new start, and a different kind of failure

T.S. Eliot, East Coker, V, 1940

Abstract: Brouwer's Continuity Principle distinguishes intuitionistic mathematics from other varieties of constructive mathematics, giving it its own flavour. We discuss the plausibility of this assumption and show how it is used. We explain how one may understand its consequences even if one hesitates to accept it as an axiom.

Mathematical Subject Classification: 03F55

1. BROUWER'S CONTINUITY PRINCIPLE

We let $\mathbb{N}$ be the set of all natural numbers. Its elements $0, 1, 2, \ldots$ are produced one by one. $\mathbb{N}$ is a never finished project that is executed step-by-step.
We let $\mathcal{N}$ be the set of all infinite sequences of natural numbers.
The acceptance of $\mathcal{N}$ as a totality has been a major step in the history of mathematical thinking, and led to the development of set theory. With Cantor's diagonal argument in mind, Brouwer probed the meaning of the words: "every *possible* infinite sequence of natural numbers" and

P. Schuster, U. Berger and H. Osswald (eds.),
Reuniting the Antipodes – Constructive and Nonstandard Views of the Continuum, 285–302.

found a way to sensibly use them.
An element α of $\mathcal{N}$ is a function from $\mathbb{N}$ to $\mathbb{N}$, $\alpha = \alpha(0), \alpha(1), \alpha(2), \ldots$ Every such element is produced step-by-step, and like the set $\mathbb{N}$ itself, it is a never finished project.
The production process may consist in the evaluation of an algorithm, like:

> The sequence $\underline{0}$ with the constant value 0:
> $\underline{0}(0) = 0$, $\underline{0}(1) = 0$, $\underline{0}(2) = 0, \ldots$

or:

> d, the decimal expansion of π:
> $d(0) = 1$, $d(1) = 4$, $d(2) = 1, \ldots$

or:

> $\mathbf{example_0}$:
> for each n, $\mathbf{example_0}(n) = 1$ if there is no $i < n$ such that for each $j < 99$, $d(i+j) = 9$ but for each $j < 99$, $d(n+j) = 9$, and $\mathbf{example_0}(n) = 0$ otherwise.

Brouwer suggested also the following non-algorithmic project,

> $\mathbf{example_1}$:
> for each n, $\mathbf{example_1}(n) = 0$ as long as we do not find a proof that for each j, $\mathbf{example_0}(j) = 0$, and $\mathbf{example_1}(n) = 1$ otherwise.

$\mathbf{example_1}$ is unusual and strange; it depends on one's future mathematical experience and requires a subdivision of (part of) the mathematical future into stages $0, 1, 2, \ldots$. Also if one doubts that this could be done in a sensible way, the example may make clear that we are prepared to recognize all kinds of projects for constructing step-by-step an infinite sequence of natural numbers.
We even give up the idea that there must be a "rule" or a "secret" guiding the development of such a sequence. It suffices that we fulfil our intention to construct a next value every time, that is, we may create a sequence by freely choosing its successive values. The result of such a free step-by-step construction is unpredictable but may be a disappointingly regular sequence like the sequence $\underline{0}$. There is no obligation to demonstrate one's freedom by avoiding every repetitive pattern of behaviour. We go so far as to believe that every infinite sequence may be imagined to be the result of a free step-by-step construction.

We now discuss the Continuity Principle.
Let $R \subseteq \mathcal{N} \times \mathbb{N}$ be a relation between infinite sequences of natural numbers and natural numbers. If the pair $\langle \alpha, m \rangle$ belongs to R, we write

αRm and say: "m is suitable for α".
Suppose we feel entitled to claim the following:

For every α in $\mathcal{N}$ there exists m in $\mathbb{N}$ such that αRm.

This is a strong claim for the following two reasons:

(i) We take seriously the expression: "for every α in $\mathcal{N}$".
We in no way want to delimit the range of this quantifier and in particular allow infinite sequences that grow step-by-step by free choices.

(ii) We take seriously the expression: "there exists m in $\mathbb{N}$".
Given any α we must be able to construct and produce a natural number m suitable for α, and a natural number is a finite object.

Strong claims need strong evidence and thus may be seen to have strong implications.
We argue as follows: if α is coming into existence step-by-step and we calculate a number m suitable for α, the construction of this number m will be completed when only finitely many values of α have been decided upon. The number m will be suitable not only for α itself but for every infinite sequence β that has these first finitely many values the same as α.
Let us summarize our conclusion, (we use $m, n, \ldots$ as variables over the set $\mathbb{N}$ of natural numbers):

1.1 *Brouwer's Continuity Principle:*
For every $R \subseteq \mathcal{N} \times \mathbb{N}$, if $\forall \alpha \in \mathcal{N}\ \exists m\ [\alpha Rm]$, then
$\forall \alpha \in \mathcal{N} \exists n \exists m \forall \beta \in \mathcal{N}$ [if for every $i < n$, $\alpha(i) = \beta(i)$, then βRm].

This axiom is called *Brouwer's principle for numbers* in Kleene and Vesley[7] 1965 and $WC - \mathbb{N}$ in Troelstra and van Dalen [8] 1988.

The first occasion where Brouwer is using the principle seems to be Brouwer [1] 1918, page 13, or Brouwer [3] 1975, page 160 (we freely render Brouwer's words):

A law that assigns to every α in $\mathcal{N}$ some m in $\mathbb{N}$ must have determined m after some initial segment of α has become known.

The Continuity Principle is not a far-fetched assumption but a natural axiom, borne out by experience. We never were in a situation in which we found reason to affirm the premiss but hesitated to uphold the conclusion.

Brouwer considers it as evident. Having mentioned the possibility of creating an infinite sequence by successive choices, he offers no further apology. In Heyting [6] 1952/53, some explication is given, not very different from the one we sketched.

Kleene and Vesley [7] 1965 prove the consistency of their formal system of intuitionistic analysis including the Continuity Principle by making use of a non-intended realizability interpretation.

The Continuity Principle changes the landscape of mathematics.

Brouwer is offering us a pair of spectacles and promising new vistas. We should not only study the spectacles but also put them on and describe what we see.

2. THE CONTINUITY OF REAL FUNCTIONS

The continuity of real functions is an immediate consequence of Brouwer's Continuity Principle, as we showed in Veldman [10] 1982. We now repeat this argument.

We first formulate a generalization of the principle that easily follows from the principle itself.

2.1 Let X be a subset of $\mathcal{N}$. X will be called a *spread* if and only if the following two conditions are satisfied:

(i) For every finite sequence $s = \langle s(0), \ldots, s(n-1) \rangle$ of natural numbers one may decide if there exists α in X such that for each $i < n$, $\alpha(i) = s(i)$.

(ii) For every α in $\mathcal{N}$, if for each n in $\mathbb{N}$ there exists β in X such that for each $i < n$, $\alpha(i) = \beta(i)$, then α itself belongs to X.

A spread is a closed subset of Baire space $\mathcal{N}$ that satisfies the classically empty condition (i).

2.2 *Brouwer's Continuity Principle, general formulation:*
For every spread $X \subseteq \mathcal{N}$, for every $R \subseteq X \times \mathbb{N}$,
if $\forall \alpha \in X\, \exists m\, [\alpha R m]$, then
$\forall \alpha \in X \exists n \exists m \forall \beta \in X$ [if for every $i < n$, $\alpha(i) = \beta(i)$, then $\beta R m$].

One may prove 2.2 from 1.1 by defining a so-called retraction of $\mathcal{N}$ onto X, that is, a continuous function r from $\mathcal{N}$ onto X such that for every α in X, $r(\alpha) = \alpha$, and then arguing straightforwardly.

We have to say something on real numbers.

2.3 Let $q_0, q_1, \ldots$ be an enumeration of the set $\mathbb{Q}$ of the rational numbers. For the purposes of this paper, an element α of $\mathcal{N}$ will be called a *canonical real number* if and only if, for each n, $|q_{\alpha(n)} - q_{\alpha(n+1)}| < \frac{1}{2^{n+1}}$. We let $\mathbb{R}$ be the set of all canonical real numbers. Observe that $\mathbb{R}$ is a spread.

2.4 We now define the binary relation of real coincidence on the set $\mathbb{R}$. Let α, β be canonical real numbers. We say that α *really coincides with* β, notation $\alpha =_{\mathbb{R}} \beta$, if and only if, for each n, $|q_{\alpha(n)} - q_{\beta(n)}| < \frac{1}{2^{n-1}}$. We don't go into the definition of the usual operations on the set $\mathbb{R}$.

2.5 A *real function* f is an effective method that associates to every canonical real number α a canonical real number $f(\alpha)$ in such a way that, for all α, β in $\mathbb{R}$, if $\alpha =_{\mathbb{R}} \beta$, then $f(\alpha) =_{\mathbb{R}} f(\beta)$.

2.6 Theorem: (*Continuity of real functions*)
Let f be a real function.
Then $\forall \alpha \in \mathbb{R} \forall m \exists n \forall \beta \in \mathbb{R}$ [if $|\alpha - \beta| < \frac{1}{2^n}$, then $|f(\alpha) - f(\beta)| < \frac{1}{2^m}$].

Proof: Let f be a real function and α a canonical real number, and m a natural number.
Applying 2.2, we calculate n, such that for every γ in $\mathbb{R}$, if for each $i < n$, $\alpha(i) = \gamma(i)$, then $(f(\alpha))(m+2) = (f(\gamma))(m+2)$, therefore for every k, $|q_{(f(\alpha))(m+2+k)} - q_{(f(\gamma))(m+2+k)}| < \frac{1}{2^{m+1}}$ and $|f(\alpha) - f(\gamma)| < \frac{1}{2^m}$.
Observe that, for every canonical real number β, if $|\alpha - \beta| < \frac{1}{2^n}$, then there exists a canonical real number γ such that for every $i < n$, $\alpha(i) = \gamma(i)$, and $\beta =_{\mathbb{R}} \gamma$, therefore $f(\beta) =_{\mathbb{R}} f(\gamma)$, and $|f(\alpha) - f(\beta)| < \frac{1}{2^m}$.
⊠

2.7 It is possible to generalize the result of Theorem 2.6.
Let D be a subset of $\mathbb{R}$. D will be called a *domain of continuity* if every real function from D to $\mathbb{R}$ is continuous, that is, for every function f from D to $\mathbb{R}$, if for all α, β in D such that $\alpha =_{\mathbb{R}} \beta$, also $f(\alpha) =_{\mathbb{R}} f(\beta)$, then $\forall \alpha \in D\ \forall m\ \exists n\ \forall \beta \in D$ [if $|\alpha - \beta| < \frac{1}{2^n}$, then $|f(\alpha) - f(\beta)| < \frac{1}{2^m}$].

2.8 We let Perhaps ($\mathbb{Q}$) be the set of all canonical real numbers α that change their value at most one time, that is, for all i, k, if $i < k$ and $\alpha(i) \neq \alpha(i+1)$, then $\alpha(i+1) = \alpha(k)$.

2.9 Let $b_0, b_1, \ldots$ be an enumeration of the set $\mathbb{Q}_2$ of all binary rational numbers.
We let Bin be the set of all canonical real numbers α such that for all i,

k, if $b_i \leq q_{\alpha(i)}$, then $b_i \leq q_{\alpha(i+k)}$, and if $q_{\alpha(i)} \leq b(i)$, then $q_{\alpha(i+k)} \leq b(i)$. Observe that every α in Bin has a binary development. Conversely, if α in $\mathbb{R}$ has a binary development, then there exists γ in Bin such that $\alpha =_{\mathbb{R}} \gamma$.

2.10 Theorem:

(i) Perhaps ($\mathbb{Q}$) is a domain of continuity.

(ii) Bin is a domain of continuity.

Proof: Observe that both Perhaps ($\mathbb{Q}$) and Bin are spreads and repeat the argument by which we established Theorem 2.6.
⊠

3. STRONG COUNTER-EXAMPLES

Brouwer's attack upon classical logic started with the observation that upon his interpretation of the logical connectives we have no reason to affirm the principle of the excluded third, $P \vee \neg P$.
For instance, we have no reason to affirm either: $\mathbf{example_0} = \underline{0}$ or: $\neg$ ($\mathbf{example_0} = \underline{0}$), therefore we have no reason to affirm: $\mathbf{example_0} = \underline{0} \vee \neg(\mathbf{example_0} = \underline{0})$.
This does not mean that the assumption $\mathbf{example_0} = \underline{0} \vee \neg(\mathbf{example_0} = \underline{0})$ leads to a contradiction; on the contrary, one may prove, in intuitionistic logic, for every proposition P, $\neg\neg(P \vee \neg P)$.
The Continuity Principle brings home to us that the assumption that we could decide, for *every* α in $\mathcal{N}$, if $\alpha = \underline{0}$ or $\neg(\alpha = \underline{0})$, is contradictory indeed.

3.1 Theorem: (*Absurdity of the Principle of Excluded Third*)
$\neg\forall\alpha \in \mathcal{N}\ [\alpha = \underline{0} \vee \neg(\alpha = \underline{0})]$.

Proof: Assume $\forall\alpha \in \mathcal{N}\ [\alpha = \underline{0} \vee \neg(\alpha = \underline{0})]$, therefore
$\forall\alpha \in \mathcal{N}\ \exists m\ [(m = 0 \wedge \alpha = \underline{0}) \vee (m = 1 \wedge \neg(\alpha = \underline{0}))]$.
Applying the Continuity Principle we find n such that
either for every α in $\mathcal{N}$, if for every $i < n$, $\alpha(i) = 0$, then $\alpha = \underline{0}$,
or for every α in $\mathcal{N}$, if for every $i < n$, $\alpha(i) = 0$, then $\neg(\alpha = \underline{0})$.
Both alternatives are absurd.
⊠

This technique can be applied more generally. Once we come to see that we sometimes are unable to take a decision of some given kind, it very often will be possible to derive a contradiction from the assumption that we should have a general method to take that kind of decision. The following Theorem is a further illustration of this remark.
For each rational number r we let r^* be the function from $\mathbb{N}$ to $\mathbb{N}$ such that, for each n, $q_{r^*(n)} = r$.
Observe that r^* belongs to $\mathbb{R}$. If confusion is improbable we sometimes write r where r^* would be more appropriate.

3.2 Theorem: (*Negative Continuity Theorem*)
There is no real function f such that, $f(0) =_{\mathbb{R}} 1$ but for each positive n, $f\left(\frac{1}{2^n}\right) =_{\mathbb{R}} 0$.

Proof: Assume there exists such a function.
Let $t : \mathbb{N} \to \mathbb{N}$ be such that for each positive n, $q_{t(n)} = \frac{1}{2^n}$. We now define a real number β as follows:
for each n, if for each $i \leq n$, $\mathbf{example_0}(i) = 0$, then $\beta(n) = t(n)$, and if there exists $i < n$ such that $\mathbf{example_0}(i) \neq 0$, then $\beta(n) = t(i_0)$ where i_0 is the least i such that $\mathbf{example_0}(i) \neq 0$.
Observe that, if $\mathbf{example_0} = \underline{0}$, then $\beta =_{\mathbb{R}} 0$ and $f(\beta) =_{\mathbb{R}} 1$ and if $\mathbf{example_0} \neq \underline{0}$, then $f(\beta) =_{\mathbb{R}} 0$.
We are unable to calculate $f(\beta)$, for we do not know how to define $\big(f(\beta)\big)(2)$. If $q_{\left(f(\beta)\right)(2)} \geq \frac{1}{2}$, then $\mathbf{example_0} = \underline{0}$ and if $q_{\left(f(\beta)\right)(2)} \leq \frac{1}{2}$,then $\mathbf{example_0} \neq \underline{0}$.
Generalizing this example, we may define, for every α in $\mathcal{N}$, a suitable canonical real number β with the property: if $\alpha = \underline{0}$, then $f(\beta) =_{\mathbb{R}} 1$ and if $\alpha \neq \underline{0}$, then $f(\beta) =_{\mathbb{R}} 0$.
Considering $\big(f(\beta)\big)(2)$ we may decide $\alpha = \underline{0}$ or $\alpha \neq \underline{0}$.
We obtain a contradiction by Theorem 3.1.
⊠

3.3 We have included Theorem 3.2 for historical reasons.
In the first Section of Brouwer [2] 1927, Brouwer proves the Negative Continuity Theorem, but he proves it in the weak sense only: there cannot be such a function because if we had one, then we could solve some problem for which we do not have a solution. Brouwer fails to distinguish carefully between this weak negation and the stronger one, which says that the negated proposition would lead to a contradiction. If one wants to prove the Negative Continuity Theorem in this strong sense,

one needs the Continuity Principle.
But why should one prove the Negative Continuity Theorem rather than the Continuity Theorem 2.6 itself, if one is prepared to use the Continuity Principle? (The Negative Continuity Theorem of course easily follows from Theorem 2.6.)
Brouwer probably felt doubtful about the argument we used for proving Theorem 2.6. He does not state a Continuity Theorem and, in Brouwer [2] 1927, goes on to prove, by a far more complicated argument, that every real function from the closed interval $[0, 1]$ to $\mathbb{R}$ must be uniformly continuous.
Something like our argument for Theorem 2.6 is used in Heyting [6] 1952/53. Charles Parsons, writing, in van Heijenoort [5] 1967, an introductory note to Brouwer [2] 1927, disputes the correctness of this reasoning, but unconvincingly so, as I explain in Veldman [10] 1982.

4. BROUWER'S FIRST APPLICATION

Brouwer [1] 1918 uses the Continuity Principle for the following application:

4.1 Theorem:
There is no injective function from $\mathcal{N}$ to $\mathbb{N}$.

Proof: Let f be a function from $\mathcal{N}$ to $\mathbb{N}$.
Determine n such that for each α in $\mathcal{N}$, if for each $i < n$, $\alpha(i) = 0$, then $f(\alpha) = f(\underline{0})$.
Obviously, f is not injective.
⊠

4.2 Cantor's argument is valid intuitionistically and does not depend on the Continuity Principle:

For every function $f : \mathbb{N} \to \mathcal{N}$ there exists α in $\mathcal{N}$ such that for each n, $f(n) \neq \alpha$. Define α as follows: for each n, $\alpha(n) := \big(f(n)\big)(n) + 1$.

This argument does not prove Theorem 4.1.

4.3 Brouwer [1] 1918 introduces the following notion:

Let X, Y be sets. X is *smaller than* Y, or: Y is *greater than* X, notation $X \sqsubset Y$, if and only there exists an injective function from X to Y and there does not exist an injective function from Y to X. We sometimes say: X *embeds into* Y if there exists an injective function from X to Y.

Brouwer observes that $\sqsubset$ is a transitive relation between sets and that $\mathbb{N}$ is smaller than $\mathcal{N}$.
There is a slight ambiguity in the notion "smaller than", as we may take the notion of an injective function in (at least) two senses, whenever a constructive inequality or apartness relation is present.
For the theorems we want to prove in this paper it does not make any difference which interpretation we choose.

4.4 We want to show that Theorem 4.1 admits of a vast extension.
We let $\mathbb{N}^*$ be the set of all finite sequences of natural numbers. We let $*$ denote the binary operation of concatenation of finite sequences and also the operation of concatenating a finite and an infinite sequence of natural numbers.
Let T be the set of all α in $\mathcal{N}$ that assume no other values than $0, 1$, and assume the value 1 at most one time.
Here is an infinite sequence of elements of T:

$$\underline{0},\ \langle 1\rangle * \underline{0},\ \langle 0,1\rangle * \underline{0},\ \langle 0,0,1\rangle * \underline{0},\ \ldots$$

This sequence is an injective function from $\mathbb{N}$ into T.
Remark that $\mathbf{example}_0$ belongs to T but we do not know where it occurs in the above sequence.
Observe that T is a spread and use the argument from the proof of Theorem 4.1 in order to see that there is no injective function from T into $\mathbb{N}$. (In particular, the above sequence is not an enumeration of T.)
So $\mathbb{N}$ is smaller than T.

4.5 For every s in $\mathbb{N}^*$, every subset X of $\mathcal{N}$ we define:
$s * X := \{s * \alpha \mid \alpha \in X\}$. For all subsets X, Y of $\mathcal{N}$ we define:
$X \oplus Y := \langle 0\rangle * X \cup \langle 1\rangle * Y$. For every subset X of $\mathcal{N}$ we define a sequence $0 \cdot X,\ 1 \cdot X,\ 2 \cdot X, \ldots$ of subsets of $\mathcal{N}$, as follows: $0 \cdot X := \emptyset$ and, for each n, $(n+1) \cdot X := (n \cdot X) \oplus X$. For each α in $\mathcal{N}$, n in $\mathbb{N}$, we define: $\overline{\alpha}(n) := \langle \alpha(0), \ldots \alpha(n-1)\rangle$. If confusion is unlikely to arise, we write $\overline{\alpha}n$ rather than $\overline{\alpha}(n)$.

4.6 Theorem:

For each n, $n \cdot T \sqsubset (n+1) \cdot T$.

Proof: We first show: $T \sqsubset T \oplus T$. Observe that T is a subset of $T \oplus T$. Assume now that f is an injective function from $T \oplus T$ into T. Observe that $f(\underline{0}) = \underline{0}$. For assume, for some m, $\big(f(\underline{0})\big)(m) = 1$. Calculate n such that for all α in T, if for all $i < n$, $\alpha(i) = 0$, then $\big(f(\alpha)\big)(m) = \big(f(\underline{0})\big)(m)$,

and $f(\alpha) = f(\underline{0})$. So f is not injective.
For similar reasons, $f(\langle 1 \rangle * \underline{0}) = \underline{0}$. So f is not injective.
Observe that, for each n, $n \cdot T$ is a subset of $(n+1) \cdot T$. Assume now that for some n, f is an injective function from $(n+1) \cdot T$ into $n \cdot T$. Observe, by repeating the above argument, that f has to map each one of the sequences $\langle 1 \rangle * \underline{0}, \ldots, \overline{\underline{0}}n * \langle 1 \rangle * \underline{0}$ onto one of the sequences $\langle 1 \rangle * \underline{0}, \ldots, \overline{\underline{0}}(n-1) * \langle 1 \rangle * \underline{0}$, and so cannot be injective.
⊠

4.7 For every subset X of $\mathcal{N}$, we let $\overline{X}$ be the set of all α in $\mathcal{N}$ such that for each n in $\mathbb{N}$ there exists β in X such that $\overline{\beta}n = \overline{\alpha}n$.
$\overline{X}$ is called the *closure of* X.
$\overline{X}$ will be a spread provided we are able to decide for each s in $\mathbb{N}^*$, if there exists α in X such that, for some n, $\overline{\alpha}n = s$.
We now define a sequence $T_0, T_1, T_2, \ldots$ of subsets of $\mathcal{N}$, as follows:
$T_0 := \{\underline{0}\}$ and for each m, $T_{m+1} := \overline{\bigcup_{n \in \mathbb{N}} \overline{\underline{0}}n * \langle 1 \rangle * T_m}$
Observe that each T_m is a spread and that T_1 coincides with the set T introduced in Section 4.4. For each m, T_m is the set of all α in $\mathcal{N}$ that assume no other values than $0, 1$ and that assume the value 1 at most m times. For each n, m, $n \cdot T_m$ is a subset of T_{m+1}.

4.8 Theorem:
For each n, m, $\quad n \cdot T_m \sqsubset (n+1) \cdot T_m \sqsubset T_{m+1}$.

Proof: Observe that, for each n, $n \cdot T_1$ is a subset of T_2, so, in view of Theorem 4.6, $n \cdot T_1 \sqsubset T_2$, because if T_2 would embed into $n \cdot T_1$, also $(n+1) \cdot T_1$ would embed into $n \cdot T_1$. We now show that T_2 is smaller than $T_2 \oplus T_2$ and leave the rest of the proof to the reader.
Observe that T_2 is a subset of $T_2 \oplus T_2$.
Assume that f is an injective function from $T_2 \oplus T_2$ into T_2. Observe that $f(\underline{0}) = \underline{0}$. For assume, for some m, $\big(f(\underline{0})\big)(m) = 1$. Calculate n such that for each α in T_2, if for all $i < n$, $\alpha(i) = 0$ then $\overline{\big(f(\alpha)\big)}(m+1) = \overline{\big(f(\underline{0})\big)}(m+1)$. Now define a mapping g from T_2 into $\mathcal{N}$ as follows: for each β in T_2, $g(\beta)$ is the sequence γ such that $f(\overline{\underline{0}}n * \beta) = \overline{f(\underline{0})}(m+1) * \gamma$.
Observe that g embeds T_2 into T_1. Contradiction.
For similar reasons $f(\langle 1 \rangle * \underline{0}) = \underline{0}$. So f is not injective.
⊠

4.9 Theorem 4.8 admits of a further refinement.
Consider the countable ordinal $\omega^{\cdot\omega}$. It consists of all polynomials

$\omega^{\cdot n_0} \cdot p_0 + \omega^{\cdot n_1} \cdot p_1 + \cdots + \omega^{\cdot n_{k-1}} \cdot p_{k-1}$ where $\langle n_0, n_1, \ldots, n_{k-1} \rangle$ is a strictly decreasing sequence of natural numbers and $\langle p_0, p_1, \ldots, p_{k-1} \rangle$ is a finite sequence of natural numbers of the same length k.
We may associate a subset of $\mathcal{N}$ to any such polynomial by defining: $\mathbb{T}\,(\omega^{\cdot n_0} \cdot p_0 + \omega^{\cdot n_1} \cdot p_1 + \cdots \; + \omega^{\cdot n_{k-1}} \cdot p_{k-1}) := p_0 \cdot T_{n_0} \oplus p_1 \cdot T_{n_1} \oplus \ldots \oplus p_{k-1} \cdot T_{n_{k-1}}$. Let $<$ denote the usual ordering on $\omega^{\cdot \omega}$.

4.10 Theorem:
For all α, β in $\omega^{\cdot \omega}$, if $\alpha < \beta$, then $\mathbb{T}(\alpha) \sqsubset \mathbb{T}(\beta)$.

Proof: We leave the straightforward proof to the reader.
⊠

4.11 These considerations may of course be extended further into the transfinite. Suppose we have a sequence $U_0, U_1, \ldots$ of spreads such that for each n, $U_n \sqsubset U_{n+1}$. Define $V := \overline{\bigcup_{n \in \mathbb{N}} \underline{\overline{0}}(n) * \langle 1 \rangle * U_n}$ and observe: for each n, $U_n \sqsubset V$ and $n \cdot V \sqsubset (n+1) \cdot V$.

5. A MODEL-THEORETIC OBSERVATION

5.1 We consider sentences in the first-order-language of equality.
Let X, Y be sets. We define: X *differs elementarily from* Y if there exists a sentence φ in the first-order-language of equality such that $\langle X \rangle \models \varphi$ and $\langle Y \rangle \models \neg\varphi$.
The sets $\mathbb{N}$ and $\mathcal{N}$ are elementarily different: consider the sentence $\forall x \forall y\, [x = y \vee \neg(x = y)]$ and recall Theorem 3.1. (When interpreting such a sentence, we use Tarski's truth definition, but of course understand connectives and quantifiers intuitionistically.)
The sets T and $\mathcal{N}$ are also elementarily different.
Observe that for all α in T, either $\alpha = \langle 1 \rangle * \underline{0}$ or not: $\alpha = \langle 1 \rangle * \underline{0}$ and that for every β in $\mathcal{N}$, not for all α in $\mathcal{N}$, either $\alpha = \beta$ or not: $\alpha = \beta$. So the sentence $\exists x\; \forall y\; [x = y \vee \neg(x = y)]$ is true in the structure $\langle T \rangle$ but not in the structure $\langle \mathcal{N} \rangle$.

5.2 Theorem:
For all α, β in $\omega^{\cdot \omega}$,
if $\alpha < \beta$, then $\mathbb{T}(\alpha)$ differs elementarily from $\mathbb{T}(\beta)$.

Proof: We define a sequence $\text{Undec}_0, \text{Undec}_1, \ldots$ of formulas in the first-order-language of equality as follows:
$\text{Undec}_0(x) := \neg\forall x_1\; [x = x_1 \vee \neg(x = x_1)]$, and for each n:

$\text{Undec}_{n+1}(x) := \text{Undec}_n(x) \wedge \neg\forall x_{n+2}\ [\text{Undec}_n(x_{n+2}) \rightarrow (x = x_{n+2} \vee \neg(x = x_{n+2}))]$.
(We assume that $x, x_0, x_1, \ldots$ are different individual variables of our language.)
For any formula $\varphi = \varphi(x)$ we use $\exists! x[\varphi(x)]$ as an abbreviation of the formula $\exists x\ [\varphi(x) \wedge \forall y\ [\varphi(y) \rightarrow y = x]]$.
Observe that the sentence $\exists! x[\text{Undec}_0(x)]$ is true in $\langle T_1 \rangle$ but not in $\langle 2 \cdot T_1 \rangle$, observe that the sentence $\exists x\ \exists y\ [\neg(x = y) \wedge \text{Undec}_0(x) \wedge \text{Undec}_0(y) \wedge \forall z\ [\text{Undec}_0(z) \rightarrow (z = x \vee z = y)]]$ is true in $\langle 2 \cdot T_1 \rangle$ but not in $\langle 3 \cdot T_1 \rangle$, observe that, for each n, the set of all α in T_{n+1} satisfying the formula Undec_0 coincides with T_n, observe that for each n, the sentence $\exists x\ [\text{Undec}_n(x)]$ is true in T_{n+1} but not in T_n and now complete the proof yourself.
⊠

The observations in Section 4 and 5 are new.
Some intuitionistic model theory is developed in Veldman and Jansen [15] 1990 and Veldman and Waaldijk [16] 1996. Both papers contain applications of the Continuity Principle.

6. BEGINNING THE BOREL HIERARCHY

6.1 Let X be a subset of the set $\mathbb{R}$ of canonical real numbers. We define:
X is *basic open* if and only if there exists q in $\mathbb{Q}$ and n in $\mathbb{N}$ such that $X = \{\alpha | \alpha \in \mathbb{R} |\ |\alpha - q| < \frac{1}{2^n}\}$.
X is *open* or $\mathbf{\Sigma}^0_1$ if and only if there exists a sequence $X_0, X_1, \ldots$ of basic open sets such that $X = \bigcup_{n \in \mathbb{N}} X_n$.
X is *closed* or $\mathbf{\Pi}^0_1$ if and only if there exists an open set Y such that $X = \mathbb{R} \backslash Y$.
X is *positively Borel* if X may be obtained from open and closed sets by means of the repeated application of the operations of countable union and countable intersection.
X is $\mathbf{\Sigma}^0_2$ if and only if there exists a sequence $X_0, X_1, \ldots$ of closed sets such that $X = \bigcup\limits_{n \in \mathbb{N}} X_n$.
X is $\mathbf{\Pi}^0_2$ if and only if there exists a sequence $X_0, X_1, \ldots$ of open sets such that $X = \bigcap\limits_{n \in \mathbb{N}} X_n$.
The study of the positively Borel sets starts with the observation that neither one of the classes $\mathbf{\Pi}^0_2$, $\mathbf{\Sigma}^0_2$ includes the other.

6.2 We let **Rat** be the set of all canonical real numbers α such that there exists a real rational number q with the property: for each n, $|q_{\alpha(n)} - q| < \frac{1}{2^n}$.
We let **PosIrr** be the set of all canonical real numbers α such that for each n, $q_{\alpha(2n)} < q_{\alpha(2n+2)} < q_{\alpha(2n+3)} < q_{\alpha(2n+1)}$ and either $q_n < q_{\alpha(2n)}$ or $q_{\alpha(2n+1)} < q_n$.
Observe that for every α in **PosIrr**, every q in $\mathbb{Q}$ there exists n such that $|\alpha - q| > \frac{1}{2^n}$, that is, α is *positively irrational.*
Conversely, every positively irrational real number really coincides with an element of **PosIrr**.
Observe that **PosIrr** is a spread.
Observe that **Rat** is $\mathbf{\Sigma}^0_2$ and that the set of all positively irrational canonical real numbers is a $\mathbf{\Pi}^0_2$ subset of $\mathbb{R}$.

6.3 Theorem:

For every sequence $G_0, G_1, \ldots$ of open sets,
if **Rat** $\subseteq \bigcap_{n\in\mathbb{N}} G_n$, there exists α in **PosIrr** such that $\alpha \in \bigcap_{n\in\mathbb{N}} G_n$.

Proof: Let $G_0, G_1, \ldots$ be a sequence of open sets such that **Rat** $\subseteq \bigcap_{n\in\mathbb{N}} G_n$. We construct step by step a canonical real number α satisfying the condition: for each n, $q_{\alpha(2n)} < q_{\alpha(2n+2)} < q_{\alpha(2n+3)} < q_{\alpha(2n+1)}$. Let n be a natural number and suppose we defined already $\alpha(0), \alpha(1), \ldots, \alpha(2n-2), \alpha(2n-1)$, such that $q_{\alpha(2n-2)} < q_{\alpha(2n-1)}$.
We then define $\alpha(2n)$ and $\alpha(2n+1)$ in such a way that $q_{\alpha(2n-2)} < q_{\alpha(2n)} < q_{\alpha(2n+1)} < q_{\alpha(2n-1)}$ *and* either $q_n < q_{\alpha(2n)}$ or $q_{\alpha(2n+1)} < q_n$ *and* the open interval $(q_{\alpha(2n)}, q_{\alpha(2n+1)})$ is a subset of G_n *and* $q_{\alpha(2n+1)} - q_{\alpha(2n)} < \frac{1}{2^{2n}}$.
It will be clear that α is positively irrational and belongs to every G_n.
⊠

6.4 Theorem:

For every sequence $F_0, F_1, \ldots$ of closed sets,
if **PosIrr** $\subseteq \bigcup_{n\in\mathbb{N}} F_n$, there exists α in **Rat** such that $\alpha \in \bigcup_{n\in\mathbb{N}} F_n$.

Proof: Let $F_0, F_1, \ldots$ be a sequence of closed sets such that **PosIrr** $\subseteq \bigcup_{n\in\mathbb{N}} F_n$. Let α_0 be some element of **PosIrr**. We recall that **PosIrr** is a spread, and apply the Continuity Principle. Determine m, n such that for all α in **PosIrr**, if $\overline{\alpha}(2m) = \overline{\alpha_0}(2m)$, then α belongs to F_n. Now every positively irrational number from the open interval

$(q_{\alpha_0(2m-2)}, q_{\alpha_0(2m-1)})$ will belong to the closed set F_n, and the closed interval $[q_{\alpha_0(2m-2)}, q_{\alpha_0(2m-1)}]$ will be a subset of F_n and F_n will contain many rational numbers.
⊠

6.5 Observe that Theorem 6.4 depends on the Continuity Principle, and that Theorem 6.3 does not but is proved by a straightforward "*Baire Category*" argument. Theorem 6.3 shows that there exist $\mathbf{\Sigma}^0_2$ sets that are not $\mathbf{\Pi}^0_2$. The Continuity Principle enables one to prove that there are easier examples. For all rational numbers r, s we let $[r, s]$ be the set of all canonical real numbers α such that, for each n, $r - \frac{1}{2^n} < q_{\alpha(n)} < s + \frac{1}{2^n}$.

6.6 Theorem:
The set $[0,1] \cup [1,2]$ is $\mathbf{\Sigma}^0_2$ but not $\mathbf{\Pi}^0_2$.

Proof: The set $[0,1] \cup [1,2]$ is obviously $\mathbf{\Sigma}^0_2$, so it suffices to show that it is not $\mathbf{\Pi}^0_2$. Let $G_0, G_1, \ldots$ be a sequence of open sets such that $[0,1] \cup [1,2] = \bigcap_{n \in \mathbb{N}} G_n$. Observe that for each n, $[0,2] \subseteq G_n$, therefore $[0,2] = [0,1] \cup [1,2]$. Therefore, for every α in $\mathbb{R}$, either, for all n, $q_{\alpha(n)} < 1 + \frac{1}{2^n}$ or, for all n, $1 - \frac{1}{2^n} < q_{\alpha(n)}$.
Consider the element α_0 of $\mathbb{R}$ defined by: for all n, $q_{\alpha_0(n)} = 1 + \left(-\frac{1}{2}\right)^{n+2}$. Applying the Continuity Principle we determine n such that either for all α in $\mathbb{R}$, if $\overline{\alpha_0}n = \overline{\alpha}n$, then, for all n, $q_{\alpha(n)} < 1 + \frac{1}{2^n}$, or for all α in $\mathbb{R}$, if $\overline{\alpha_0}n = \overline{\alpha}n$, then, for all n, $1 - \frac{1}{2^n} < q_{\alpha(n)}$.
Both alternatives are absurd.
⊠

6.7 The problem of establishing the second level of the Borel hierarchy was considered by Brouwer around 1924, see Brouwer [4] 1991, and Veldman [14] 1999B. One of his proofs is somewhat circuitous and not wholly correct.

7. BOREL HIERARCHY THEOREMS

7.1 We now consider Baire space $\mathcal{N}$. A subset X of $\mathcal{N}$ is called *basic open* if and only if there exists s in $\mathbb{N}^*$ such that X is the set of all α in $\mathcal{N}$ with the property: for some n, $\overline{\alpha}n = s$. We define *open, closed* and *positively Borel* subsets of $\mathcal{N}$ like the corresponding subsets of $\mathbb{R}$ in Section 6.1.

7.2 Let X, Y be subsets of $\mathcal{N}$. We define: X *reduces to* Y if and only if there exists a continuous function f from $\mathcal{N}$ to $\mathcal{N}$ such that for every α, α belongs to X if and only if $f(\alpha)$ belongs to Y.
We also define: X is *strongly irreducible to* Y if and only if for every continuous function f from $\mathcal{N}$ to $\mathcal{N}$, if f maps X into Y, then f also maps some element of $\mathcal{N} \setminus X$ into Y.
We now state without proof:

7.3 Theorem: (*Intuitionistic Borel Hierarchy Theorem*)
For every positively Borel set P there exists a positively Borel set Q such that Q is strongly irreducible to P.

7.4 The Continuity Principle plays a key rôle in the proof of Theorem 7.3. We may elucidate this rôle as follows.
For very many positively Borel set P one may construct a continuous function h from $\mathcal{N}$ to $\mathcal{N}$ such that $P = \mathrm{Ran}(h)$.
Suppose we are given Borel sets P, Q and continuous functions h_0, h_1 from $\mathcal{N}$ to $\mathcal{N}$, such that $P = \mathrm{Ran}(h_0)$ and $Q = \mathrm{Ran}(h_1)$, and also a continuous function f from $\mathcal{N}$ to $\mathcal{N}$ mapping P into Q.

$$\begin{array}{ccc} \mathcal{N} & \overset{g}{\cdots\rightarrow} & \mathcal{N} \\ \downarrow h_0 & & \downarrow h_1 \\ \mathcal{N} & \underset{f}{\longrightarrow} & \mathcal{N} \end{array}$$

Observe that for every α there exists β such that $f\big(h_0(\alpha)\big) = h_1(\beta)$. The Continuity Principle leads one to suspect: there exists a continuous function g from $\mathcal{N}$ to $\mathcal{N}$ such that $h_1 \circ g = f \circ h_0$. There is a stronger form of the Continuity Principle that gives exactly this conclusion, but in the actual proof, our version of the Principle suffices. In any case, the fact that f maps P into Q is taken very seriously: we require strong evidence for it.
The above picture may help one to see the meaning of the Intuitionistic Borel Hierarchy Theorem, which is very different from the meaning of the Hierarchy Theorem proved by Borel and Lebesgue.

7.5 Cantor space $\mathcal{C}$ is the set of all α in $\mathcal{N}$ that assume no other values than $0, 1$. We introduce two subsets of Cantor space, **Finite** and **Almostfinite**. We studied these sets in Veldman 1995. **Finite** is the set of all α in $\mathcal{C}$ such that $\exists n \, \forall m > n \, [\alpha(m) = 0]$. **Almostfinite** is the set of all α in $\mathcal{C}$ such that for every strictly increasing sequence γ of natural numbers there exists n such $\alpha(\gamma(n)) = 0$. Observe that α belongs to **Finite** if and only if α is the characteristic function of a finite subset of $\mathbb{N}$.

7.6 Theorem: (*Intuitionistic Borel Hierarchy Theorem, Second Version*)
For every positively Borel set P there exists a positively Borel set Q such that Q is strongly irreducible to P and **Finite** $\subseteq Q \subseteq$ **Almostfinite**.

From a classical point of view, the result is surprising as, with classical logic, the sets **Finite** and **Almostfinite** coincide.
Almostfinite is a subset of **Finite**$^{\neg\neg}$, that is, the set of all α in $\mathcal{C}$ such that $\neg\neg(\alpha \in$ **Finite**), and the statement that these two sets coincide is equivalent with the (implausible) generalized Markov principle.
One may prove that the set **Almostfinite** is not positively Borel.
The Theorem establishes the Borel hierarchy and at the same time shows in a spectacular way the expressiveness of intuitionistic logic.

7.7 Let α, β belong to Cantor space $\mathcal{C}$. We define: α *admits* β if and only if, for each n, $\alpha(\overline{\beta}n) = 0$. (We are assuming that finite sequences of natural numbers are coded by natural numbers and do not distinguish between a finite sequence of natural numbers and its code number.)
As in Veldman 1999, we let **Share**(T) be the set of all α in $\mathcal{C}$ that admit a member of T, and in this sense, *share* a member with T. Observe that $\overline{\mathbf{Share}(T)}$ is a spread, as for every α in $\mathcal{C}$, α belongs to $\overline{\mathbf{Share}(T)}$ if and only if for each n there exists γ in T such that $\alpha(\overline{\gamma}n) = 0$. Observe that for every α in $\overline{\mathbf{Share}(T)}$, if, for some n, $\alpha(\overline{\underline{0}}n) \neq 0$, then $\neg\neg\exists j < n[\alpha$ admits $\overline{\underline{0}}j * \langle 1 \rangle * \underline{0}]$, and if $\neg\exists n\ [\alpha(\overline{\underline{0}}n) \neq 0]$, then α admits $\underline{0}$ and belongs to **Share**(T), so in any case, $\neg\neg(\alpha \in$ **Share**$(T))$, therefore $\overline{\mathbf{Share}(T)}$ coincides with **Share**$(T)^{\neg\neg}$, the double complement of **Share**(T) in $\mathcal{C}$.

7.8 Theorem: (*Intuitionistic Borel Hierarchy Theorem, Third Version*)
For every positively Borel set P there exists a positively Borel set Q such that Q is irreducible to P and **Share**$(T) \subseteq Q \subseteq \overline{\mathbf{Share}(T)}$.

Observe that we are a bit more careful than in Theorem 7.6, as we do not require that Q is strongly irreducible to P. One may prove that the set **Share**(T) is not positively Borel. Observe that, classically,

Almostfinite is $\mathbf{\Sigma}^0_2$ and **Share**(T) is $\mathbf{\Pi}^0_1$.
Theorems 7.6 and 7.8 are fascinating consequences of the Continuity Principle. We intend to prove them in a future paper. Theorem 7.3 is proved in Veldman 1999B.

References

[1] L.E.J. Brouwer (1918), *Begründung der Mengenlehre unabhängig vom logischen Satz vom ausgeschlossenen Dritten.* Erster Teil: Allgemeine Mengenlehre. Koninklijke Nederlandsche Akademie van Wetenschappen, Verhandelingen, 1e sectie 12 no. 5, 43 p. Also: Brouwer 1975, pp. 150-190, with footnotes, pp. 573-585.

[2] L.E.J. Brouwer (1927), *Über Definitionsbereiche von Funktionen,* Math. Annalen 97, pp. 60-75. (English translation of §§1-3 in: van Heijenoort 1967, pp. 446-463, with an introduction by Charles Parsons, see especially footnote g, page 448).

[3] L.E.J. Brouwer (1975), *Collected Works*, Vol. I. *Philosophy and Foundations of Mathematics*, ed. A. Heyting, (North-Holland Publ. Co., Amsterdam etc.).

[4] L.E.J. Brouwer (1991), *Intuitionismus*, herausgegeben, eingeleitet und kommentiert von D. van Dalen, (Wissenschaftsverlag Mannheim etc.)

[5] J. van Heijenoort (1967), *From Frege to Gödel, A Source Book in Mathematical Logic*, (Harvard University Press, Cambridge Mass.)

[6] A. Heyting (1952/53), *Inleiding tot de Intuïtionistische Wiskunde*, College uitgewerkt door J.J. de Iongh, (Introduction to Intuitionistic Mathematics, Course Notes elaborated by J.J. de Iongh), s.n.

[7] S.C. Kleene and R.E. Vesley (1965), *The Foundations of Intuitionistic Mathematics, especially in relation to Recursive Functions*, (North-Holland Publ. Co., Amsterdam etc.) .

[8] A.S. Troelstra and D. van Dalen (1988), *Constructivism in Mathematics*, volume I, (North-Holland Publ. Co., Amsterdam etc.).

[9] W. Veldman (1981), *Investigations in Intuitionistic Hierarchy Theory*, Ph.D. Thesis, Katholieke Universiteit Nijmegen.

[10] W. Veldman (1982) *On the continuity of functions in intuitionistic real analysis*, Report 8210, Department of Mathematics, Katholieke Universiteit Nijmegen.

[11] W. Veldman (1990), A survey of intuitionistic descriptive set theory in P.P. Petkov (ed.), *Mathematical Logic*, Proceedings of the Heyting Conference 1988, (Plenum Press, New York and London), pp. 155-174.

[12] W. Veldman (1995), *Some intuitionistic variations on the notion of a finite set of natural numbers*, in: H.C.M. de Swart, L.J.M. Bergmans (ed.), *Perspectives on Negation, essays in honour of Johan J. de Iongh on the occasion of his 80th birthday*, (Tilburg University Press, Tilburg), pp. 177-202.

[13] W. Veldman (1999A), *On sets enclosed between a set and its double complement*, in: A. Cantini e.a. (ed.), *Logic and Foundations of Mathematics*, Proceedings Xth International Congress on Logic, Methodology and Philosophy of Science, Florence 1995, Volume III, (Kluwer Academic Publishers, Dordrecht etc.), pp. 143-154.

[14] W. Veldman (1999B), *The Borel hierarchy in intuitionistic mathematics*, Report 9930, Department of Mathematics, Katholieke Universiteit Nijmegen, submitted for publication to: Annals of Pure and Applied Logic.

[15] W. Veldman and M. Jansen (1990), *Some observations on intuitionistically elementary properties of linear orderings*, Archive for Mathematical Logic, vol. 29, pp. 171-187.

[16] W. Veldman and F. Waaldijk (1996), *Some elementary results in intuitionistic model theory*, Journal of Symbolic Logic, vol. 61, pp. 745-767

PEIRCE AND THE CONTINUUM FROM A PHILOSOPHICAL POINT OF VIEW

Julia Zink
Pontificia Università Gregoriana, Roma
since May 2000: Ludwig-Maximilians-Universität München; Graduiertenkolleg SIL CIS, Oettingenstrasse 67, D-80538 München
julia.zink@lrz.uni-muenchen.de

Keywords: Continuum, Peirce, Tertium non Datur

Mathematical Subject Classification: 00A30, 01A60, 03A05, 03B60, 26A15

Abstract: Peirce used the concept of continuity for the description of generality. His continuum is of greater "cardinality" than any set and contains infinitesimals. Arguing in continuous contexts demands different logical laws from those in discrete contexts. It is not clear what Peirce meant by the Principle of Excluded Middle. Peirce developed a three-valued logic. We have good reasons to assume that the Principle of Bivalence is not valid for statements about the whole continuum.

Introduction

Charles Sanders Santiago Peirce was an American philosopher, mathematician and chemist who lived from 1839 to 1914. The favorite field of Peirce was logic, but he did also a great deal of work in the area of measurement and chemistry. From 1879 to 1884 he was lecturer for logic at the John Hopkins University. The last thirty years of his life he lived in the countryside and could hardly earn a living. During his whole life, especially in the last period, he was extremely productive. He published about 12000 pages, about 80000 handwritten pages remained unpublished. Till today a great part of his writings is accessible only on microfiche [8].

In this paper characteristics of the theory of continuity of Peirce will be presented. It is necessary for the understanding of the Peircean view of the continuum to look at his philosophy and to be aware of the changes in his philosophy and logic, especially in his theory of continuity. The

P. Schuster, U. Berger and H. Osswald (eds.),
Reuniting the Antipodes – Constructive and Nonstandard Views of the Continuum, 303–316.

changes in the Peircean view of the continuum are described by Potter and Shields who distinguish four different periods [19], of which only the Kantistic and the post-Cantorian periods will be discussed here. In this paper the concepts and definitions will be used as Peirce formulates them in *Reasoning and the Logic of Things* [16], if not noted otherwise. As a term for distinguishable things we will normally use the word "individual", in geometric contexts the word "point" and in other mathematical contexts the word "element".

For the Peircean philosophy the three categories Firstness, Secondness and Thirdness are fundamental. These three categories and the concept of generality will be described. Peirce finds the categories by a characterization of that which is present in mind. Furthermore, Peirce claims that reality can be described in terms of these three categories. Afterwards, some mathematical properties of the Peircean continuum will be discussed. Then the connection between generality and continuity will be explored. The Peircean continuum is vast; it contains all sets and, therefore, all possibilities. Finally, it will be examined, considering his notes about three-valued logic, whether the Principle of Excluded Middle is valid for statements about the whole continuum.

1. THE THREE PEIRCEAN CATEGORIES

Peirce recognizes three states of the human mind that make our knowledge possible and that correspond to the Peircean categories Firstness, Secondness and Thirdness [18]. He claims that it is possible to describe the whole of reality in terms of the three categories [16]. First, in interpreting Peirce, every category will be characterized through the corresponding state of mind [18]. Second, by giving a literal citation, it will be shown how this category can be found in reality [16].

1.1. Firstness

We have the experience of *Firstness* in our first perception, before we react and before we begin to argue. To understand this state of mind imagine a person in a dream state who sees nothing else, but the color red. The person is neither thinking about it nor reasoning about it nor answering some kind of question nor reacting to it [18].

Peirce describes this category as follows ([16], p 147):

> *Firstness* may be defined as follows: It is the mode in which anything would be for itself, irrespective of anything else, so that it would not

> make any difference though nothing else existed, or ever had existed, or could exist.

Firstness is something absolutely simple, it is found for example in sense-qualities like red, or warmth, it is feeling.

1.2. Secondness

Secondness corresponds to reaction. If the person in his dream-state sees that the red changes rapidly into a very glaring light, he may react with puzzlement. Another person who touches a hot stove will react by pulling back his hand without thinking about that. Each of these reactions indicates that there are two things in his mind. These two things are connected. The one is that what the person has perceived, and the other one is the reaction to the perception [18].

Secondness can be defined as a modification of the being of one subject by another subject. Peirce writes ([16], pp 147-148):

> ..., Secondness is that in each of two absolutely severed and remote subjects which pairs it with the other, not for my mind nor for, or by, any mediating subject or circumstance whatsoever, but in those two subjects alone; so that it would be just the same if nothing else existed, or ever had existed, or could exist....It is that a blind reaction takes place between the two objects.

1.3. Thirdness

The third state of mind is that of thinking. When someone thinks, he produces a representation of the perception and maybe also of the reaction and reflects about the perception and the reaction by using the representation. The person is aware of learning. While he is aware of learning, he is going through a process by which he finds the perceived phenomenon to be governed by a rule. If we have a mediatory interest in something, which means that we want to get an idea about something, we will represent the thing in a sign. Signs can express a Thirdness. If a sign expresses a law, it is general. The sentence "Biological species can reproduce themselves" is a general sign; it is a law ([18]; [16]).

For signs it is characteristic that they need a triadic relation [18].

Peirce writes about Thirdness ([16], p 148):

> It is a modification of the being of one subject which is a mode of a second so far as it is a modification of a third. It might be called an inherent reason.

It is important to know that not all signs express a Thirdness. A simple example for a sign that expresses Firstness is the exclamation "Red!". This exclamation expresses just the perception of the color red, which is a Firstness. Peirce uses the sentence "Enoch is a man" as an example of a sign that expresses Firstness. The sentence "Enoch is a man" expresses a link between Enoch and the quality of being a man. This link seems to be regarded by Peirce as absolutely simple. He considers this link as absolutely simple because it has no reason. Peirce writes ([16], p 148):

> There is no reason for it; such is Enoch's nature,—that is all.

On the other hand for sentences like "Peter is bald" we can find an explanation; maybe Peter has had an illness which made him lose his hair.

Furthermore Peirce describes how Firstness, by the operation of Thirdness, brings forth Secondness ([16], p 148):

> On the other hand, the result, that Enoch dies like other men, as result or effect, expresses a Secondness. The necessity of the conclusion is just the brute force of this Secondness.

In interpreting Peirce, it can be said that Enoch is an example of a man who dies. He is an example who fulfills the general rule "Man is mortal", and, thereby, actualizes the rule.

2. MATHEMATICAL PROPERTIES OF THE PEIRCEAN CONTINUUM

In order to describe the continuum Peirce argues about cardinals. The continuum is in a certain sense so big that we cannot give it a cardinality. For continuous sets, it makes no sense to compare their multitude.

From each set we can build the power set, which is of greater cardinality than the original set, as it was proved by Cantor (see [4], p 511), and probably independently of him, by Peirce (see [16]; [7], 3.548; [4], p 511). Peirce assumes the General Continuum Hypothesis as true [14], he tries to prove the Continuum Hypothesis [14]. Peirce assumes that all possible infinite sets have the cardinality

$$\aleph_0 \text{ or } 2^{\aleph_0} \text{ or } 2^{2^{\aleph_0}} \text{ or...or } 2^{2^{\cdot^{\cdot^{\cdot^{2^{\aleph_0}}}}}}$$

He tries to prove that the power set of a collection which exceeds all these cardinals is not bigger than the set itself, but this proof is not valid. Peirce himself also was not sure about his proof (see [14]; [2], p 625; [4], pp 516-517).

Peirce wants his continuum to be of greater power than any of these sets. From his claim that all infinite sets have the cardinalities of power sets of power sets ... of the natural numbers he deduces that his continuum cannot be a set. The bigger a set is, the more difficult it is to distinguish its elements [14], but the elements remain always distinct. For example, take the natural numbers and the real numbers on a line. The difficulty to distinguish irrational numbers derives from the fact, that every irrational number is determined by an infinite number of elements, because it has infinitely many decimal places. To distinguish elements from a set of the cardinality $2^{2^{\aleph_0}}$ it would be necessary to compare uncountable many elements. Therefore these are even more difficult to distinguish than real numbers. Another cause for the difficulty to distinguish two numbers, can be density. Two rational number which are close together seem to be more difficult to distinguish, because there is no minimal distance between the rational numbers. Therefore it one can ask if the difficulty to distinguish elements derives from their arrangement. Peirce discusses whether the potential points are not distinct, because they are arranged in a particular way (see [14], p 97). But for all sets, the elements remain distinct [14], even if it is difficult to distinguish them. To distinguish to elements of the Peircean continuum it would be necessary to compare an amount of elements which exceeds any cardinality. This feature distinguishes the continuum from any set. For Peirce, the distinguishability seems to be an important characteristic of sets. His continuum is not a set anymore. What distinguishes the continuum from a set is that the continuum has no distinct elements anymore; the elements become welded together. They can be actualized, that means, they can be marked. Unless they are marked, for example the numbers one and two do not exist in an actual mode and so they can neither be distinguished, nor are they distinct. Peirce calls this kind of existence potential. His continuum is a collection of potential elements. Potential elements are "*indeterminate yet capable of determination in any special case*"[17].

Peirce wants his continuum to contain every possible set potentially [17]. That is, he insists that every element of every set that exists or could exist can be actualized in the continuum. That means, not that the continuum should contain all sets, for example the set of all apples, but that the continuum should contain potentially a set of every cardinality. Furthermore, he wants to arrange the continuum on a line, or better on a circle, to avoid boundary points. As the continuum is not a set, it has to be interpreted; we need to discover what Peirce probably meant by this. Peirce assumes that sets of every cardinality could be arranged on a line [16]. As Zermelo proved (Peirce was probably unaware of it), it follows from the Axiom of Choice that every set can be well-ordered (see [21]; [4], p 530). If every set can be well-ordered, every set can be arranged on a line. So we can arrange one set, for example, the set of natural numbers on a line. Then we can go on building the power set and arranging this set on a line. So we get an infinite process, where all the obtained sets are arranged on a line. Peirce probably means that, if we take the limit of this process, we get the continuum where the elements are merged together and where these potential elements are arranged on a line. If we actualize one of these elements, this element is in some of the ordered sets. In this sense the continuum is ordered.

About the melting together of the points in the continuum, Peirce writes ([16], p 159):

> We, therefore, find that we have now reached a multitude so vast that the individuals of such a collection melt into one another and loose their distinct identities. Such a collection is *continuous*.

Peirce says that, even if his conclusions about the uncountable infinite were false, he would retain that the continuum is of so vast a multitude that the elements cannot remain individuals, but become welded together [16].

Peirce illustrates how the potential points behave when they are actualized, that is, when one of them is marked. This mark is a discontinuity, and Peirce says that this point is made different by marking it. Peirce says that when the continuum is arranged on a circle [16] and the circle is cut at the marked point, this marked point becomes devided into two halfs. Points have no absolute identity [15]. If the circle was put together again at the two points, they would become one single point again. If the mark ceased to have any distinguishing character, there would be no distinct point anymore; the discontinuity would disappear.

Peirce makes a second thought-experiment [16]. Take the same circle as before, cut at one point. If one point at one end of the cut would burst away, this point could burst into a multitude of points as big as the irrational numbers between zero and one [16]. So we can say that if we mark a point in a continuum, and, thereby, get a discontinuity, uncountably many potential points are hidden behind this point. Probably Peirce meant, that if we actualize one point, this point contains also all points infinitesimally close to it, and these are uncountably many. Peirce requires these properties mentioned in the two thought-experiments from his notion of continuum.

Myrvold developed a model of the continuum (see [4], pp 535-537), based on the Peircean ideas. The model has the following properties:

- it can be arranged on a line,
- it contains all sets,
- hidden behind any point on the line, there are uncountably many points.

In the model of Myrvold, the Axiom of Choice is assumed to be true.

According to Myrvold it is possible to add infinitesimals to the real numbers by taking a fresh constant ε and by supposing that $0 < \varepsilon < x$ for all real $x > 0$. Let the x_i range over the real numbers. Let the rule $ab < ac$ if $b < c$ and $0 < a$ hold also for an infinitesimal b.

Let $\mathbf{R}_1 := \mathbf{R}$. We can define a new set $\mathbf{R}_2$, which has elements of the form: $x = x_0 + \varepsilon x_1$. Similarly, the new set $\mathbf{R}_\omega$ shall have elements of the form:

$$x = x_0 + \varepsilon x_1 + ... + \varepsilon^n x_n + ...;$$

hence $\mathbf{R}_{\omega+1}$ can be defined by adding another fresh constant ε^ω supposing that $0 < \varepsilon^\omega < \varepsilon^n$ for all natural numbers n. Its elements are supposed to have the form:

$$x = x_0 + \varepsilon x_1 + ... + \varepsilon^n x_n + ... + \varepsilon^\omega x_\omega.$$

Assuming that the $\mathbf{R}_i$ for $i < a$ are already defined we can go on defining $\mathbf{R}_a$ for each ordinal a. If $a - 1$ is a limit-number a fresh constant ε^{a-1} with $0 < \varepsilon^{a-1} < \varepsilon^i$ for all ordinals $i < a - 1$ has to be added as it was done for $\mathbf{R}_{\omega+1}$. If $a - 1$ is not a limit-number suppose that ε^{a-1} is $\varepsilon\varepsilon^{a-2}$. So $\mathbf{R}_a$ is a collection of elements of the form: $\sum_{i<a} \varepsilon^i x_i$. Therefore every $\mathbf{R}_a$ corresponds therefore to a collection of well-ordered sequences $(x_i : i < a)$, each of which corresponds to one of the series $\sum_{i<a} \varepsilon^i x_i$ mentioned above. For the ordinal a there are $(2^{\aleph_0})^{card(a)}$ elements in $\mathbf{R}_a$. From the well-ordering theorem it follows that for any cardinal k there is

an $\mathbf{R}_a$ whose cardinality is at least k. Each $\mathbf{R}_a$ can be lexicographically ordered: for any $x, y \in \mathbf{R}_a$,

$$x < y \Leftrightarrow \exists i < a \quad (x_i < y_i \wedge \forall j < i \quad (x_j = y_j)) .$$

None of these sets is a continuum in the Peircean sense. However, the class that includes all $\mathbf{R}_a$ is the model presented by Myrvold as a faithfull representation of the Peircean continuum. Peirce would say here, furthermore, that in the class which contains all $\mathbf{R}_a$ the points are merged together and distinguishable only by actualizing them.

3. GENERALITY AND CONTINUITY

In order to understand the connection between generality and continuity we have to go into the meaning of some Peircean concepts. The concepts are vagueness, possibility, generality, potentiality and necessity. They are to be connected with the categories and with modal logic. Vagueness will only be explained as far as it is necessary to understand generality.

Vagueness and possibility as well as generality and necessity are connected; the connection is of a similar mode. The explanation of this connection will be concluded with an appropriate literal citation. Next the concepts of possibility and potentiality will be compared. Afterwards, the concepts of potentiality and continuity, necessity and generality will be connected.

Peirce distinguishes so-called vague contexts from general contexts. Vague contexts are those to which the Principle of Contradiction does not apply. General contexts are those to which the Principle of Excluded Middle does not apply (see [7], 5.505).

This seems to be a unequivocal definition, but it is not entirely clear what Peirce means by Principle of Contradiction and by the Principle of Excluded Middle.

When we express a possibility, we make a vague statement, and possibilities are in the category of Firstness. In this case, the right of giving a precise interpretation of the statement is left to the speaker (see [9]). So, if the speaker says "A man is tall", it is not clear which man he means. So there are different possibilities to interpret this sentence. It can be interpreted referring to a man who is really tall or it can be interpreted referring to a man who is not tall. In the first case, the statement is true, in the second case the statement is not true. So it can be said that the undetermined sentence is both true and false in the sense that it has an interpretation which is true, and an interpretation which is false. Peirce says: "*For, as we have seen above, that which characterizes and defines an assertion of Possibility is its emancipation from the Principle of Contradiction, while it remains subject to the Principle of Excluded Third*" ([12] cit in: [1], p 79).

When we make a necessary statement, we make a general statement. In necessary statements we express a rule and rules are in the category of Thirdness. Peirce says that necessity is the support of actuality by reason (see [10], cit in: [1], p 78). Furthermore he says, that the interpretation of those kinds of statements is left to the interpreter (see [9]). He can apply the rule to any situation which it concerns. As these situations are unknown at the moment, and as Peirce means that even the laws of nature are in evolution, he claims, that general statements are neither true nor false, because they are indeterminate. They have the same kind of indeterninacy as habits. Peirce says that the Principle of Excluded Middle is not valid for general assertions (see [12] cit in: [1], p 79). As Noble writes, the difference between a mere possibility and generality is the difference between Firstness and Thirdness (see [5], p 170).

When we express a Secondness, we talk about something that is actual, and to these contexts the Principle of Contradiction and the Principle of Excluded Middle do apply.

To resume the above, it will be explained what Peirce said (see [12] cit in: [1], p 79). Assertions of actuality are in the middle of assertions of possibility and assertions of necessity. The assertion of actuality "*acknowledges allegiance to both formulae*", that is the Principle of Contradiction and the Principle of Excluded Middle. On the contrary, this is not valid for assertions of possibility and assertions of necessity: for the assertions of possibility the Principle of Contradiction is not valid, "*while it remains subject to the Principle of Excluded Third*". An assertion of necessity "*remains subject to the Principle of Contradiction, but throws off the yoke of the Principle of Excluded Third*".

Potentiality is connected with generality and necessity. Rules are general. Something that is *potential* is "*indeterminate yet capable of determination in any special case*" [17].

If we say, for example, "Man is mortal" and if we want to express a rule with it, every death of a man is a determination of the rule "Man is mortal" in a special case.

It is a characteristic feature of general contexts that a general claim makes a statement about the whole of potential individuals. Peirce describes the relation between possible and potential by saying that the possible is undetermined, whereas the potential is undetermined, but capable of determination in any special case. So the potential is possible but not entirely vague. If we affirm a rule, this rule should range over all appropriate possibilities, and these possibilities are potential as well. Possible situations which a rule would apply to can be converted to actual ones by them happening. This might be considered as universal quantification in the Peircean sense. So the continuum exceeds any cardinality and is a whole of potential individuals; therefore Peirce says that it is the true universal which he needs for his Logic of Relatives (see [16]).

The idea of laws and generality are important to understand Thirdness. The category of Thirdness is the category of signs. The meaning of signs is given by a habit of the interpreters, because signs have to be repeatable. Each sign is embedded in a continuity of potential interpretations and their relations, these interpretaions are themselves signs. Therefore each sign is vague. Furthermore, all the interpretations are related to the object which they represent by a rule which embeds them into a continuum. Similarly, a law of nature connects the potential situations in which it would apply to a continuum. Both, habits and laws of nature, are considered by Peirce to be in evolution and indeterminate (see[6], p 121; [13], p 165).

Peirce tries to find the general, in order to describe the kind of generality needed in his Logic of Relatives . He compares cardinalities of sets. Then he looks for the collection that includes all sets potentially. Since the continuum contains all sets potentially, it includes all possibilities potentially. That is the reason why Peirce says that continuity is the true universal [16].

4. ARGUING IN CONTINUOUS OR GENERAL CONTEXTS

The Principle of Excluded Middle does apply in contexts where we talk about individuals. If we argue in continuous contexts, we do not talk about individuals, but about some potential points, or about all potential points. If we make statements about the whole continuum or, in other words, about all potential points in the continuum, we make a general statement. Furthermore, it is a characteristic of general statements that they make an assertion about a whole of potential individuals, and therefore about a continuum (see [7] 1.218, 1.420). Peirce claims that the

Principle of Excluded Middle does not apply to general assertions, a circumstance by which he defines and characterizes them. But he uses two formulations of the Principle of Excluded Middle, as was made evident by Lane [3]. He comes to the conclusion that Peirce characterizes general assertions by the invalidity of the first formulation of the Principle of Excluded Middle, which can be written as (S is P) $\vee$ (S is $\sim P$). The second formulation today is often called Principle of Bivalence: for any proposition p, either p is true or else p is false. What shall be shown here is that in general contexts, Peirce rejects the validity of the Principle of Bivalence.

To this end, attention will be given first to some of the contexts in which the principle (S is P) $\vee$ (S is $\sim P$) is called Principle of Excluded Middle by Peirce. For example, Peirce says that by the Principle of Excluded Middle *"is always meant the principle that no pair of mutually contradictory predicates is false for any individual subject"*([11], cit in [3]). Peirce writes immediately after the above cited that *"the principle of excluded middle only applies to an individual"*. He gives the reason for this, with the example "Any man is wise". He says that it *"is not true that "Any man is wise" nor that "Any man is not wise"."* According to this citation the principle can be written as: (S is P) $\vee$ (S is $\sim P$). It is important here that Peirce does not say that this principle is not valid in general contexts, but he just makes the weaker claim, that this principle is valid for *each individual.* Other similar citations can be found in [3].

Now it will be examined in which contexts this principle fails. Peirce says that general sentences (see [7] 2.289, 2.324, 2.328, 2.357) can often be recognized by words like "any" or "every". These words indicate that the interpretation is left to the interpreter who can choose whatever individual he wants (see [9]). Therefore "Any man is wise" means that whatever man you choose will be wise. Many of the Peircean general assertions have the form: "S is P", where S is an general subject term and P is a predicate. So he predicates "wise"or "not wise"of the general subject term "Any man". If we write down the first principle (S is P) $\vee$ (S is $\sim P$) for S being the general subject-term *Any* Q, we get:

$$(\mathit{Any}\ Q \text{ is } P) \vee (\ \mathit{Any}\ Q \text{ is } \sim P)\,.$$

It is not true, that:

$$(\mathit{Any} \quad x \in \{2,3\} \quad \text{is even}) \vee (\mathit{Any} \quad x \in \{2,3\} \quad \text{is } \sim \text{even})\,.$$

So this principle may fail already if we assign one property to more than one individual. It seems, however that this proposition is not a general proposition in the Peircean sense, because it does not express a rule, but something about a special set. In order to describe such contexts, we do not need a continuum. Lane probably sustains this principle because he hardly considers the connection between generality, continuity and modality (see [3], p 701). Furthermore, Lane does not want to accept the statement that the assertion "Man is mortal" is neither true nor false (see [3], p 681).

Peirce gave also another formulation to the Principle of Excluded Middle: In the section about *N-Valued Logic* in *The New Elements of Mathematics,* Peirce defines the Principle of Excluded Middle in the same way as we define today the Principle of Bivalence ([15], p 750):

> Every proposition is either true or false.

Furthermore Peirce began to develop a three-valued logic in February 1909. This includes a rejection of the Principle of Bivalence. This fact is mentioned also by Lane (see [3], p 701). The principle $(S \text{ is } P) \vee (S \text{ is } \sim P)$ may fail already for more than one individual. The continuum contains no individuals, but potential individuals, which are merged together, that is they are neither distinguishable, nor distinct. So arguing in general form about the continuum can not be characterized in the same way as arguing about more than one individual.

In the following citation about triadic logic, it will be seen, that the formulation of Peirce is somehow confusing or imprecise. He connects there the *neither being true nor false* of an assertion of the form "S is P" with the S *being neither* P *nor not* P. It seems that he considers both as being just the same. The reason for this may be that for individuals the principle $(S \text{ is } P) \vee (S \text{ is } \sim P)$ is at least very similar to the Principle of Bivalence for individuals. Peirce writes ([1], p 75; [8]):

> Triadic Logic is that Logic which, though not rejecting entirely the Principle of Excluded Middle, nevertheless recognizes that every proposition, S is P is either true, or false, or else has a lower mode of being such that it can neither be determinately P, nor determinately not P, but is at the limit between P and not P.[1]

So it seems that there are good reasons to say that from the Peircean view after 1909 the Principle of Bivalence, "For any proposition p, either

[1]What seems strange is, that Peirce developed his triadic logic so late, this is discussed in [1], p 82.

p is true or else p is false" does not apply in general contexts and for the same reasons also not in contexts where we make assertions about the whole continuum.

Acknowledgments

I want to express my gratitude to U. Berger and P. Schuster, who invited me to give a talk and to publish this article. I want to thank J. Bell, J. Moschovakis who made me see interesting open questions, and also the referee. Finally I want to thank A. Labella and J. Ehrat SJ for discussions about this article.

References

[1] Fisch, M, Turquette, A. (1966). "Peirce's Triadic Logic", Transactions [2] of the Charles S. Peirce Society 2 (2), 71-85.

[2] Herron, T. (1997). "Charles Sanders Peirce's Theories of Infinitesimals", TCSPS 33 (3), 590 - 645.

[3] Lane, R. (1997). " Peirce Entanglement with the Principle of Excluded Middle and Contradiction", TCSPS 33 (3), 680-793.

[4] Myrvold, W.(1995)." Peirce on Cantor's Paradox and the Continuum", TCSPS 31 (3), 508-541.

[5] Noble, N.(1989)." Peirce's Definitions of Continuity and the Concept of Possibility", TCSPS 25 (2), 149-174.

[6] Oehler, K. (1993). *Charles Sanders Peirce,* München: Beck'sche Reihe.

[7] Peirce, C. (1931-1958). *Collected Papers*, (eds. Hartshorne C. et al.), Cambridge, MA: Harvard University Press [3].

[8] Peirce, C. (1966). " Logic Notebook (MS 339)", 344 r, in: The Charles S. Peirce Papers, microfilm edition, Cambridge, MA: Harvard University Library, Photographic Service [4].

[9] Peirce, C. (1966). "MS 9", 1903, in: The Charles S. Peirce Papers, microfilm edition, Cambridge, MA: Harvard University Library, Photographic Service.

[2] Transactions of he Charles S. Peirce Society will in the following be abbreviated TCSPS.

[3] These writings are cited using the usual decimal numeration. Before the point the volume is given, after the point the paragraph number.

[4] MS 339, February 23, 1909. The manuscript number MS 339 is that which is given by [20]. The handwritten notes are printed in [1], pp 73-75.

[10] Peirce, C.(1966). "Prescott Book (MS 277)", January 1908, in: The Charles S. Peirce Papers, microfilm edition, Cambridge, MA: Harvard University Library, Photographic Service.

[11] Peirce, C. (1966). "MS 611", 1908, 13 in: The Charles S. Peirce Papers, microfilm edition, Cambridge, MA: Harvard University Library, Photographic Service.

[12] Peirce, C.(1966). "The Art of Reasoning Elucidated (MS 678)", 34-35, in: The Charles S. Peirce Papers, microfilm edition, Cambridge, MA: Harvard University Library, Photographic Service.

[13] Peirce, C. (1983). *Phänomen und Logik der Zeichen*, (ed. Pape, H.), Frankfurt am Main: Suhrkamp.

[14] Peirce, C.(1976)"Multitude and Continuity", 84-88, 97-100, in: The New Elements of Mathematics, Vol III, (ed. Eisele, C.), The Hague, Paris: Mouton Publishers.

[15] Peirce, C. (1976). "N-Valued Logic", 748, 750, in: The New Elements of Mathematics, Vol III, (ed. Eisele, C.), The Hague, Paris: Mouton Publishers.

[16] Peirce, C. (1992). "The Logic of Relatives", 147-148, 157-160, in: Reasoning and the Logic of Things, (ed. Ketner, K.), Cambridge, MA: Library of Congress.

[17] Peirce, C. (1992). "The Logic of Continuity", 247, in: Reasoning and the Logic of Things, (ed. Ketner, K.), Cambridge, MA: Library of Congress.

[18] Peirce, C. (1998). "What is a Sign", 4-5, in: The essential Peirce, Volume II (ed. by the Peirce Edition Project), Bloomington, Indiana: Indiana University Press.

[19] Potter, V; Shields, P. (1977). "Peirce's Definitions of Continuity", TCSPS 13 (1), 20-34.

[20] Robin, R. (1967). *Annotated Catalogue of the Papers of Charles S. Peirce*, Amherst: University of Massachusetts Press.

[21] Zermelo, E.(1904)."Beweis, daß jede Menge wohlgeordnet werden kann", Mathematische Annalen 59, 514 - 516.

1. J. M. Bochénski, *A Precis of Mathematical Logic*. Translated from French and German by O. Bird. 1959 ISBN 90-277-0073-7
2. P. Guiraud, *Problèmes et méthodes de la statistique linguistique*. 1959 ISBN 90-277-0025-7
3. H. Freudenthal (ed.), *The Concept and the Role of the Model in Mathematics and Natural and Social Sciences*. 1961 ISBN 90-277-0017-6
4. E. W. Beth, *Formal Methods*. An Introduction to Symbolic Logic and to the Study of Effective Operations in Arithmetic and Logic. 1962 ISBN 90-277-0069-9
5. B. H. Kazemier and D. Vuysje (eds.), *Logic and Language*. Studies dedicated to Professor Rudolf Carnap on the Occasion of His 70th Birthday. 1962 ISBN 90-277-0019-2
6. M. W. Wartofsky (ed.), *Proceedings of the Boston Colloquium for the Philosophy of Science, 1961–1962*. [Boston Studies in the Philosophy of Science, Vol. I] 1963 ISBN 90-277-0021-4
7. A. A. Zinov'ev, *Philosophical Problems of Many-valued Logic*. A revised edition, edited and translated (from Russian) by G. Küng and D.D. Comey. 1963 ISBN 90-277-0091-5
8. G. Gurvitch, *The Spectrum of Social Time*. Translated from French and edited by M. Korenbaum and P. Bosserman. 1964 ISBN 90-277-0006-0
9. P. Lorenzen, *Formal Logic*. Translated from German by F.J. Crosson. 1965 ISBN 90-277-0080-X
10. R. S. Cohen and M. W. Wartofsky (eds.), *Proceedings of the Boston Colloquium for the Philosophy of Science, 1962–1964*. In Honor of Philipp Frank. [Boston Studies in the Philosophy of Science, Vol. II] 1965 ISBN 90-277-9004-0
11. E. W. Beth, *Mathematical Thought*. An Introduction to the Philosophy of Mathematics. 1965 ISBN 90-277-0070-2
12. E. W. Beth and J. Piaget, *Mathematical Epistemology and Psychology*. Translated from French by W. Mays. 1966 ISBN 90-277-0071-0
13. G. Küng, *Ontology and the Logistic Analysis of Language*. An Enquiry into the Contemporary Views on Universals. Revised ed., translated from German. 1967 ISBN 90-277-0028-1
14. R. S. Cohen and M. W. Wartofsky (eds.), *Proceedings of the Boston Colloquium for the Philosophy of Sciences, 1964–1966*. In Memory of Norwood Russell Hanson. [Boston Studies in the Philosophy of Science, Vol. III] 1967 ISBN 90-277-0013-3
15. C. D. Broad, *Induction, Probability, and Causation*. Selected Papers. 1968 ISBN 90-277-0012-5
16. G. Patzig, *Aristotle's Theory of the Syllogism*. A Logical-philosophical Study of *Book A* of the *Prior Analytics*. Translated from German by J. Barnes. 1968 ISBN 90-277-0030-3
17. N. Rescher, *Topics in Philosophical Logic*. 1968 ISBN 90-277-0084-2
18. R. S. Cohen and M. W. Wartofsky (eds.), *Proceedings of the Boston Colloquium for the Philosophy of Science, 1966–1968, Part I*. [Boston Studies in the Philosophy of Science, Vol. IV] 1969 ISBN 90-277-0014-1
19. R. S. Cohen and M. W. Wartofsky (eds.), *Proceedings of the Boston Colloquium for the Philosophy of Science, 1966–1968, Part II*. [Boston Studies in the Philosophy of Science, Vol. V] 1969 ISBN 90-277-0015-X
20. J. W. Davis, D. J. Hockney and W. K. Wilson (eds.), *Philosophical Logic*. 1969 ISBN 90-277-0075-3
21. D. Davidson and J. Hintikka (eds.), *Words and Objections*. Essays on the Work of W. V. Quine. 1969, rev. ed. 1975 ISBN 90-277-0074-5; Pb 90-277-0602-6
22. P. Suppes, *Studies in the Methodology and Foundations of Science. Selected Papers from 1951 to 1969*. 1969 ISBN 90-277-0020-6
23. J. Hintikka, *Models for Modalities*. Selected Essays. 1969 ISBN 90-277-0078-8; Pb 90-277-0598-4

24. N. Rescher *et al.* (eds.), *Essays in Honor of Carl G. Hempel.* A Tribute on the Occasion of His 65th Birthday. 1969 ISBN 90-277-0085-0
25. P. V. Tavanec (ed.), *Problems of the Logic of Scientific Knowledge.* Translated from Russian. 1970 ISBN 90-277-0087-7
26. M. Swain (ed.), *Induction, Acceptance, and Rational Belief.* 1970 ISBN 90-277-0086-9
27. R. S. Cohen and R. J. Seeger (eds.), *Ernst Mach: Physicist and Philosopher.* [Boston Studies in the Philosophy of Science, Vol. VI]. 1970 ISBN 90-277-0016-8
28. J. Hintikka and P. Suppes, *Information and Inference.* 1970 ISBN 90-277-0155-5
29. K. Lambert, *Philosophical Problems in Logic.* Some Recent Developments. 1970 ISBN 90-277-0079-6
30. R. A. Eberle, *Nominalistic Systems.* 1970 ISBN 90-277-0161-X
31. P. Weingartner and G. Zecha (eds.), *Induction, Physics, and Ethics.* 1970 ISBN 90-277-0158-X
32. E. W. Beth, *Aspects of Modern Logic.* Translated from Dutch. 1970 ISBN 90-277-0173-3
33. R. Hilpinen (ed.), *Deontic Logic.* Introductory and Systematic Readings. 1971 *See also* No. 152. ISBN Pb (1981 rev.) 90-277-1302-2
34. J.-L. Krivine, *Introduction to Axiomatic Set Theory.* Translated from French. 1971 ISBN 90-277-0169-5; Pb 90-277-0411-2
35. J. D. Sneed, *The Logical Structure of Mathematical Physics.* 2nd rev. ed., 1979 ISBN 90-277-1056-2; Pb 90-277-1059-7
36. C. R. Kordig, *The Justification of Scientific Change.* 1971 ISBN 90-277-0181-4; Pb 90-277-0475-9
37. M. Čapek, *Bergson and Modern Physics.* A Reinterpretation and Re-evaluation. [Boston Studies in the Philosophy of Science, Vol. VII] 1971 ISBN 90-277-0186-5
38. N. R. Hanson, *What I Do Not Believe, and Other Essays.* Ed. by S. Toulmin and H. Woolf. 1971 ISBN 90-277-0191-1
39. R. C. Buck and R. S. Cohen (eds.), *PSA 1970.* Proceedings of the Second Biennial Meeting of the Philosophy of Science Association, Boston, Fall 1970. In Memory of Rudolf Carnap. [Boston Studies in the Philosophy of Science, Vol. VIII] 1971 ISBN 90-277-0187-3; Pb 90-277-0309-4
40. D. Davidson and G. Harman (eds.), *Semantics of Natural Language.* 1972 ISBN 90-277-0304-3; Pb 90-277-0310-8
41. Y. Bar-Hillel (ed.), *Pragmatics of Natural Languages.* 1971 ISBN 90-277-0194-6; Pb 90-277-0599-2
42. S. Stenlund, *Combinators, γ Terms and Proof Theory.* 1972 ISBN 90-277-0305-1
43. M. Strauss, *Modern Physics and Its Philosophy.* Selected Paper in the Logic, History, and Philosophy of Science. 1972 ISBN 90-277-0230-6
44. M. Bunge, *Method, Model and Matter.* 1973 ISBN 90-277-0252-7
45. M. Bunge, *Philosophy of Physics.* 1973 ISBN 90-277-0253-5
46. A. A. Zinov'ev, *Foundations of the Logical Theory of Scientific Knowledge (Complex Logic).* Revised and enlarged English edition with an appendix by G. A. Smirnov, E. A. Sidorenka, A. M. Fedina and L. A. Bobrova. [Boston Studies in the Philosophy of Science, Vol. IX] 1973 ISBN 90-277-0193-8; Pb 90-277-0324-8
47. L. Tondl, *Scientific Procedures.* A Contribution concerning the Methodological Problems of Scientific Concepts and Scientific Explanation. Translated from Czech by D. Short. Edited by R.S. Cohen and M.W. Wartofsky. [Boston Studies in the Philosophy of Science, Vol. X] 1973 ISBN 90-277-0147-4; Pb 90-277-0323-X
48. N. R. Hanson, *Constellations and Conjectures.* 1973 ISBN 90-277-0192-X

49. K. J. J. Hintikka, J. M. E. Moravcsik and P. Suppes (eds.), *Approaches to Natural Language.* 1973 ISBN 90-277-0220-9; Pb 90-277-0233-0
50. M. Bunge (ed.), *Exact Philosophy.* Problems, Tools and Goals. 1973 ISBN 90-277-0251-9
51. R. J. Bogdan and I. Niiniluoto (eds.), *Logic, Language and Probability.* 1973 ISBN 90-277-0312-4
52. G. Pearce and P. Maynard (eds.), *Conceptual Change.* 1973 ISBN 90-277-0287-X; Pb 90-277-0339-6
53. I. Niiniluoto and R. Tuomela, *Theoretical Concepts and Hypothetico-inductive Inference.* 1973 ISBN 90-277-0343-4
54. R. Fraissé, *Course of Mathematical Logic* – Volume 1: *Relation and Logical Formula.* Translated from French. 1973 ISBN 90-277-0268-3; Pb 90-277-0403-1 (For *Volume 2* see under No. 69).
55. A. Grünbaum, *Philosophical Problems of Space and Time.* Edited by R.S. Cohen and M.W. Wartofsky. 2nd enlarged ed. [Boston Studies in the Philosophy of Science, Vol. XII] 1973 ISBN 90-277-0357-4; Pb 90-277-0358-2
56. P. Suppes (ed.), *Space, Time and Geometry.* 1973 ISBN 90-277-0386-8; Pb 90-277-0442-2
57. H. Kelsen, *Essays in Legal and Moral Philosophy.* Selected and introduced by O. Weinberger. Translated from German by P. Heath. 1973 ISBN 90-277-0388-4
58. R. J. Seeger and R. S. Cohen (eds.), *Philosophical Foundations of Science.* [Boston Studies in the Philosophy of Science, Vol. XI] 1974 ISBN 90-277-0390-6; Pb 90-277-0376-0
59. R. S. Cohen and M. W. Wartofsky (eds.), *Logical and Epistemological Studies in Contemporary Physics.* [Boston Studies in the Philosophy of Science, Vol. XIII] 1973 ISBN 90-277-0391-4; Pb 90-277-0377-9
60. R. S. Cohen and M. W. Wartofsky (eds.), *Methodological and Historical Essays in the Natural and Social Sciences. Proceedings of the Boston Colloquium for the Philosophy of Science, 1969–1972.* [Boston Studies in the Philosophy of Science, Vol. XIV] 1974 ISBN 90-277-0392-2; Pb 90-277-0378-7
61. R. S. Cohen, J. J. Stachel and M. W. Wartofsky (eds.), *For Dirk Struik. Scientific, Historical and Political Essays.* [Boston Studies in the Philosophy of Science, Vol. XV] 1974 ISBN 90-277-0393-0; Pb 90-277-0379-5
62. K. Ajdukiewicz, *Pragmatic Logic.* Translated from Polish by O. Wojtasiewicz. 1974 ISBN 90-277-0326-4
63. S. Stenlund (ed.), *Logical Theory and Semantic Analysis.* Essays dedicated to Stig Kanger on His 50th Birthday. 1974 ISBN 90-277-0438-4
64. K. F. Schaffner and R. S. Cohen (eds.), *PSA 1972. Proceedings of the Third Biennial Meeting of the Philosophy of Science Association.* [Boston Studies in the Philosophy of Science, Vol. XX] 1974 ISBN 90-277-0408-2; Pb 90-277-0409-0
65. H. E. Kyburg, Jr., *The Logical Foundations of Statistical Inference.* 1974 ISBN 90-277-0330-2; Pb 90-277-0430-9
66. M. Grene, *The Understanding of Nature.* Essays in the Philosophy of Biology. [Boston Studies in the Philosophy of Science, Vol. XXIII] 1974 ISBN 90-277-0462-7; Pb 90-277-0463-5
67. J. M. Broekman, *Structuralism: Moscow, Prague, Paris.* Translated from German. 1974 ISBN 90-277-0478-3
68. N. Geschwind, *Selected Papers on Language and the Brain.* [Boston Studies in the Philosophy of Science, Vol. XVI] 1974 ISBN 90-277-0262-4; Pb 90-277-0263-2
69. R. Fraissé, *Course of Mathematical Logic* – Volume 2: *Model Theory.* Translated from French. 1974 ISBN 90-277-0269-1; Pb 90-277-0510-0 (For *Volume 1* see under No. 54)

70. A. Grzegorczyk, *An Outline of Mathematical Logic.* Fundamental Results and Notions explained with all Details. Translated from Polish. 1974 ISBN 90-277-0359-0; Pb 90-277-0447-3
71. F. von Kutschera, *Philosophy of Language.* 1975 ISBN 90-277-0591-7
72. J. Manninen and R. Tuomela (eds.), *Essays on Explanation and Understanding.* Studies in the Foundations of Humanities and Social Sciences. 1976 ISBN 90-277-0592-5
73. J. Hintikka (ed.), *Rudolf Carnap, Logical Empiricist.* Materials and Perspectives. 1975 ISBN 90-277-0583-6
74. M. Čapek (ed.), *The Concepts of Space and Time.* Their Structure and Their Development. [Boston Studies in the Philosophy of Science, Vol. XXII] 1976 ISBN 90-277-0355-8; Pb 90-277-0375-2
75. J. Hintikka and U. Remes, *The Method of Analysis.* Its Geometrical Origin and Its General Significance. [Boston Studies in the Philosophy of Science, Vol. XXV] 1974 ISBN 90-277-0532-1; Pb 90-277-0543-7
76. J. E. Murdoch and E. D. Sylla (eds.), *The Cultural Context of Medieval Learning.* [Boston Studies in the Philosophy of Science, Vol. XXVI] 1975 ISBN 90-277-0560-7; Pb 90-277-0587-9
77. S. Amsterdamski, *Between Experience and Metaphysics.* Philosophical Problems of the Evolution of Science. [Boston Studies in the Philosophy of Science, Vol. XXXV] 1975 ISBN 90-277-0568-2; Pb 90-277-0580-1
78. P. Suppes (ed.), *Logic and Probability in Quantum Mechanics.* 1976 ISBN 90-277-0570-4; Pb 90-277-1200-X
79. H. von Helmholtz: *Epistemological Writings. The Paul Hertz / Moritz Schlick Centenary Edition of 1921 with Notes and Commentary by the Editors.* Newly translated from German by M. F. Lowe. Edited, with an Introduction and Bibliography, by R. S. Cohen and Y. Elkana. [Boston Studies in the Philosophy of Science, Vol. XXXVII] 1975 ISBN 90-277-0290-X; Pb 90-277-0582-8
80. J. Agassi, *Science in Flux.* [Boston Studies in the Philosophy of Science, Vol. XXVIII] 1975 ISBN 90-277-0584-4; Pb 90-277-0612-2
81. S. G. Harding (ed.), *Can Theories Be Refuted?* Essays on the Duhem-Quine Thesis. 1976 ISBN 90-277-0629-8; Pb 90-277-0630-1
82. S. Nowak, *Methodology of Sociological Research.* General Problems. 1977 ISBN 90-277-0486-4
83. J. Piaget, J.-B. Grize, A. Szeminśska and V. Bang, *Epistemology and Psychology of Functions.* Translated from French. 1977 ISBN 90-277-0804-5
84. M. Grene and E. Mendelsohn (eds.), *Topics in the Philosophy of Biology.* [Boston Studies in the Philosophy of Science, Vol. XXVII] 1976 ISBN 90-277-0595-X; Pb 90-277-0596-8
85. E. Fischbein, *The Intuitive Sources of Probabilistic Thinking in Children.* 1975 ISBN 90-277-0626-3; Pb 90-277-1190-9
86. E. W. Adams, *The Logic of Conditionals.* An Application of Probability to Deductive Logic. 1975 ISBN 90-277-0631-X
87. M. Przełęcki and R. Wójcicki (eds.), *Twenty-Five Years of Logical Methodology in Poland.* Translated from Polish. 1976 ISBN 90-277-0601-8
88. J. Topolski, *The Methodology of History.* Translated from Polish by O. Wojtasiewicz. 1976 ISBN 90-277-0550-X
89. A. Kasher (ed.), *Language in Focus: Foundations, Methods and Systems.* Essays dedicated to Yehoshua Bar-Hillel. [Boston Studies in the Philosophy of Science, Vol. XLIII] 1976 ISBN 90-277-0644-1; Pb 90-277-0645-X

90. J. Hintikka, *The Intentions of Intentionality and Other New Models for Modalities.* 1975
ISBN 90-277-0633-6; Pb 90-277-0634-4
91. W. Stegmüller, *Collected Papers on Epistemology, Philosophy of Science and History of Philosophy.* 2 Volumes. 1977 Set ISBN 90-277-0767-7
92. D. M. Gabbay, *Investigations in Modal and Tense Logics with Applications to Problems in Philosophy and Linguistics.* 1976 ISBN 90-277-0656-5
93. R. J. Bogdan, *Local Induction.* 1976 ISBN 90-277-0649-2
94. S. Nowak, *Understanding and Prediction.* Essays in the Methodology of Social and Behavioral Theories. 1976 ISBN 90-277-0558-5; Pb 90-277-1199-2
95. P. Mittelstaedt, *Philosophical Problems of Modern Physics.* [Boston Studies in the Philosophy of Science, Vol. XVIII] 1976 ISBN 90-277-0285-3; Pb 90-277-0506-2
96. G. Holton and W. A. Blanpied (eds.), *Science and Its Public: The Changing Relationship.* [Boston Studies in the Philosophy of Science, Vol. XXXIII] 1976
ISBN 90-277-0657-3; Pb 90-277-0658-1
97. M. Brand and D. Walton (eds.), *Action Theory.* 1976 ISBN 90-277-0671-9
98. P. Gochet, *Outline of a Nominalist Theory of Propositions.* An Essay in the Theory of Meaning and in the Philosophy of Logic. 1980 ISBN 90-277-1031-7
99. R. S. Cohen, P. K. Feyerabend, and M. W. Wartofsky (eds.), *Essays in Memory of Imre Lakatos.* [Boston Studies in the Philosophy of Science, Vol. XXXIX] 1976
ISBN 90-277-0654-9; Pb 90-277-0655-7
100. R. S. Cohen and J. J. Stachel (eds.), *Selected Papers of Léon Rosenfield.* [Boston Studies in the Philosophy of Science, Vol. XXI] 1979 ISBN 90-277-0651-4; Pb 90-277-0652-2
101. R. S. Cohen, C. A. Hooker, A. C. Michalos and J. W. van Evra (eds.), *PSA 1974. Proceedings of the 1974 Biennial Meeting of the Philosophy of Science Association.* [Boston Studies in the Philosophy of Science, Vol. XXXII] 1976 ISBN 90-277-0647-6; Pb 90-277-0648-4
102. Y. Fried and J. Agassi, *Paranoia.* A Study in Diagnosis. [Boston Studies in the Philosophy of Science, Vol. L] 1976 ISBN 90-277-0704-9; Pb 90-277-0705-7
103. M. Przełęcki, K. Szaniawski and R. Wójcicki (eds.), *Formal Methods in the Methodology of Empirical Sciences.* 1976 ISBN 90-277-0698-0
104. J. M. Vickers, *Belief and Probability.* 1976 ISBN 90-277-0744-8
105. K. H. Wolff, *Surrender and Catch.* Experience and Inquiry Today. [Boston Studies in the Philosophy of Science, Vol. LI] 1976 ISBN 90-277-0758-8; Pb 90-277-0765-0
106. K. Kosík, *Dialectics of the Concrete.* A Study on Problems of Man and World. [Boston Studies in the Philosophy of Science, Vol. LII] 1976 ISBN 90-277-0761-8; Pb 90-277-0764-2
107. N. Goodman, *The Structure of Appearance.* 3rd ed. with an Introduction by G. Hellman. [Boston Studies in the Philosophy of Science, Vol. LIII] 1977
ISBN 90-277-0773-1; Pb 90-277-0774-X
108. K. Ajdukiewicz, *The Scientific World-Perspective and Other Essays, 1931-1963.* Translated from Polish. Edited and with an Introduction by J. Giedymin. 1978 ISBN 90-277-0527-5
109. R. L. Causey, *Unity of Science.* 1977 ISBN 90-277-0779-0
110. R. E. Grandy, *Advanced Logic for Applications.* 1977 ISBN 90-277-0781-2
111. R. P. McArthur, *Tense Logic.* 1976 ISBN 90-277-0697-2
112. L. Lindahl, *Position and Change.* A Study in Law and Logic. Translated from Swedish by P. Needham. 1977 ISBN 90-277-0787-1
113. R. Tuomela, *Dispositions.* 1978 ISBN 90-277-0810-X
114. H. A. Simon, *Models of Discovery and Other Topics in the Methods of Science.* [Boston Studies in the Philosophy of Science, Vol. LIV] 1977 ISBN 90-277-0812-6; Pb 90-277-0858-4

115. R. D. Rosenkrantz, *Inference, Method and Decision.* Towards a Bayesian Philosophy of Science. 1977 ISBN 90-277-0817-7; Pb 90-277-0818-5
116. R. Tuomela, *Human Action and Its Explanation.* A Study on the Philosophical Foundations of Psychology. 1977 ISBN 90-277-0824-X
117. M. Lazerowitz, *The Language of Philosophy.* Freud and Wittgenstein. [Boston Studies in the Philosophy of Science, Vol. LV] 1977 ISBN 90-277-0826-6; Pb 90-277-0862-2
118. Not published 119. J. Pelc (ed.), *Semiotics in Poland, 1894–1969.* Translated from Polish. 1979 ISBN 90-277-0811-8
120. I. Pörn, *Action Theory and Social Science.* Some Formal Models. 1977 ISBN 90-277-0846-0
121. J. Margolis, *Persons and Mind.* The Prospects of Nonreductive Materialism. [Boston Studies in the Philosophy of Science, Vol. LVII] 1977 ISBN 90-277-0854-1; Pb 90-277-0863-0
122. J. Hintikka, I. Niiniluoto, and E. Saarinen (eds.), *Essays on Mathematical and Philosophical Logic.* 1979 ISBN 90-277-0879-7
123. T. A. F. Kuipers, *Studies in Inductive Probability and Rational Expectation.* 1978 ISBN 90-277-0882-7
124. E. Saarinen, R. Hilpinen, I. Niiniluoto and M. P. Hintikka (eds.), *Essays in Honour of Jaakko Hintikka on the Occasion of His 50th Birthday.* 1979 ISBN 90-277-0916-5
125. G. Radnitzky and G. Andersson (eds.), *Progress and Rationality in Science.* [Boston Studies in the Philosophy of Science, Vol. LVIII] 1978 ISBN 90-277-0921-1; Pb 90-277-0922-X
126. P. Mittelstaedt, *Quantum Logic.* 1978 ISBN 90-277-0925-4
127. K. A. Bowen, *Model Theory for Modal Logic.* Kripke Models for Modal Predicate Calculi. 1979 ISBN 90-277-0929-7
128. H. A. Bursen, *Dismantling the Memory Machine.* A Philosophical Investigation of Machine Theories of Memory. 1978 ISBN 90-277-0933-5
129. M. W. Wartofsky, *Models.* Representation and the Scientific Understanding. [Boston Studies in the Philosophy of Science, Vol. XLVIII] 1979 ISBN 90-277-0736-7; Pb 90-277-0947-5
130. D. Ihde, *Technics and Praxis.* A Philosophy of Technology. [Boston Studies in the Philosophy of Science, Vol. XXIV] 1979 ISBN 90-277-0953-X; Pb 90-277-0954-8
131. J. J. Wiatr (ed.), *Polish Essays in the Methodology of the Social Sciences.* [Boston Studies in the Philosophy of Science, Vol. XXIX] 1979 ISBN 90-277-0723-5; Pb 90-277-0956-4
132. W. C. Salmon (ed.), *Hans Reichenbach: Logical Empiricist.* 1979 ISBN 90-277-0958-0
133. P. Bieri, R.-P. Horstmann and L. Krüger (eds.), *Transcendental Arguments in Science.* Essays in Epistemology. 1979 ISBN 90-277-0963-7; Pb 90-277-0964-5
134. M. Marković and G. Petrović (eds.), *Praxis.* Yugoslav Essays in the Philosophy and Methodology of the Social Sciences. [Boston Studies in the Philosophy of Science, Vol. XXXVI] 1979 ISBN 90-277-0727-8; Pb 90-277-0968-8
135. R. Wójcicki, *Topics in the Formal Methodology of Empirical Sciences.* Translated from Polish. 1979 ISBN 90-277-1004-X
136. G. Radnitzky and G. Andersson (eds.), *The Structure and Development of Science.* [Boston Studies in the Philosophy of Science, Vol. LIX] 1979 ISBN 90-277-0994-7; Pb 90-277-0995-5
137. J. C. Webb, *Mechanism, Mentalism and Metamathematics.* An Essay on Finitism. 1980 ISBN 90-277-1046-5
138. D. F. Gustafson and B. L. Tapscott (eds.), *Body, Mind and Method.* Essays in Honor of Virgil C. Aldrich. 1979 ISBN 90-277-1013-9
139. L. Nowak, *The Structure of Idealization.* Towards a Systematic Interpretation of the Marxian Idea of Science. 1980 ISBN 90-277-1014-7

140. C. Perelman, *The New Rhetoric and the Humanities.* Essays on Rhetoric and Its Applications. Translated from French and German. With an Introduction by H. Zyskind. 1979 ISBN 90-277-1018-X; Pb 90-277-1019-8
141. W. Rabinowicz, *Universalizability.* A Study in Morals and Metaphysics. 1979 ISBN 90-277-1020-2
142. C. Perelman, *Justice, Law and Argument.* Essays on Moral and Legal Reasoning. Translated from French and German. With an Introduction by H.J. Berman. 1980 ISBN 90-277-1089-9; Pb 90-277-1090-2
143. S. Kanger and S. Öhman (eds.), *Philosophy and Grammar.* Papers on the Occasion of the Quincentennial of Uppsala University. 1981 ISBN 90-277-1091-0
144. T. Pawlowski, *Concept Formation in the Humanities and the Social Sciences.* 1980 ISBN 90-277-1096-1
145. J. Hintikka, D. Gruender and E. Agazzi (eds.), *Theory Change, Ancient Axiomatics and Galileo's Methodology.* Proceedings of the 1978 Pisa Conference on the History and Philosophy of Science, Volume I. 1981 ISBN 90-277-1126-7
146. J. Hintikka, D. Gruender and E. Agazzi (eds.), *Probabilistic Thinking, Thermodynamics, and the Interaction of the History and Philosophy of Science.* Proceedings of the 1978 Pisa Conference on the History and Philosophy of Science, Volume II. 1981 ISBN 90-277-1127-5
147. U. Mönnich (ed.), *Aspects of Philosophical Logic.* Some Logical Forays into Central Notions of Linguistics and Philosophy. 1981 ISBN 90-277-1201-8
148. D. M. Gabbay, *Semantical Investigations in Heyting's Intuitionistic Logic.* 1981 ISBN 90-277-1202-6
149. E. Agazzi (ed.), *Modern Logic – A Survey.* Historical, Philosophical, and Mathematical Aspects of Modern Logic and Its Applications. 1981 ISBN 90-277-1137-2
150. A. F. Parker-Rhodes, *The Theory of Indistinguishables.* A Search for Explanatory Principles below the Level of Physics. 1981 ISBN 90-277-1214-X
151. J. C. Pitt, *Pictures, Images, and Conceptual Change.* An Analysis of Wilfrid Sellars' Philosophy of Science. 1981 ISBN 90-277-1276-X; Pb 90-277-1277-8
152. R. Hilpinen (ed.), *New Studies in Deontic Logic.* Norms, Actions, and the Foundations of Ethics. 1981 ISBN 90-277-1278-6; Pb 90-277-1346-4
153. C. Dilworth, *Scientific Progress.* A Study Concerning the Nature of the Relation between Successive Scientific Theories. 3rd rev. ed., 1994 ISBN 0-7923-2487-0; Pb 0-7923-2488-9
154. D. Woodruff Smith and R. McIntyre, *Husserl and Intentionality.* A Study of Mind, Meaning, and Language. 1982 ISBN 90-277-1392-8; Pb 90-277-1730-3
155. R. J. Nelson, *The Logic of Mind.* 2nd. ed., 1989 ISBN 90-277-2819-4; Pb 90-277-2822-4
156. J. F. A. K. van Benthem, *The Logic of Time.* A Model-Theoretic Investigation into the Varieties of Temporal Ontology, and Temporal Discourse. 1983; 2nd ed., 1991 ISBN 0-7923-1081-0
157. R. Swinburne (ed.), *Space, Time and Causality.* 1983 ISBN 90-277-1437-1
158. E. T. Jaynes, *Papers on Probability, Statistics and Statistical Physics.* Ed. by R. D. Rozenkrantz. 1983 ISBN 90-277-1448-7; Pb (1989) 0-7923-0213-3
159. T. Chapman, *Time: A Philosophical Analysis.* 1982 ISBN 90-277-1465-7
160. E. N. Zalta, *Abstract Objects.* An Introduction to Axiomatic Metaphysics. 1983 ISBN 90-277-1474-6
161. S. Harding and M. B. Hintikka (eds.), *Discovering Reality.* Feminist Perspectives on Epistemology, Metaphysics, Methodology, and Philosophy of Science. 1983 ISBN 90-277-1496-7; Pb 90-277-1538-6
162. M. A. Stewart (ed.), *Law, Morality and Rights.* 1983 ISBN 90-277-1519-X

163. D. Mayr and G. Süssmann (eds.), *Space, Time, and Mechanics.* Basic Structures of a Physical Theory. 1983 ISBN 90-277-1525-4
164. D. Gabbay and F. Guenthner (eds.), *Handbook of Philosophical Logic.* Vol. I: Elements of Classical Logic. 1983 ISBN 90-277-1542-4
165. D. Gabbay and F. Guenthner (eds.), *Handbook of Philosophical Logic.* Vol. II: Extensions of Classical Logic. 1984 ISBN 90-277-1604-8
166. D. Gabbay and F. Guenthner (eds.), *Handbook of Philosophical Logic.* Vol. III: Alternative to Classical Logic. 1986 ISBN 90-277-1605-6
167. D. Gabbay and F. Guenthner (eds.), *Handbook of Philosophical Logic.* Vol. IV: Topics in the Philosophy of Language. 1989 ISBN 90-277-1606-4
168. A. J. I. Jones, *Communication and Meaning.* An Essay in Applied Modal Logic. 1983 ISBN 90-277-1543-2
169. M. Fitting, *Proof Methods for Modal and Intuitionistic Logics.* 1983 ISBN 90-277-1573-4
170. J. Margolis, *Culture and Cultural Entities.* Toward a New Unity of Science. 1984 ISBN 90-277-1574-2
171. R. Tuomela, *A Theory of Social Action.* 1984 ISBN 90-277-1703-6
172. J. J. E. Gracia, E. Rabossi, E. Villanueva and M. Dascal (eds.), *Philosophical Analysis in Latin America.* 1984 ISBN 90-277-1749-4
173. P. Ziff, *Epistemic Analysis.* A Coherence Theory of Knowledge. 1984 ISBN 90-277-1751-7
174. P. Ziff, *Antiaesthetics.* An Appreciation of the Cow with the Subtile Nose. 1984 ISBN 90-277-1773-7
175. W. Balzer, D. A. Pearce, and H.-J. Schmidt (eds.), *Reduction in Science.* Structure, Examples, Philosophical Problems. 1984 ISBN 90-277-1811-3
176. A. Peczenik, L. Lindahl and B. van Roermund (eds.), *Theory of Legal Science.* Proceedings of the Conference on Legal Theory and Philosophy of Science (Lund, Sweden, December 1983). 1984 ISBN 90-277-1834-2
177. I. Niiniluoto, *Is Science Progressive?* 1984 ISBN 90-277-1835-0
178. B. K. Matilal and J. L. Shaw (eds.), *Analytical Philosophy in Comparative Perspective.* Exploratory Essays in Current Theories and Classical Indian Theories of Meaning and Reference. 1985 ISBN 90-277-1870-9
179. P. Kroes, *Time: Its Structure and Role in Physical Theories.* 1985 ISBN 90-277-1894-6
180. J. H. Fetzer, *Sociobiology and Epistemology.* 1985 ISBN 90-277-2005-3; Pb 90-277-2006-1
181. L. Haaparanta and J. Hintikka (eds.), *Frege Synthesized.* Essays on the Philosophical and Foundational Work of Gottlob Frege. 1986 ISBN 90-277-2126-2
182. M. Detlefsen, *Hilbert's Program.* An Essay on Mathematical Instrumentalism. 1986 ISBN 90-277-2151-3
183. J. L. Golden and J. J. Pilotta (eds.), *Practical Reasoning in Human Affairs.* Studies in Honor of Chaim Perelman. 1986 ISBN 90-277-2255-2
184. H. Zandvoort, *Models of Scientific Development and the Case of Nuclear Magnetic Resonance.* 1986 ISBN 90-277-2351-6
185. I. Niiniluoto, *Truthlikeness.* 1987 ISBN 90-277-2354-0
186. W. Balzer, C. U. Moulines and J. D. Sneed, *An Architectonic for Science.* The Structuralist Program. 1987 ISBN 90-277-2403-2
187. D. Pearce, *Roads to Commensurability.* 1987 ISBN 90-277-2414-8
188. L. M. Vaina (ed.), *Matters of Intelligence.* Conceptual Structures in Cognitive Neuroscience. 1987 ISBN 90-277-2460-1

189. H. Siegel, *Relativism Refuted.* A Critique of Contemporary Epistemological Relativism. 1987 ISBN 90-277-2469-5
190. W. Callebaut and R. Pinxten, *Evolutionary Epistemology.* A Multiparadigm Program, with a Complete Evolutionary Epistemology Bibliograph. 1987 ISBN 90-277-2582-9
191. J. Kmita, *Problems in Historical Epistemology.* 1988 ISBN 90-277-2199-8
192. J. H. Fetzer (ed.), *Probability and Causality.* Essays in Honor of Wesley C. Salmon, with an Annotated Bibliography. 1988 ISBN 90-277-2607-8; Pb 1-5560-8052-2
193. A. Donovan, L. Laudan and R. Laudan (eds.), *Scrutinizing Science.* Empirical Studies of Scientific Change. 1988 ISBN 90-277-2608-6
194. H.R. Otto and J.A. Tuedio (eds.), *Perspectives on Mind.* 1988 ISBN 90-277-2640-X
195. D. Batens and J.P. van Bendegem (eds.), *Theory and Experiment.* Recent Insights and New Perspectives on Their Relation. 1988 ISBN 90-277-2645-0
196. J. Österberg, *Self and Others.* A Study of Ethical Egoism. 1988 ISBN 90-277-2648-5
197. D.H. Helman (ed.), *Analogical Reasoning.* Perspectives of Artificial Intelligence, Cognitive Science, and Philosophy. 1988 ISBN 90-277-2711-2
198. J. Woleński, *Logic and Philosophy in the Lvov-Warsaw School.* 1989 ISBN 90-277-2749-X
199. R. Wójcicki, *Theory of Logical Calculi.* Basic Theory of Consequence Operations. 1988 ISBN 90-277-2785-6
200. J. Hintikka and M.B. Hintikka, *The Logic of Epistemology and the Epistemology of Logic.* Selected Essays. 1989 ISBN 0-7923-0040-8; Pb 0-7923-0041-6
201. E. Agazzi (ed.), *Probability in the Sciences.* 1988 ISBN 90-277-2808-9
202. M. Meyer (ed.), *From Metaphysics to Rhetoric.* 1989 ISBN 90-277-2814-3
203. R.L. Tieszen, *Mathematical Intuition.* Phenomenology and Mathematical Knowledge. 1989 ISBN 0-7923-0131-5
204. A. Melnick, *Space, Time, and Thought in Kant.* 1989 ISBN 0-7923-0135-8
205. D.W. Smith, *The Circle of Acquaintance.* Perception, Consciousness, and Empathy. 1989 ISBN 0-7923-0252-4
206. M.H. Salmon (ed.), *The Philosophy of Logical Mechanism.* Essays in Honor of Arthur W. Burks. With his Responses, and with a Bibliography of Burk's Work. 1990 ISBN 0-7923-0325-3
207. M. Kusch, *Language as Calculus vs. Language as Universal Medium.* A Study in Husserl, Heidegger, and Gadamer. 1989 ISBN 0-7923-0333-4
208. T.C. Meyering, *Historical Roots of Cognitive Science.* The Rise of a Cognitive Theory of Perception from Antiquity to the Nineteenth Century. 1989 ISBN 0-7923-0349-0
209. P. Kosso, *Observability and Observation in Physical Science.* 1989 ISBN 0-7923-0389-X
210. J. Kmita, *Essays on the Theory of Scientific Cognition.* 1990 ISBN 0-7923-0441-1
211. W. Sieg (ed.), *Acting and Reflecting.* The Interdisciplinary Turn in Philosophy. 1990 ISBN 0-7923-0512-4
212. J. Karpiński, *Causality in Sociological Research.* 1990 ISBN 0-7923-0546-9
213. H.A. Lewis (ed.), *Peter Geach: Philosophical Encounters.* 1991 ISBN 0-7923-0823-9
214. M. Ter Hark, *Beyond the Inner and the Outer.* Wittgenstein's Philosophy of Psychology. 1990 ISBN 0-7923-0850-6
215. M. Gosselin, *Nominalism and Contemporary Nominalism.* Ontological and Epistemological Implications of the Work of W.V.O. Quine and of N. Goodman. 1990 ISBN 0-7923-0904-9
216. J.H. Fetzer, D. Shatz and G. Schlesinger (eds.), *Definitions and Definability.* Philosophical Perspectives. 1991 ISBN 0-7923-1046-2
217. E. Agazzi and A. Cordero (eds.), *Philosophy and the Origin and Evolution of the Universe.* 1991 ISBN 0-7923-1322-4

218. M. Kusch, *Foucault's Strata and Fields*. An Investigation into Archaeological and Genealogical Science Studies. 1991 ISBN 0-7923-1462-X
219. C.J. Posy, *Kant's Philosophy of Mathematics*. Modern Essays. 1992 ISBN 0-7923-1495-6
220. G. Van de Vijver, *New Perspectives on Cybernetics*. Self-Organization, Autonomy and Connectionism. 1992 ISBN 0-7923-1519-7
221. J.C. Nyíri, *Tradition and Individuality*. Essays. 1992 ISBN 0-7923-1566-9
222. R. Howell, *Kant's Transcendental Deduction*. An Analysis of Main Themes in His Critical Philosophy. 1992 ISBN 0-7923-1571-5
223. A. García de la Sienra, *The Logical Foundations of the Marxian Theory of Value*. 1992 ISBN 0-7923-1778-5
224. D.S. Shwayder, *Statement and Referent*. An Inquiry into the Foundations of Our Conceptual Order. 1992 ISBN 0-7923-1803-X
225. M. Rosen, *Problems of the Hegelian Dialectic*. Dialectic Reconstructed as a Logic of Human Reality. 1993 ISBN 0-7923-2047-6
226. P. Suppes, *Models and Methods in the Philosophy of Science: Selected Essays*. 1993 ISBN 0-7923-2211-8
227. R. M. Dancy (ed.), *Kant and Critique: New Essays in Honor of W. H. Werkmeister*. 1993 ISBN 0-7923-2244-4
228. J. Woleński (ed.), *Philosophical Logic in Poland*. 1993 ISBN 0-7923-2293-2
229. M. De Rijke (ed.), *Diamonds and Defaults*. Studies in Pure and Applied Intensional Logic. 1993 ISBN 0-7923-2342-4
230. B.K. Matilal and A. Chakrabarti (eds.), *Knowing from Words*. Western and Indian Philosophical Analysis of Understanding and Testimony. 1994 ISBN 0-7923-2345-9
231. S.A. Kleiner, *The Logic of Discovery*. A Theory of the Rationality of Scientific Research. 1993 ISBN 0-7923-2371-8
232. R. Festa, *Optimum Inductive Methods*. A Study in Inductive Probability, Bayesian Statistics, and Verisimilitude. 1993 ISBN 0-7923-2460-9
233. P. Humphreys (ed.), *Patrick Suppes: Scientific Philosopher*. Vol. 1: Probability and Probabilistic Causality. 1994 ISBN 0-7923-2552-4
234. P. Humphreys (ed.), *Patrick Suppes: Scientific Philosopher*. Vol. 2: Philosophy of Physics, Theory Structure, and Measurement Theory. 1994 ISBN 0-7923-2553-2
235. P. Humphreys (ed.), *Patrick Suppes: Scientific Philosopher*. Vol. 3: Language, Logic, and Psychology. 1994 ISBN 0-7923-2862-0
Set ISBN (Vols 233–235) 0-7923-2554-0
236. D. Prawitz and D. Westerståhl (eds.), *Logic and Philosophy of Science in Uppsala*. Papers from the 9th International Congress of Logic, Methodology, and Philosophy of Science. 1994 ISBN 0-7923-2702-0
237. L. Haaparanta (ed.), *Mind, Meaning and Mathematics*. Essays on the Philosophical Views of Husserl and Frege. 1994 ISBN 0-7923-2703-9
238. J. Hintikka (ed.), *Aspects of Metaphor*. 1994 ISBN 0-7923-2786-1
239. B. McGuinness and G. Oliveri (eds.), *The Philosophy of Michael Dummett*. With Replies from Michael Dummett. 1994 ISBN 0-7923-2804-3
240. D. Jamieson (ed.), *Language, Mind, and Art*. Essays in Appreciation and Analysis, In Honor of Paul Ziff. 1994 ISBN 0-7923-2810-8
241. G. Preyer, F. Siebelt and A. Ulfig (eds.), *Language, Mind and Epistemology*. On Donald Davidson's Philosophy. 1994 ISBN 0-7923-2811-6
242. P. Ehrlich (ed.), *Real Numbers, Generalizations of the Reals, and Theories of Continua*. 1994 ISBN 0-7923-2689-X

243. G. Debrock and M. Hulswit (eds.), *Living Doubt*. Essays concerning the epistemology of Charles Sanders Peirce. 1994 ISBN 0-7923-2898-1
244. J. Srzednicki, *To Know or Not to Know*. Beyond Realism and Anti-Realism. 1994 ISBN 0-7923-2909-0
245. R. Egidi (ed.), *Wittgenstein: Mind and Language*. 1995 ISBN 0-7923-3171-0
246. A. Hyslop, *Other Minds*. 1995 ISBN 0-7923-3245-8
247. L. Pólos and M. Masuch (eds.), *Applied Logic: How, What and Why*. Logical Approaches to Natural Language. 1995 ISBN 0-7923-3432-9
248. M. Krynicki, M. Mostowski and L.M. Szczerba (eds.), *Quantifiers: Logics, Models and Computation*. Volume One: Surveys. 1995 ISBN 0-7923-3448-5
249. M. Krynicki, M. Mostowski and L.M. Szczerba (eds.), *Quantifiers: Logics, Models and Computation*. Volume Two: Contributions. 1995 ISBN 0-7923-3449-3
Set ISBN (Vols 248 + 249) 0-7923-3450-7
250. R.A. Watson, *Representational Ideas from Plato to Patricia Churchland*. 1995 ISBN 0-7923-3453-1
251. J. Hintikka (ed.), *From Dedekind to Gödel*. Essays on the Development of the Foundations of Mathematics. 1995 ISBN 0-7923-3484-1
252. A. Wiśniewski, *The Posing of Questions*. Logical Foundations of Erotetic Inferences. 1995 ISBN 0-7923-3637-2
253. J. Peregrin, *Doing Worlds with Words*. Formal Semantics without Formal Metaphysics. 1995 ISBN 0-7923-3742-5
254. I.A. Kieseppä, *Truthlikeness for Multidimensional, Quantitative Cognitive Problems*. 1996 ISBN 0-7923-4005-1
255. P. Hugly and C. Sayward: *Intensionality and Truth*. An Essay on the Philosophy of A.N. Prior. 1996 ISBN 0-7923-4119-8
256. L. Hankinson Nelson and J. Nelson (eds.): *Feminism, Science, and the Philosophy of Science*. 1997 ISBN 0-7923-4162-7
257. P.I. Bystrov and V.N. Sadovsky (eds.): *Philosophical Logic and Logical Philosophy*. Essays in Honour of Vladimir A. Smirnov. 1996 ISBN 0-7923-4270-4
258. Å.E. Andersson and N-E. Sahlin (eds.): *The Complexity of Creativity*. 1996 ISBN 0-7923-4346-8
259. M.L. Dalla Chiara, K. Doets, D. Mundici and J. van Benthem (eds.): *Logic and Scientific Methods*. Volume One of the Tenth International Congress of Logic, Methodology and Philosophy of Science, Florence, August 1995. 1997 ISBN 0-7923-4383-2
260. M.L. Dalla Chiara, K. Doets, D. Mundici and J. van Benthem (eds.): *Structures and Norms in Science*. Volume Two of the Tenth International Congress of Logic, Methodology and Philosophy of Science, Florence, August 1995. 1997 ISBN 0-7923-4384-0
Set ISBN (Vols 259 + 260) 0-7923-4385-9
261. A. Chakrabarti: *Denying Existence*. The Logic, Epistemology and Pragmatics of Negative Existentials and Fictional Discourse. 1997 ISBN 0-7923-4388-3
262. A. Biletzki: *Talking Wolves*. Thomas Hobbes on the Language of Politics and the Politics of Language. 1997 ISBN 0-7923-4425-1
263. D. Nute (ed.): *Defeasible Deontic Logic*. 1997 ISBN 0-7923-4630-0
264. U. Meixner: *Axiomatic Formal Ontology*. 1997 ISBN 0-7923-4747-X
265. I. Brinck: *The Indexical 'I'*. The First Person in Thought and Language. 1997 ISBN 0-7923-4741-2
266. G. Hölmström-Hintikka and R. Tuomela (eds.): *Contemporary Action Theory*. Volume 1: Individual Action. 1997 ISBN 0-7923-4753-6; Set: 0-7923-4754-4

267. G. Hölmström-Hintikka and R. Tuomela (eds.): *Contemporary Action Theory.* Volume 2: Social Action. 1997 ISBN 0-7923-4752-8; Set: 0-7923-4754-4
268. B.-C. Park: *Phenomenological Aspects of Wittgenstein's Philosophy.* 1998 ISBN 0-7923-4813-3
269. J. Paśniczek: *The Logic of Intentional Objects.* A Meinongian Version of Classical Logic. 1998 Hb ISBN 0-7923-4880-X; Pb ISBN 0-7923-5578-4
270. P.W. Humphreys and J.H. Fetzer (eds.): *The New Theory of Reference.* Kripke, Marcus, and Its Origins. 1998 ISBN 0-7923-4898-2
271. K. Szaniawski, A. Chmielewski and J. Woleński (eds.): *On Science, Inference, Information and Decision Making.* Selected Essays in the Philosophy of Science. 1998 ISBN 0-7923-4922-9
272. G.H. von Wright: *In the Shadow of Descartes.* Essays in the Philosophy of Mind. 1998 ISBN 0-7923-4992-X
273. K. Kijania-Placek and J. Woleński (eds.): *The Lvov–Warsaw School and Contemporary Philosophy.* 1998 ISBN 0-7923-5105-3
274. D. Dedrick: *Naming the Rainbow.* Colour Language, Colour Science, and Culture. 1998 ISBN 0-7923-5239-4
275. L. Albertazzi (ed.): *Shapes of Forms.* From Gestalt Psychology and Phenomenology to Ontology and Mathematics. 1999 ISBN 0-7923-5246-7
276. P. Fletcher: *Truth, Proof and Infinity.* A Theory of Constructions and Constructive Reasoning. 1998 ISBN 0-7923-5262-9
277. M. Fitting and R.L. Mendelsohn (eds.): *First-Order Modal Logic.* 1998 Hb ISBN 0-7923-5334-X; Pb ISBN 0-7923-5335-8
278. J.N. Mohanty: *Logic, Truth and the Modalities from a Phenomenological Perspective.* 1999 ISBN 0-7923-5550-4
279. T. Placek: *Mathematical Intiutionism and Intersubjectivity.* A Critical Exposition of Arguments for Intuitionism. 1999 ISBN 0-7923-5630-6
280. A. Cantini, E. Casari and P. Minari (eds.): *Logic and Foundations of Mathematics.* 1999 ISBN 0-7923-5659-4 set ISBN 0-7923-5867-8
281. M.L. Dalla Chiara, R. Giuntini and F. Laudisa (eds.): *Language, Quantum, Music.* 1999 ISBN 0-7923-5727-2; set ISBN 0-7923-5867-8
282. R. Egidi (ed.): *In Search of a New Humanism.* The Philosophy of Georg Hendrik von Wright. 1999 ISBN 0-7923-5810-4
283. F. Vollmer: *Agent Causality.* 1999 ISBN 0-7923-5848-1
284. J. Peregrin (ed.): *Truth and Its Nature (if Any).* 1999 ISBN 0-7923-5865-1
285. M. De Caro (ed.): *Interpretations and Causes.* New Perspectives on Donald Davidson's Philosophy. 1999 ISBN 0-7923-5869-4
286. R. Murawski: *Recursive Functions and Metamathematics.* Problems of Completeness and Decidability, Gödel's Theorems. 1999 ISBN 0-7923-5904-6
287. T.A.F. Kuipers: *From Instrumentalism to Constructive Realism.* On Some Relations between Confirmation, Empirical Progress, and Truth Approximation. 2000 ISBN 0-7923-6086-9
288. G. Holmström-Hintikka (ed.): *Medieval Philosophy and Modern Times.* 2000 ISBN 0-7923-6102-4
289. E. Grosholz and H. Breger (eds.): *The Growth of Mathematical Knowledge.* 2000 ISBN 0-7923-6151-2

290. G. Sommaruga: *History and Philosophy of Constructive Type Theory*. 2000 ISBN 0-7923-6180-6
291. J. Gasser (ed.): *A Boole Anthology*. Recent and Classical Studies in the Logic of George Boole. 2000 ISBN 0-7923-6380-9
292. V.F. Hendricks, S.A. Pedersen and K.F. Jørgensen (eds.): *Proof Theory*. History and Philosophical Significance. 2000 ISBN 0-7923-6544-5
293. W.L. Craig: *The Tensed Theory of Time*. A Critical Examination. 2000 ISBN 0-7923-6634-4
294. W.L. Craig: *The Tenseless Theory of Time*. A Critical Examination. 2000 ISBN 0-7923-6635-2
295. L. Albertazzi (ed.): *The Dawn of Cognitive Science*. Early European Contributors. 2001 ISBN 0-7923-6799-5
296. G. Forrai: *Reference, Truth and Conceptual Schemes*. A Defense of Internal Realism. 2001 ISBN 0-7923-6885-1
297. V.F. Hendricks, S.A. Pedersen and K.F. Jørgensen (eds.): *Probability Theory*. Philosophy, Recent History and Relations to Science. 2001 ISBN 0-7923-6952-1
298. M. Esfeld: *Holism in Philosophy of Mind and Philosophy of Physics*. 2001 ISBN 0-7923-7003-1
299. E.C. Steinhart: *The Logic of Metaphor*. Analogous Parts of Possible Worlds. 2001 ISBN 0-7923-7004-X
300. To be published.
301. T.A.F. Kuipers: *Structures in Science Heuristic Patterns Based on Cognitive Structures*. An Advanced Textbook in Neo-Classical Philosophy of Science. 2001 ISBN 0-7923-7117-8
302. G. Hon and S.S. Rakover (eds.): *Explanation*. Theoretical Approaches and Applications. 2001 ISBN 1-4020-0017-0
303. G. Holmström-Hintikka, S. Lindström and R. Sliwinski (eds.): *Collected Papers of Stig Kanger with Essays on his Life and Work*. Vol. I. 2001 ISBN 1-4020-0021-9; Pb ISBN 1-4020-0022-7
304. G. Holmström-Hintikka, S. Lindström and R. Sliwinski (eds.): *Collected Papers of Stig Kanger with Essays on his Life and Work*. Vol. II. 2001 ISBN 1-4020-0111-8; Pb ISBN 1-4020-0112-6
305. C.A. Anderson and M. Zelëny (eds.): *Logic, Meaning and Computation*. Essays in Memory of Alonzo Church. 2001 ISBN 1-4020-0141-X
306. P. Schuster, U. Berger and H. Osswald (eds.): *Reuniting the Antipodes – Constructive and Nonstandard Views of the Continuum*. 2001 ISBN 1-4020-0152-5

Previous volumes are still available.

KLUWER ACADEMIC PUBLISHERS – DORDRECHT / BOSTON / LONDON